Cross-cultural Perspectives on Well-Being and Sustainability in Organizations

Annamaria Di Fabio
Editor

Cross-cultural Perspectives on Well-Being and Sustainability in Organizations

Springer

Editor
Annamaria Di Fabio
Department of Education, Languages, Intercultures, Literatures and Psychology
University of Florence
Florence, Italy

ISBN 978-3-030-86708-9 ISBN 978-3-030-86709-6 (eBook)
https://doi.org/10.1007/978-3-030-86709-6

This Springer imprint is published by the registered company Springer Nature Switzerland AG
The registered company address is: Gewerbestrasse 11, 6330 Cham, Switzerland

This book has been independently peer-reviewed by international academics who are experts in the field. The reviews included the entire text that appears in the published book and peer reviews were exceptionally positive.

Foreword

Sustainability is an important challenge for humanity and is receiving a wide attention from international bodies, scientific disciplines, governments, NGOs and society in general during the last half a century. It has been briefly characterized as the capacity for continuance into the long-term future (Martin, 2003) although often different disciplines have provided their own definitions (ecology, economics, biology and sociology) focusing on specific facets. A comprehensive view of sustainability is lacking and the concept is becoming fuzzy and confusing (Faber et al. 2005). In an effort to provide a systematic and comprehensive view, the Brundtland Commission pointed out that, beyond the physical needs and economics, "sustainability requires views of human needs and well-being that incorporate such non-economic variables as education and health enjoyed for their own sake, clean air and water, and the protection of natural beauty. It must also work to remove disabilities from disadvantaged groups, many of whom live in ecologically vulnerable areas" (WCED, 1987).

Sustainable development is considered the process to achieve sustainability. Thus, the Brundtland Commission characterizes it as the "development that meets the needs of the present without compromising the ability of future generations to meet their own needs" (WCDE, 1987). Three main pillars of sustainable development are considered: economic growth, environmental protection and social equality. In practice, conceptualizations of sustainable development cover even a broader spectrum. Lozano (2008) has distinguished five types of conceptualizations: the conventional economists' perspective; the non-environmental degradation perspective; the integrational perspective, i.e. encompassing the economic, environmental and social aspects; (4) the inter-generational perspective; and (5) the holistic perspective (Lozano 2008). More recently, Bolis, Morioka and Sznelwar (2014) developed a model with an axiological perspective. It proposes that "sustainable development can be seen as development aimed at improving the well-being of society as a whole (including future generations), enabled by an axiological perspective in decision-making processes, considering the limitations of environmental resources. The model explicitly includes a value-based mindset in the concept of sustainable development" (p. 7). It highlights the relation of sustainability with people's well-being, considering also future generations, and emphasizes the decision-making processes with the tension between individual and functionalistic drivers on one end and the axiological collective driver on the other one. Then, the model considers subjective and emotional components and values such as cooperation, equity and equality, including a collective, intergenerational and axiological approach.

During the las decade, psychology has paid growing attention and awareness to the processes of sustainable development. In this way, the *European Journal of Sustainable Development Research*, in an editorial article recently published (Di Fabio and Rosen, 2018), declares a new frontier consisting in "opening the black box of Psychological Processes in the science of sustainable development".

The UN General Assembly approved, on 12 September, 2015, the Sustainable Development Goals and the agenda 2030. The goals aim to stimulate action in the five critical areas: People, Planet, Prosperity, Peace and Partnership. The declaration included 17 goals with 169 targets establishing, for the first time, shared goals concerning all the countries around the world and not just the developing ones, as it was the case for the Millennial Goals. This has been seen as a novel approach of global governance by goal setting (Biermann, Kanie & Kim, 2017). They are integrated and balance the three dimensions of sustainable development. The goals are primarily included in one of the five "P" areas: *People* (goal 1: No poverty; 2: Zero hunger; 3: Good health and well-being; 4: Quality education and 5: Gender equality), *Planet* (goals 6: Clean water and sanitation; 12: Responsible consumption and production; 13: Climate action; 14: Life below water; and 15: Life on land), *Prosperity* (goals 8: Decent work and economic growth; 9: Industry innovation and infrastructure; 10: Reduced inequalities and 11: Sustainable cities and communities), *Peace* (goal 16: peace, justice and strong institutions) and *Partnership* (goal 17: partnership for the goals).

Organizations and work are then explicitly dealt with in goal 8 and are mentioned in several targets of different goals. The goal 8 aims to "promote sustained, inclusive and sustainable economic growth full and productive employment and decent work for all". Its targets include issues such as "*decent job creation, entrepreneurship, creativity and innovation, and encourage the formalization and growth of micro-, small- and medium-sized enterprises*". They also aim to "*achieve full and productive employment and decent work for all women and men, including for young people and persons with disabilities, and equal pay for work of equal value*". Moreover, they pursue to "*reduce the proportion of youth not in employment, education or training*", to "*eradicate forced labour, and modern slavery and human trafficking and secure the prohibition and elimination of the worst forms of child labour, including recruitment and use of child soldiers, and by 2025 end child labour in all its forms*" Finally, they target "*protect labour rights and promote safe and secure working environments for all workers, including migrant workers, in particular women migrants, and those in precarious employment*".

In addition to this array of important issues, other goals also include issues concerning work and/or organizations. In fact, goal 3 includes "*training and retention of the health workforce*" and goal 4 aims to "*increase the number of youth and adults who have relevant skills, including technical and vocational skills, for employment, decent jobs and entrepreneurship*, and "*supply of qualified teachers, including through international cooperation for teacher training in developing countries*". Goal 5 includes targets related to work and organizations such as "*recognize and value unpaid care and domestic work*" and "*ensure women's full and effective participation and equal opportunities for leadership at all levels of decision-making*". Finally, referring to companies, goal 12 sets as a target to "*encourage companies to adopt sustainable practices and to integrate sustainability information into their reporting cycle*".

Organizations are important social formations in modern societies and their cooperation and functioning are critical to achieve the sustainable goals. This is the reason why the UN global compact has been promoted (https://www.unglobalcompact.org/) aiming "to mobilize a global movement of sustainable companies and stakeholders to create the world we want". It supports companies to: (1) "Do business responsibly by aligning their strategies and operations with Ten Principles on human rights, labour, environment and anti-corruption; and (2) Take strategic actions to advance broader societal goals, such as the UN Sustainable Development Goals, with an emphasis on collaboration and innovation". In fact, corporations and organizations may play an important role to promote Sustainable Development Goals (Foster & Viale, 2020).

Taking into consideration the wide array of issues and challenges concerning organizations work and well-being of people, the contribution of work and organizational psychology, in cooperation with other disciplines and actors, is a must in this immense endeavour. It is becoming plausible that the psychological lens is critical to promote sustainable development, and more specifically, the contributions of work and organizational psychology are urgently needed. There is growing awareness about the wide array of issues where the contributions of psychology can be beneficial for sustainable development. Organizational psychology can be extremely helpful in promoting organizational behaviours at the collective level that may join efforts towards sustainable development in many different issues concerning people (employees' well-being, decent work, gender equality, etc.), the planet (promoting green jobs, contributing to climate change adaptation, and caring about environment and ecology) and sustainable prosperity (clean industry, responsible production, etc.). All these efforts need to consider not just improving and protecting the current situation but also take into account the legacy for future generations.

Moreover, organizations need to care about their own sustainability in the broader context of sustainable development. Work and organizational psychology can be important pillars for their own sustainability and may contribute to make organizations healthier, safer, more effective and productive. Organizational justice, and participation, management processes and practices to assure dignity at work and strategies that promote corporate social responsibility are important components of the organizational sustainability. This is mainly relevant when psychology contributes to improve quality of life and dignity of the most vulnerable workers, especially in informal work. Carr (2021) has recently presented an overview of important contributions through humanitarian work and organizational psychology.

In this context, the present book is a milestone on the study of organizational well-being and sustainability. With a psychological lens, the different chapters present topics that are clear examples of the contributions that work and organizational psychology can make to the challenges of promoting sustainability and well-being. It provides an important conceptualization and framework for the role of work and organizational psychology in this field (Chap. 1) and also offers relevant keys about the methodological approaches to research from this perspective (Chap. 2). Then, a number of relevant issues are dealt with, such as training of professionals to better promote well-being (Chap. 3), the employees' biological stress markers and their dynamics (Chap. 4) and a better understanding of the role of emotional intelligence and positive emotions (Chap. 5). It is also interesting the conceptualization of the pillars of sustainability and the ways to promote them (Chap. 6), the assessment of leadership for sustainable human capital in organizations (Chap.7), and the study of the key features of leadership in indigenous communities (Chap. 8).

In the second part of the book, special attention is paid to innovation both from theory and empirical studies. Chap. 9 deals with the trait emotional intelligence through which the positive and negative affect influence the meaning of work. Social identity is analysed as a challenge for sustainable management (Chap. 10), and Chap. 11 studies sleep during the first days after giving birth. Other important topics included in this book are: career indecision (Chap. 12), acceptance of change for sustainability in organizations (Chap. 13), competitiveness during the transition towards a market economy (Chap. 14), analysis of the care programme changes in South Asian countries (Chap. 15), and work-life balance and well-being of Tunisian mangers. Finally, the book closes with an analysis and reflection on balance between innovation, health and wealth, which is a critical issue of sustainability from a psychological perspective (Chap. 17).

As the title points out, this book includes a rich bundle of *perspectives* from different cultures when considering well-being and sustainability in organizations. In fact, the authors come from a broad scope of countries in different regions of the world: Europe (Italy, Spain, Hungary), North America (USA), Asia (United Arab Emirates, India, Bangladesh, Malaysia, Japan), Africa (South Africa, Tunisia) and Oceania (Australia). This rich array offers an interesting and somewhat kaleidoscopic view of sustainability and well-being in work and organizations.

These brief considerations are just a quick overview of a rich collection of contributions that provide excellent perspectives on a number of issues concerning organizations, paying special attention to the focus of sustainability and well-being and using the lens of psychological sciences, especially the organizational and work psychology one. Before, closing this foreword, I would like to encourage the readers to go through the chapters of the book because they will benefit from a rich, complex and informative picture that emerges on this really challenging issues for the psychology of the twenty-first century.

University of Valencia
Valencia, Spain

José M. Peiró

References

Biermann, F., Kanie, N., & Kim, R. E. (2017). Global governance by goal-setting: the novel approach of the UN Sustainable Development Goals. Current Opinion in Environmental Sustainability, 26, 26–31

Bolis, I., Morioka, S.N. & Sznelwar, L. (2014): When sustainable development risks losing its meaning. Delimiting the concept with a comprehensive literature review and a conceptual model *Journal of cleaner Production*; u3, 2014, 7–20

Carr, S. Humanitarian Work and Organizational Psychology. *Oxford Research Encyclopedia of Psychology*. Retrieved 8 Sep. 2021, from https://oxfordre.com/psychology/view/10.1093/acrefore/9780190236557.001.0001/acrefore-9780190236557-e-828

Di Fabio, A. and Rosen, M. A. (2018). Opening the Black Box of Psychological Processes in the Science of Sustainable Development: A New Frontier. *European Journal of Sustainable Development Research*, 2(4), 47. https://doi.org/10.20897/ejosdr/3933

Faber, N., Jorna, R., Van Engelen, J., (2005). The sustainability of "sustainability". A study into the conceptual foundations of the notion of "sustainability". *J. Environ. Assess. Policy Manag.* 7, 1e33.

Foster, L. L., & Viale, T. (2020). The changing nature of work and organizations in sustainable development. In B. J. Hoffman, M. K. Shoss, & L. A. Wegman (Eds.), *The Cambridge handbook of the changing nature of work* (pp. 583–618). Cambridge University Press

Lozano, R. (2008). Envisioning sustainability three-dimensionally. *Journal of cleaner production*, *16*(17), 1838–1846

Martin S. (2003). *Sustainability, systems thinking and professional practice*. Worcester: University College, Worcester.

WCED (1987) *Report of the World Commission on Environment and Development: Our Common Future*. Transmitted to the General Assembly as an Annex to document A/42/427 – Development and International Co-operation: Environment. Downloaded from https://sustainabledevelopment.un.org/content/documents/5987our-common-future.pdf

Preface

The background for this book is the new research area of the psychology of sustainability and sustainable development (Di Fabio, 2017a, b; Di Fabio & Rosen, 2018) that represents a relatively new research area in the field of Sustainability Science. It contributes to sustainability issues by introducing a psychological perspective and promoting the trans-disciplinary framework at the basis of Sustainability Science. This perspective could also permit the advancement of psychological contributions to the concrete answers to each of the 17 UN Sustainable Development Goals (United Nations, 2020), which extend to 2030 and which are of vital importance for the world and humanity.

This volume presents cross-cultural contributions, studying in depth the dark and the bright sides (first part) and theoretical approaches and empirical research, widening the perspectives for innovation in the prevention science for sustainable organizations and sustainability in organizations (second part).

The first part, Studying in Depth the Dark and the Bright Sides: Cross-Cultural Contributions, includes eight chapters. The first contribution, by Annamaria Di Fabio, delineates the theoretical perspective of the psychology of sustainability and sustainable development that constitutes the framework of reference for this volume. In particular, the critical contribution of this perspective for establishing healthy organizations and healthy business is underlined, emphasizing its promising value for research and interventions and also considering strength-based prevention perspectives (Di Fabio & Saklofske, 2021), as well as a primary prevention perspective (Di Fabio & Kenny, 2015, 2016; Hage et al., 2007). The second chapter, by Lynette H. Bikos, Kaitlin Patton, Thomas Pankau, and Lauren Hirsch, offers a contribution in relation to methodological approaches for research on sustainability themes in cross-cultural perspectives. The chapter reviews different kinds of quantitative analyses to offer a view of the status of quantitative research that has examined the role of time in international adaptation. The authors present different statistical theory from the paradigmatic, computational algorithm, and statistic-specific levels with a specific focus on the advantages of multi-level modeling (MLM). MLM permits the simultaneous evaluation of within-person and between-person indicators in line with the concept of harmonization (Di Fabio & Tsuda,

2018) that highlights interpersonal, intrapersonal, and contextual factors including temporal and spatial dimensions.

The third chapter, by Mickey Stein, Larry E. Beutler, Satoko Kimpara, Nancy A. Haug, Hannah Brunet, Kathleen Someah, Christopher J. Edwards, and Sandra Macias, presents a study on the effectiveness of Systematic Treatment Selection (STS)-assisted supervision (SAS). The study showed that patients whose trainee therapist received SAS presented more positive outcomes than patients whose trainee therapist received usual supervision. STS permits helpers to choose the most appropriate treatment for each patient, offering a valuable contribution in the process of promoting health and well-being considering the perspective of the psychology of sustainability and sustainable development. The following chapter, by Hisayoshi Okamura, Jumpei Yajima, and Kengo Mihara Akira Tsuda, is focused on psychobiological responses to work-related stress in Japanese female workers in the framework of the psychology of sustainability and sustainable development and the psychology of harmonization, underlining the relevance to guaranteeing healthy lives and to enhancing well-being in different environments. Studies on work-related stress of workers represent a fundamental area of research for healthy organizations (Di Fabio et al., 2020). The fifth chapter, by José-Antonio Gutiérrez-Carrasco, Gabriela Topa, and Juan-Carlos Pérez-González, presents the relationships among trait emotional intelligence and the six rings of positive self-capital (soft skills or noncognitive factors, career adaptability, covitality, psychological strengths, positive psychological capital, and high potential) for optimal performance and sustainability. The chapter also focuses on EI training programs for adults, considering emotional intelligence as a key factor to promote sustainability in organizations. The following chapter, by Alpana Vaidya, offers a contribution regarding sustainability in Indian organizations. After defining sustainability and delineating its three pillars (economic, environmental, and social), the chapter focuses on the challenges that Indian companies and Indian government are facing regarding sustainability and sustainable development. The seventh contribution, by Pan Lee Ching, Chua Bee Seok, and Rosnah Ismail, moves to the bright side, presenting a study that analyzes the psychometric proprieties of the Malaysian version of the Human Capital Sustainability Leadership Scale (HCSLS; Di Fabio & Peiró, 2018). The Malaysian version of the scale showed adequate psychometric properties offering future opportunities in cross-cultural perspectives. The last chapter of this first section, by Christopher Pitt, Kerrie E. Doyle, Elias Mpofu, Paul Saunders, Robert William Doyle, and Lauren Parkinson Zarb, presents a model of leadership for sustainable organizational health and well-being within indigenous community perspectives. This model is focused on sustainable, inclusive approaches which include diversities of race/ethnicity, culture of origin, and cultural heritage. The authors highlight that the application of indigenist-oriented models of leadership could contribute to improve organizational cultural inclusivity, and higher employee organizational citizenship, favoring sustainable organizational competitiveness.

The second part, "Widening the Perspectives for Innovation: Theoretical Approaches and Empirical Research," includes nine chapters. The ninth chapter of this book, by Annamaria Di Fabio and Maureen E. Kenny, discussing strength-based prevention perspectives and the psychology of sustainability and sustainable development, is focused on emotional intelligence, a promising amenable resource in relation to well-being. The results of the study presented in this chapter show the contribution of trait emotional intelligence as a mediator in the relationship between both positive and negative affect and meaningful work, underlining the value of emotional intelligence as a primary prevention resource in organizations for sustainable and positive human capital development.

The following chapter, by Samuel Fernández-Salinero and Gabriela Topa, offers new challenges of social identity in the organizational field for sustainable management in today's organizations. The authors emphasize the relevance of positive and sustainable management processes according to the research area of the psychology of sustainability and sustainable development. In particular, the chapter offers a definition of social identity and shows its direct and indirect associations with organizational citizenship behavior, perceived conflict, and interactional justice. Organizational identity is seen as a possibility to develop sustainable organizations. The 11th chapter, by Nami Fujoka, Chikako Hata, and Akira Tsuda, includes a study on the actual situation of primiparas' sleep acquired by the mobile sleep diary for one month after giving birth in Japan. The results offer promising perspectives for sustainable parenting, also reaching well-being, one of the Sustainable Development Goals. The psychology of harmonization (Di Fabio & Tsuda, 2018) calls for a prevention perspective, underlining the value of harmonically recomposing different levels of internal and external complexity included in parenting. The following chapter, by Kobus Maree, presents a case study of a postgraduate psychology student that shows how conscious knowledge could be connected with subconscious advice through career construction counseling to solve career choice indecision. The opportunities for constructing a sustainable career path are delineated. The thirteenth contribution, by Alessio Gori, Andrea Svicher, Letizia Palazzeschi, and Annamaria Di Fabio, presents a study that shows the contribution of insight in mediating the relationship between trait emotional intelligence in Italian workers, underlining that insight could represent a promising primary prevention resource in relation to acceptance of change in organizations in the framework of the psychology of sustainability and sustainable development.

The fourteenth contribution, by Márta Fülöp, presents research on the role of competition in Hungarian business life, conducted three decades after the regime change, with the aim to examine how leaders adapt to socio-economic-cultural change. The chapter also discusses the possibility to conciliate competitive attitudes with new leadership styles as human capital sustainability leadership (Di Fabio & Peiró, 2018).

The following chapter, by Anwarul Hasan Sufi, describes sustainability and changing care programs in the South Asian countries with a focus on happiness considering disadvantaged people such as older workers and people with disabilities. This chapter also connects sustainability and changing care programs in the South Asian countries with the United Nations' 17 Sustainable Development Goals.

The sixteenth chapter, by Alma Hafsi, Amira Laroui, and Takwa Chammem, is focused on the theme of conciliation, centrality, and the meaning of work among

Tunisian managers, showing the effects of such measures used by some public and private companies on the mental health of this group of workers. The issue of work-life balance in relation to well-being is particularly relevant in the framework of the psychology of sustainability and sustainable development. The last chapter, by Yomna Sameer and Nada Rostom, presents a study that analyzes work-life balance and its associations with positive outcomes from a macro perspective. This study, which involved 48 countries with 143 observations (combining developing and developed countries) in the period from 2016 to 2018, showed that work-life balance is a predictor of health, wealth, and innovation as mediated by happiness. The hope is that this book can help to enhance cross-cultural perspectives for well-being and sustainability in organizations, honoring the twenty-first century and the new awareness we have on the basis of the results of empirical research.

University of Florence
Florence, Italy

Annamaria Di Fabio

References

Di Fabio, A. (2017a). Positive healthy organizations: Promoting well-being, meaningfulness, and sustainability in organizations. *Frontiers in Psychology. Organizational Psychology, 8*, 1938. https://doi.org/10.3389/fpsyg.2017.01938.

Di Fabio, A. (2017b). The psychology of sustainability and sustainable development for well-being in organizations. *Frontiers in Psychology. Organizational Psychology, 8*, 1534. https://doi.org/10.3389/fpsyg.2017.01534.

Di Fabio, A., Cheung, F., & Peiró, J.-M. (2020). Editorial special issue personality and individual differences and healthy organizations. *Personality and Individual Differences, 166*. https://doi.org/10.1016/j.paid.2020.110196.

Di Fabio, A., & Kenny, M. E. (2015). The contributions of emotional intelligence and social support for adaptive career progress among Italian youth. *Journal of Career Development, 42*, 48–59. https://doi.org/10.1177/2F0894845314533420.

Di Fabio, A., & Kenny, E. M. (2016). From decent work to decent lives: Positive Self and Relational Management (PS&RM) in the twenty-first century. *Frontiers in Psychology. Section Organizational Psychology, 7*, 361. https://doi.org/10.3389/fpsyg.2016.00361.

Di Fabio, A., & Peiró, J. M. (2018). Human capital sustainability leadership to promote sustainable development and healthy organizations. *Sustainability MDPI, 10*(7), 2413. https://doi.org/10.3390/su10072413.

Di Fabio, A., & Rosen, M. A. (2018). Opening the black box of psychological processes in the science of sustainable development: A new frontier. *European Journal of Sustainable Development Research, 2*(4), 47. https://doi.org/10.20897/ejosdr/3933.

Di Fabio, A., & Tsuda, A. (2018). The psychology of harmony and harmonization: Advancing the perspectives for the psychology of sustainability and sustainable development. *Sustainability, 10*(12), 4726. https://doi.org/10.3390/su10124726.

Hage, S. M., Romano, J. L., Conyne, R. K., Kenny, M., Matthews, C., Schwartz, J. P., & Waldo, M. (2007). Best practice guidelines on prevention practice, research, training, and social advocacy for psychologists. *The Counseling Psychologist, 35*(4), 493–566. https://doi.org/10.1177/0011000006291411.

United Nations (2020). *Sustainable development: The 17 goals*. https://sdgs.un.org/goals. Accessed 10 Nov 2020.

Introduction. Well-Being and Sustainability in Organizations Through the Flourishing of Talents: The Contribution of the Psychology of Sustainability and Sustainable Development

Human Resources Management literature introduced the expression "war for talent" (Michaels et al., 2001) in terms of both strategic business challenge and critical driver of corporate performance. If talent management as strategic investment (Davies & Davies, 2010) asks for talent definition, talent identification, and talent culture, HRM literature has shown a lack of clarity in definitions, scope, and aim of talent (Lewis & Heckman, 2006). On this track, Nijs et al. (2014) underline the value of a conceptual framework of talent including definition, operationalization, and measurement of talent and its association with excellent performance. They also underlined the importance of talent management as strategic investments in terms of talent-identification, selection, development, planning, and retention (Nijs et al., 2014). At the core of their perspective, the concept of talent for excellence (Nijs et al., 2014) emerged, including interpersonal excellence as performing better than others, and intrapersonal excellence as performing at one's personal best. Advancing in this perspective, Di Fabio (2016) deepens the concept of organizational excellence and the following related question: Which operationalization of organizational excellence is the most appropriate and will offer greater benefit for the organization? To answer this question, Di Fabio (2016) articulates the reflection starting from the entitarian or incremental nature of talent (from Dweck & Leggett, 1988 to Di Fabio, 2016), continuing with talent as predominantly an innate construct or mostly acquired (Meyers et al., 2013), discussing talents as "object" and talents as "subject" (Gallardo-Gallardo et al., 2013), talents as strengths (Clifton & Harter, 2003), and talents as capital (Farndale et al., 2010), introducing talent as constructive nonconformity (Gino, 2016), until reaching an integrated/advanced vision of talent/s linked to the critical issue of sustainability of talent/s.

Proceeding with order, theory of intelligence distinguished between entity theory and incremental theory (Dweck & Leggett, 1988). Entity theory considers abilities as fixed traits, centered on performance, including positive judgment/avoid negative judgments of competence. When individuals perceive high present ability, the behavior pattern is mastery oriented, focusing on seek challenge and high persistence; when individuals perceive low present ability, the behavior pattern is helpless, focusing on avoiding risk and low persistence. On the contrary, incremental theory sees abilities as malleable qualities, centered on learning/development with the aim to increase competence.

Meyers et al. (2013) wonder if talent is principally an innate construct or if it is mainly acquired. They argue that talent can be located on a continuum from innate talent through nature-nurture interaction to acquired talent (Meyers et al., 2013). From their perspective emerged the value of nurturing talent in organizations (Di Fabio, 2016, 2018), distinguishing two kinds of organizational factors: micro-organizational factors and macro-organizational (McDonnell et al., 2017). Micro-organizational factors refer to significant others, mentors, peers, friends, and other people nearest to gifted individuals at the beginning of their careers (Feldman, 1986; Gardner, 1983; Howe, 1993); macro-organizational factors include cultural, political, and historical conditions of the organizations that affect exceptional achievements (e.g., society/organization as a whole, culture, and the specific historical period) (Csikszentmihalyi & Robinson, 1986; Simonton, 1988; Sternberg & Lubart, 1995).

Traditional approaches to talent considered talents as "object" (talent conceived in terms of natural ability; mastery; commitment; fit) and talents as "subject" (talent as all people; talent as some people in terms of both high performers and as high potentials) (Gallardo-Gallardo et al., 2013); talents as strengths turned out to be centered on virtues instead of deficits, involving identification of talent by the individual, learning about one's own talents, and using talents to change behavior) (Clifton & Harter, 2003); talents as capital means to recognize the global competition context, and a vision of talent as human, social, intellectual, and political capital (Farndale et al., 2010). Di Fabio (2016) adds an important contribution linking talent/s to the value of constructive nonconformity and the six strategies to promote it (Gino, 2016): giving employees opportunities to be themselves; encouraging employees to bring out their signature strengths; questioning the status quo and encouraging employees to do the same; creating challenging experiences; fostering broader perspectives; and giving voice and encouraging dissenting views to reach better and deeper awareness.

Moving on to the perspective of Gallardo-Gallardo et al. (2013) and Nijs, Gallardo-Gallardo, Dries, and Sels (2014), Di Fabio (2016, 2018) proposes a revised/advanced/integrated perspective including not only objective talent/s and subjective talent/s but also sustainable talent/s, valorizing aspects of meaning, authenticity, and purposefulness following the framework of psychology of sustainability and sustainable development (Di Fabio, 2017a, b; Di Fabio & Rosen, 2018). This current perspective includes the different psychological frameworks previously recognized regarding talent/s (Dries, 2013), integrates them, and makes them evolve by adding the theme of sustainability of talent/s: educational psychology, work and organizational psychology, social psychology, positive psychology, and vocational and counseling psychology, and finally including the current psychology of sustainability and sustainable development (Di Fabio, 2017a, b; Di Fabio & Rosen, 2018).

According to educational psychology, talent is seen as giftedness (Gagné, 2000, 2004, 2009, 2010; Meyers, et al., 2013). This perspective considers excellent performer in particular domains (e.g., music, sports, arts, mathematics) and the main reference is the Differentiated Model of Giftedness and Talent (DMGT; Gagné, 2000, 2004, 2009, 2010). Giftedness is defined as having and using "outstanding natural abilities, called aptitudes, in at least one ability domain, to a degree that places an individual at least among the top 10% of age peers" (Gagné, 2000, p. 67). Talent is "the outstanding mastery of systematically developed abilities, called competencies (knowledge and skills), in at least one field of human activity to a degree that places an individual at least among the top 10% of age peers who are or have been active in that field" (Gagné, 2000, p. 67). The talent development process is relative to the progressive transformation of gifts into talents. Talented individuals have special gifts that permit them to show excellent skills in a specific field. Giftedness is considered rare and involves only a few individuals. Research on giftedness is predominantly carried out with children or adolescents aiming at explaining why giftedness emerges early on and how the learning of gifted individuals can be enhanced to develop talents.

Work and organizational psychology considers talent both as individual differences (Hough & Oswald, 2000) and as competence (Boyatzis, 2008). Talent as individual differences includes cognitive abilities, knowledge, and personality (Hough & Oswald, 2000). Talent as a competence for an effective performance (Boyatzis, 2008) focuses on behavioral manifestations of talent that allow superior performance of an individual compared to that of colleagues. Regarding talent as individual differences (Hough & Oswald, 2000), the principal aim was to develop methodologies able to appropriately predict the future job performance of workers on the basis of the psychometric paradigm. The guidelines of this psychological trend are personnel selection, promotion systems, and performance evaluation. Regarding talent as competence (Boyatzis, 2008), maximum performance appears when the individual's talent corresponds to the necessity of the job demands and the organizational environment. This approach (Boyatzis, 2008) is a behavioral approach to talent, underlining that human talent can be developed during adulthood.

Social psychology stresses talent in terms of perception of talent (Dominick & Gabriel, 2009), adequately considering what is considered talent rather than centering the talent itself. This perspective facilitates individuals in the capitalization of their resources to obtain success, also emphasizing how external perceptions, biases, and social expectations outline and determine the concept of talent: the talent exists only if it is recognized by others.

Positive psychology introduced the concept of talent as strengths, in terms of creative strengths, politeness, wisdom, thankfulness, and justice, together with positive feelings able to invigorate, energize, intrinsically motivate, sustain authenticity, and provide self-fulfillment (Peterson & Seligman, 2004). These strengths are seen as individuals' qualities that permit them to have good performance or to perform at their personal best. In the framework of positive psychology (Seligman &

Csikszentmihalyi, 2000), strengths have been considered "potentials for excellence" (Biswas-Diener et al., 2011). They are usually defined as trait-like constructs that are partially innate but can also be somewhat enhanced. Each person has determinate strengths and use them complemented by positive feelings for reaching self-actualization in terms of the realization of one's own highest potential (Peterson & Seligman, 2004).

Vocational psychology focuses on talent in terms of identity and self-concept crystallization, in particular as lasting goals, interest, and talents. It also underlines the vocational self-concept as dynamic and evolving. Regarding the self-concept, possible selves and actual selves, ideal selves, and ought-to selves are considered (Whitty, 2002). Actual self refers to how you (or another) actually believe you are; ideal self refers to how you or another would like to see yourself, including hopes and wishes for you; ought-to self refers to features you (or another) believe you should possess (Whitty, 2002). Di Fabio (2016, 2018) has enriched this perspective on the basis of counseling psychology too, likewise based on the construct of talent as identity (Di Fabio, 2014c; Di Fabio & Bernaud, 2014; Whitty, 2002) but deepening the reflection by also introducing the contributions of Guichard (2004, 2009). Regarding the self-concept, also the concepts of the Subjective Identity Form (SIF, Guichard, 2010), Core SIF, Aspired SIF, and the Subjective Identity Form System (SIFS, Guichard, 2010) are highly valuable. The Subjective Identity Form (SIF, Guichard, 2010) involves a specific social role and includes the way in which an individual perceives themself in such a role and how an individual describes themself in that role. The Subjective Identity Form is the "ways of being, acting, and interacting in relation to a certain view of oneself in a given context" (Guichard, 2013). The Core SIF encompasses what "the person wants to achieve a need for self-actualization. It corresponds to a life domain in which the person wants to attain a certain state of excellence as it gives his/her life meaning" (Guichard, 2013). The Aspired SIF refers to what "individuals want to try to achieve, as full of meaning for them in the construction of their possible future Selves" (Guichard, 2013). The Subjective Identity Form System (SIFS, Guichard, 2010) pays attention to individual identity as multiple and constantly changing. The talent/s as identity is based on plural identity (Guichard, 2008, 2010, 2013) and the Subjective Identity Form System (SIFS): from one's own Subjective Identity Form (SIF), to Aspired SIF that the individual wants to engage to realize the possible future Self to reach identitarian purposeful awareness (Di Fabio, 2014c) through the two critical key meta-competencies (Di Fabio, 2014a): adaptability (Savickas, 2001; Savickas & Porfeli, 2012) and identity (Guichard, 2004, 2010).

The psychology of sustainability and sustainable development (Di Fabio, 2017a, b; Di Fabio & Rosen, 2018) enlarges the reflection by introducing the concept of sustainable talent/s. In this perspective, talent is closely connected to authentic self (Di Fabio, 2014c), and to more in-depth and significant interest in and value for individuals. Self-attunement represents the key for developing and flourishing talents, based on objective talent/s and potential talent/s differently understood from the traditional perspective in terms of self-evaluation of "What I am able to?" and subjective talent/s and potential talent/s differently understood from the traditional perspective in terms of self-evaluation of "What energizes me? What motivates me?" considering the process of "go between the concepts" (Guichard, 2013) for greatest performance through meaningful goals (Di Fabio, 2014c). Purposeful identitarian awareness (Di Fabio, 2014c) is a fundamental concept in this perspective referring to the "contact with the most authentic aspects of one's one Self and one's professional life and lives' deepest purposes as an aid in the construction of work and life meaningfulness" (Di Fabio, 2014c, p. 158). The concept of sustainable talent/s underlined as developing talent/s is anchored in the actual zone of proximal development linked to the individual's meaning (Di Fabio, 2017b). The psychology of sustainability and sustainable development (Di Fabio, 2017a, b; Di Fabio & Rosen, 2018) offers a contribution to the sustainability and sustainable development of environment/s, individual/s, and group/s, favoring the flourishing of both interpersonal and intrapersonal talents and focusing on establishing a current open-mind culture of talent/s. A talent is a true talent if individuals perceive the uniqueness, authenticity, and purposefulness of this talent (Di Fabio, 2017b). A sustainable talent also calls for new positive and regenerating resources for nurturing talent (Di Fabio, 2017b). Intrapreneurial Self-Capital (ISC, Di Fabio, 2014b) can represent one of these resources, and can be increased through specific training (Di Fabio & Van Esbroeck, 2016), allowing one to "deal with the frequent changes and transitions by creating innovative solutions when confronted with constraints of the environment to turn constraints into resources" (Di Fabio, 2014b, p. 100). Sustainability of talent/s (Di Fabio, 2017b; Di Fabio & Maree, 2016) calls for a sustainable personal/life project. The value of meaning as a fundamental aspect of a sustainable construction of authentic aspects of self for individuals is stressed, underlining a change from motivational paradigm to meaning paradigm (Di Fabio & Blustein, 2016), also including purpose and connection. The Psychology of Working Framework (PWF; Blustein, 2006) and in the Psychology of Working Theory (PWT; Duffy et al., 2016) present the value of work meaning for the self-determination and well-being of individuals (Blustein, 2006, 2008; Duffy et al., 2016). Work meaning (Steger et al., 2012) refers to work as a subjectively meaningful experience characterized by positive experiences of meaning at work. Work meaning represents a critical indicator of eudaimonic well-being at work.

A sustainable personal/life project brings with it sustainable development respecting the talent/s of people for decent work and decent lives (Di Fabio & Blustein, 2016). Furthermore, recent research results (Villajos et al., 2019) permitted the further advancement of the HRM literature by enhancing the comprehension of links among HR practices, well-being, and performance. In detail, sustainable HRM asks for paying attention to both performance and well-being aspects strictly linked to sustainability, recognizing the mediator role of eudaimonic well-being between HR practices and creative performance (Villajos et al., 2019).

The perspective of the psychology of sustainability and sustainable development (Di Fabio, 2017a, b; Di Fabio & Rosen, 2018) pays attention to promoting and valorizing talent/s with a focus on flourishing of talent/s, internal and external harmonization (Di Fabio & Tsuda, 2018), and eudaimonic well-being, introducing innovative possibilities for HR management and development of talent/s in organizations. In the framework of healthy organizations (Di Fabio, 2017a; Di Fabio et al., 2020) a strength-based prevention perspective (Di Fabio & Saklofske, 2021) for talent/s could be carefully introduced, focusing on promotion of talent/s of workers as thriving individuals, leading to a primary prevention perspective (Di Fabio & Kenny, 2015, 2016; Hage et al., 2007) for talent/s. It calls for an organizational culture of talent/s, permeated with awareness of current empirical research findings and centered on promoting generative management and human capital sustainability leadership (Di Fabio & Peiró, 2018). Also the well-being movement (Johnson, Robertson, & Cooper, 2018; Robertson & Cooper, 2010) stresses the vision of supporting both performance and productivity for positive outcomes and health at work. The time has come for creating a virtuous circle of respecting, protecting, and enhancing talent/s for healthy business, healthy workers, and healthy organizations (Di Fabio et al., 2020).

University of Florence
Florence, Italy

Annamaria Di Fabio

References

Becker, B. E., & Huselid, M. A. (2006). Strategic human resources management: where do we go from here? *Journal of Management*, *32*(6), 898–925. https://doi.org/10.1177/0149206306293668.

Biswas-Diener, R., Kashdan, T. B., & Minhas, G. (2011). A dynamic approach to psychological strength development and intervention. *The Journal of Positive Psychology*, *6*(2), 106–118. https://doi.org/10.1080/17439760.2010.545429.

Blustein, D. L. (2006). The psychology of working: A new perspective for career development, counseling, and public policy. Erlbaum.

Blustein, D. L. (2008). The role of work in psychological health and well-being: A conceptual, historical, and public policy perspective. *American Psychologist*, *63*(4), 228–237. https://doi.org/10.1037/0003-066X.63.4.228.

Boyatzis, R. E. (2008). Competencies in the 21st century. *Journal of Management Development*, *27*(1), 5–12. https://doi.org/10.1108/02621710810840730.

Clifton, D. O., & Harter, J. K. (2003). Investing in strengths. *Positive organizational scholarship: Foundations of a new discipline*, 111–121.

Csikszentmihalyi, M., & Robinson, R. E. (1986). Culture, time, and the development of talent. In R. J. Sternberg & J. E. Davidson (Eds.), *Conceptions of giftedness* (pp. 264–284). Cambridge University Press.

Davies, B., & Davies, B. J. (2010). Talent management in academies. *International Journal of Educational Management*. *24*(5), 418–426. https://doi.org/10.1108/09513541011055983.

Di Fabio, A. (2014a). Constructing my future purposeful life: A new life construction intervention. In A. Di Fabio & J.-L. Bernaud (Eds.), *The construction of the identity in 21st century: A Festschrift for Jean Guichard* (pp. 219–239). New York: Nova Science Publishers.

Di Fabio, A. (2014b). Intrapreneurial self-capital: A new construct for the 21st century. *Journal of Employment Counseling*, *51*, 98–111. https://doi.org/10.1002/j.2161-1920.2014.00045.x.

Di Fabio, A. (2014c). The new purposeful identitarian awareness for the twenty-first century: Valorize themselves in the Life Construction from youth to adulthood and late adulthood. In A. Di Fabio & J.-L. Bernaud (Eds.), *The Construction of the Identity in 21st Century: A Festschrift for Jean Guichard* (pp. 157–168). Nova Science Publishers.

Di Fabio, A. (2016, October). How can career guidance and counseling support the development of talents? The challenge of project sustainability in the construction and managing of the personal project? Invited keynote speaker at the seminar "Supporting and Developing Talents" organized by the Euroguidance Centre Slovakia at the Slovak Academic Association for International Cooperation, Bratislava, Slovakia, October 27, 2016.

Di Fabio, A. (2017a). Positive Healthy Organizations: Promoting well-being, meaningfulness, and sustainability in organizations. *Frontiers in Psychology. Organizational Psychology, 8*, 1938. https://doi.org/10.3389/fpsyg.2017.01938.

Di Fabio, A. (2017b). The psychology of sustainability and sustainable development for well-being in organizations. *Frontiers in Psychology. Organizational Psychology, 8*, 1534. https://doi.org/10.3389/fpsyg.2017.01534.

Di Fabio, A. (2018, August). Closing Session – Sustaining intrapersonal and interpersonal talents for healthy organizations: The Psychology of Sustainability and Sustainable Development. Closing session at the First Summer School "Healthy organizations: From research to application" organized by the Department of Education and Psychology, University of Florence, Florence, Italy, August 27–30, 2018.

Di Fabio, A., & Bernaud, J.-L. (Eds.). (2014). The construction of the identity in 21st century: A Festschrift for Jean Guichard. New York: Nova Science Publishers.

Di Fabio, A., Cheung, F., & Peiró, J.-M. (2020). Editorial Special Issue Personality and individual differences and healthy organizations. *Personality and Individual Differences, 166*. https://doi.org/10.1016/j.paid.2020.110196.

Di Fabio, A., & Kenny, M. E. (2015). The contributions of emotional intelligence and social support for adaptive career progress among Italian youth. *Journal of Career Development, 42*, 48–59. https://doi.org/10.1177/0894845314533420.

Di Fabio, A., & Kenny, E. M. (2016). From decent work to decent lives: Positive Self and Relational Management (PS&RM) in the twenty-first century. *Frontiers in Psychology. Section Organizational Psychology, 7*, 361. https://doi.org/10.3389/fpsyg.2016.00361.

Di Fabio, A., & Maree, J. G. (2016). A psychological perspective on the future of work: promoting sustainable projects and meaning-making through grounded reflexivity. *Counseling. Giornale Italiano di Ricerca e Applicazioni, 9*(3). https://doi.org/10.14605/CS931619.

Di Fabio, A., & Rosen, M. A. (2018). Opening the Black Box of Psychological Processes in the Science of Sustainable Development: A New Frontier. *European Journal of Sustainable Development Research, 2*(4), 47. https://doi.org/10.20897/ejosdr/3933.

Di Fabio, A., & Saklofske, D. H. (2021). The relationship of compassion and self-compassion with personality and emotional intelligence. *PAID 40th Anniversary Special Issue. Personality and Individual Differences, 157*. https://doi.org/10.1016/j.paid.2020.110109.

Di Fabio, A. & Tsuda, A. (2018). The psychology of harmony and harmonization: Advancing the perspectives for the psychology of sustainability and sustainable development. *Sustainability, 10*(12), 4726. https://doi.org/10.3390/su10124726.

Di Fabio, A., & Van Esbroeck, R. (2016). Intrapreneurial Self-Capital: A concept fitting a life-designing intervention. *Counseling. Giornale Italiano di Ricerca e Applicazioni, 9*(2). https://doi.org/10.14605/CS921611.

Dominick, P. G., & Gabriel, A. S. (2009). Two sides to the story: An interactionist perspective on identifying potential. *Industrial and Organizational Psychology, 2*(4), 430–433.

Dries, N. (2013). The psychology of talent management: A review and research agenda. *Human Resource Management Review, 23*(4), 272–285. https://doi.org/10.1016/j.hrmr.2013.05.001.

Duffy, R. D., Blustein, D. L., Diemer, M. A., & Autin, K. L. (2016). The psychology of working theory. *Journal of Counseling Psychology, 63*(2), 127–148. https://doi.org/10.1037/cou0000140.

Dweck, C. S., & Leggett, E. L. (1988). A social-cognitive approach to motivation and personality. *Psychological Review, 95*(2), 256–273. https://doi.org/10.1037/0033-295X.95.2.256.

Feldman, D. H. (1986). Giftedness as a developmentalist sees it. In R. J. Sternberg & J. E. Davidson (Eds.), *Conceptions of giftedness* (pp. 285–305). Cambridge University Press.

Farndale, E., Scullion, H., & Sparrow, P. (2010). The role of the corporate HR function in global talent management. *Journal of World Business, 45*(2), 161–168. https://doi.org/10.1016/j.jwb.2009.09.012.

Gagné, F. (2000). Understanding the complex choreography of talent development through DMGT-based analysis. *International handbook of giftedness and talent*, *2*, 67–79.

Gagné, F. (2004). Transforming gifts into talents: The DMGT as a developmental theory. *High Ability Studies*, *15*(2), 119–147. https://doi.org/10.1080/1359813042000314682.

Gagné, M. (2009). A model of knowledge-sharing motivation. Human Resource Management: Published in Cooperation with the School of Business Administration, The University of Michigan and in alliance with the Society of Human Resources Management, 48(4), 571–589.

Gagné, F. (2010). Motivation within the DMGT 2.0 framework. *High Ability Studies*, *21*(2), 81–99. https://doi.org/10.1080/13598139.2010.525341.

Gallardo-Gallardo, E., Dries, N., & González-Cruz, T. F. (2013). What is the meaning of 'talent' in the world of work? *Human Resource Management Review*, *23*(4), 290–300. https://doi.org/10.1016/j.jwb.2013.11.002.

Gardner, H. (1983). Frames of mind: The theory of multiple intelligences. New York: Basic Books.

Gino, F. (2016). Rebel talent. *Harvard Business Review*. Retrieved from https://www.hbs.edu/ris/Publication%20Files/Let%20your%20workers%20rebel_b87d0da9-de68-45be-a026-22dee862e6e4.pdf.

Guichard, J. (2004). Se faire soi. *L'Orientation Scolaire et Professionnelle*, *33*(4), 499–533. https://doi.org/10.4000/osp.226.

Guichard, J. (2008). Proposition d'un schéma d'entretien constructiviste de conseil en orientation pour des adolescents ou de jeunes adultes. *L'Orientation Scolaire et Professionnelle*, *37*, 413–440. https://doi.org/10.4000/osp.9715.

Guichard, J. (2009). Self-constructing. *Journal of Vocational Behavior*, *75*(3), 251–258. https://doi.org/10.1016/j.jvb.2009.03.004.

Guichard, J. (2010, March). Les théories de la construction des parcours professionnels et de la construction de soi: Deux approches de la construction de la vie individuelle. Paper presented at Colloque International INETOP, Paris, France.

Guichard, J. (2013, June). Forms of reflexivity and transformations of systems of subjective
identity forms during life designing dialogues. Lecture presented at the SIO International Conference —Life Designing and career counseling: Building hope and resilience, Padova, Italy.

Hage, S. M., Romano, J. L., Conyne, R. K., Kenny, M., Matthews, C., Schwartz, J. P., & Waldo, M. (2007). Best practice guidelines on prevention practice, research, training, and social advocacy for psychologists. *The Counseling Psychologist*, *35*(4), 493–566. https://doi.org/10.1177/0011000006291411.

Hough, L. M., & Oswald, F. L. (2000). Personnel selection: Looking toward the future--Remembering the past. *Annual Review of Psychology*, *51*(1), 631–664. https://doi.org/10.1146/annurev.psych.51.1.631.

Howe, M. J. A. (1993). The early lives of child prodigies. In G. R. Bock & K. Ackrill (Eds.), *The origins and development of high ability* (pp. 85–105). John Wiley & Sons.

Huselid, M. A., & Becker, B. E. (2011). Bridging micro and macro domains: Workforce differentiation and strategic human resource management. *Journal of Management*, *37*(2), 421–428. https://doi.org/10.1177/0149206310373400.

Johnson, S., Robertson, I., & Cooper, C.L. (2018). *Wellbeing: Productivity and Happiness at*
Work. (2nd ed.). London: Palgrave Macmillan.

Lepak, D. P., & Snell, S. A. (1999). The human resource architecture: Toward a theory of human capital allocation and development. *Academy of Management Review*, *24*(1), 31–48. https://doi.org/10.5465/amr.1999.1580439.

Lewis, R. E., & Heckman, R. J. (2006). Talent management: A critical review. *Human Resource Management Review*, *16*(2), 139–154. https://doi.org/10.1016/j.hrmr.2006.03.001.

McDonnell, A., Collings, D. G., Mellahi, K., & Schuler, R. (2017). Talent management: a systematic review and future prospects. *European Journal of International Management*, *11*(1), 86–128. https://doi.org/10.1504/EJIM.2017.081253.

Meyers, M. C., Van Woerkom, M., & Dries, N. (2013). Talent—Innate or acquired? Theoretical considerations and their implications for talent management. *Human Resource Management Review*, *23*(4), 305–321. https://doi.org/10.1016/j.hrmr.2013.05.003.

Michaels, E., Handfield-Jones, H., & Axelrod, B. (2001). *The war for talent*. Harvard Business Press.

Nijs, S., Gallardo-Gallardo, E., Dries, N., & Sels, L. (2014). A multidisciplinary review into the definition, operationalization, and measurement of talent. *Journal of World Business*, *49*(2), 180–191. https://doi.org/10.1016/j.jwb.2013.11.002.

Peterson, C., & Seligman, M. E. (2004). *Character strengths and virtues: A handbook and classification* (Vol. 1). Oxford University Press.

Robertson, I., & Cooper, C.L. (2010). *Wellbeing: Productivity and happiness at work*. Palgrave Macmillan.

Savickas, M. L. (2001). Toward a comprehensive theory of career development: Disposition, concerns and narratives. In F. T. L. Leong & A. Barak (Eds.), *Contemporary models in vocational psychology: A volume in honor of Samuel H. Osipow* (pp. 295–320). Lawrence Erlbaum.

Savickas, M. L., & Porfeli, E. J. (2012). Career Adapt-Abilities Scale: Construction, reliability, and measurement equivalence across 13 countries. *Journal of Vocational Behavior*, *80*(3), 661–673. https://doi.org/10.1016/j.jvb.2012.01.011.

Seligman, M.E.P., & Csikszentmihalyi, M. (2000). Positive psychology: An introduction. *American Psychologist*, *55*(1), 5–14. https://doi.org/10.1037/0003-066X.55.1.5.

Silzer, R., & Dowell, B. E. (2010). Strategic talent management matters. In R. Silzer & B. E. Dowell (Eds.), *Strategy-driven talent management: A leadership imperative* (pp. 3–72). O'Reilly online learning.

Simonton, D. K. (1988). Presidential style: Personality, biography, and performance. *Journal of Personality and Social Psychology*, *55*, 928–936.

Steger, M. F., Dik, B. J., & Duffy, R. D. (2012). Measuring meaningful work: The work and meaning inventory (WAMI). *Journal of Career Assessment*, *20*(3), 322–337. https://doi.org/0.1177/1069072711436160.

Sternberg, R. J., & Lubart, T. I. (1995). Defying the crowd: Cultivating creativity in a culture of conformity. Free Press.

Ulrich, D., & Smallwood, N. (2012). What is talent? Leader to leader, 2012(63), 55–61. https://doi.org/10.1002/ltl.20011.

Villajos, E., Tordera, N., & Peiro, J. M. (2019). Human resource practices, eudaimonic well-being, and creative performance: the mediating role of idiosyncratic deals for sustainable human resource management. *Sustainability*, *11*(24), 6933. https://doi.org/10.3390/su11246933.

Whitty, M. (2002). Possible selves: An exploration of the utility of a narrative approach. *Identity: An International Journal of Theory and Research*, *2*(3), 211–228. https://doi.org/10.1207/S1532706XID0203_02.

Contents

Part I
Studying in Depth the Dark and the Bright Sides: Cross-Cultural Contributions

Chapter 1
The Psychology of Sustainability in Organizations: A New Scenario for Healthy Organizations, Healthy Business, Harmonization, and Decent Work

Annamaria Di Fabio

1.1 The Psychology of Sustainability and Sustainable Development

The psychology of sustainability and sustainable development (Di Fabio, 2017a, b; Di Fabio & Rosen, 2018) represents a relatively new research area within the domain of sustainability science (Rosen, 2017). Sustainability science offers a transdisciplinary viewpoint with contributions from several different disciplines, including natural and applied sciences as well as social sciences and humanities (Dincer & Rosen, 2013). The focus is on an in-depth study of the various and complex factors at play in the degradation of many systems (human, environmental, and engineered) in order to protect the earth from damage and also to promote human health and well-being. Sustainability science addresses sustainable development according to the Sustainable Development Goals (United Nations, 2020), integrating perspectives and contributions offered by various disciplines.

In this transdisciplinary framework, the psychology of sustainability and sustainable development was born, contributing a psychological lens to the field of sustainability science. Linking the black box of psychological processes (Di Fabio & Rosen, 2018) to sustainability is an exciting challenge for facilitating sustainable development in different environments. The authors recognize that psychological processes are embedded in decisions and behaviors concerning the natural environment and that these processes are critical for the success of positive outcomes in protecting the natural environment. Psychological processes are also essential in relation to other kinds of environments to enhance well-being; this perspective considers not only the natural environment but also different kind of environments—from natural

A. Di Fabio (✉)
University of Florence, Florence, Italy
e-mail: annamaria.difabio@unifi.it

A. Di Fabio (ed.), *Cross-cultural Perspectives on Well-Being and Sustainability in Organizations*, https://doi.org/10.1007/978-3-030-86709-6_1

to personal, social, organizational, community, digital, cross-cultural, and ultimately global environments (Di Fabio & Rosen, 2018).

In summary, the psychology of sustainability and sustainable development offers a look at problems from a psychological perspective, advancing the traditional sustainability perspective according to the "three Es" of environment, economy, and equity (Brundtland Report, 1987; Harris, 2003). This innovative framework invigorates and advances the more traditional ecological, economic, and social approaches (Di Fabio & Rosen, 2018). Furthermore, the contribution offered by the psychological lens changes the connotation of the narrative on sustainability and sustainable development—from a more traditional narrative based on damage and threat and the avoidance of exploitation and depletion to a narrative that emphasizes opportunity and challenge as well as growth, flexibility, and regeneration of resources. The viewpoint shifts from difficulties, scarcity, and danger toward constructive challenges and meaningful opportunities.

The focus on the psychological dynamics involved in behaviors and outcomes opens up promising possibilities for tangible advancements in sustainability and sustainable development. The framework of the psychology of sustainability and sustainable development also highlights the value of acting within a primary prevention perspective (Di Fabio & Kenny, 2015, 2016; Hage et al., 2007), as well as within a strength-based prevention perspective (Di Fabio & Saklofske, 2021). The challenge is not only to remedy the urgent damage but also to foster a culture of sustainability and sustainable development by instilling interest, motivation, and self-efficacy in people (Di Fabio & Rosen, 2020). This will help to generate positive and concrete outcomes in the short, medium, and long term.

1.2 Psychology of Sustainability and Sustainable Development for Healthy Organizations

Healthy organizations are defined by their ability to promote worker health and safety as well as to enhance their organizational effectiveness (Di Fabio, 2017a; Di Fabio et al., 2020; Di Fabio & Peiró, 2018; Peiró & Rodríguez, 2008; Tetrick & Peiró, 2012). The virtuous circuit of healthy organizations is founded on worker well-being and organizational productivity (De Smet et al., 2007; Di Fabio, 2017a; Grawitch & Ballard, 2016). Healthy organizations emphasize the relevance of worker health and performance for organizational success, effectiveness, and safety; the well-being of workers and the well-being of the organization are thoroughly associated (Di Fabio, 2017a; Peiró, 2008; Peiró & Tetrick, 2011).

The critical shift from an ill-health perspective to a positive-health perspective is underlined by Tetrick and Peiró's (2012) research on well-being in organizations, focusing on the enhancement of health, well-being, and flourishing (Schaufeli, 2004). This perspective promotes healthy workplaces and contexts for healthy workers, enhancing talents for improved performance, satisfaction, and well-being

(Tetrick & Peiró, 2012). In this way, healthy organizations introduce a positive perspective that focuses on achievement, excellence, and strengths, equilibrating and centering on positive perspectives at different levels: individual, group, organization, and inter-organization (Di Fabio, 2017a; Henry, 2005; Tetrick & Peiró, 2012).

The positive psychology perspective calls for strengthening individuals and organizations (Di Fabio, 2017a; Di Fabio & Kenny, 2016) to face the complexity of the current scenario (Blustein et al., 2019) and to promote the well-being of workers, groups, and organizations. Considering a prevention framework (Di Fabio & Kenny, 2015, 2016; Peiró, 2008; Tetrick & Peiró, 2012), the positive psychology perspective seeks cross-level interactions between individual and collective perspectives and also between short- and long-term objectives, introducing an anticipatory and proactive approach to improve well-being in organizations. There is a shift from a more traditional perspective, focused on mitigating risks to worker safety and health, toward a newer perspective, focused on enhancing and developing positive experiences and narratives (Di Fabio, 2017a). The current emphasis on promoting safer and healthier working environments (Tetrick & Peiró, 2012) within a preventative framework (Di Fabio & Kenny, 2015, 2016; Di Fabio & Saklofske, 2021) focuses on building individuals' psychological resources and strengthening individuals, groups, and organizations. This perspective is at the basis of occupational health psychology (Cox et al., 2000; Raymond et al., 1990; Tetrick & Peiró, 2012), which stresses the value of prevention, underlining the value of positive work experiences and the importance of building strengths (Di Fabio, 2017a).

There is a need in the current changeable and competitive marketplace for further research into organizational practices that lead to positive work experiences. Following the perspectives offered by occupational health psychology—together with organizational health psychology (Antoniou & Cooper, 2005), the positive psychology perspective, and the importance of focusing on organizational practices—it is possible to introduce the positive organizational health psychology (Di Fabio, 2017a; Di Fabio & Peiró, 2017, 2018). This perspective considers a positive organizational approach—not a focus on fixing deficiency and failure, but one that promotes well-being at many levels: individual, group, organization, and inter-organization (Di Fabio, 2017a). This perspective is aligned with the well-being movement (Robertson & Cooper, 2010; Johnson et al., 2018), aimed at enhancing well-being at work by helping leaders understand the importance of facilitating environments where workers can experience trust, valorization, autonomy, and work–life balance (Hart & Cooper, 2001). The importance of well-being in organizations is valorized by a sustainable framework. The psychology of sustainability and sustainable development makes a promising contribution to organizational well-being and also supports the framework of positive organizational health psychology. It means offering opportunities for research and intervention from a prevention perspective—both a primary prevention perspective and strength-based prevention perspectives. The focus is on essential psychological factors that promote sustainability and sustainable development through organizational well-being at all the various organizational levels: workers, groups, organizations, and inter-organizational processes.

The psychology of sustainability and sustainable development, in relation to healthy organizations, considers both hedonic and eudaimonic well-being, stressing the particular value of the latter. Since meaning is essential for sustainability (Di Fabio, 2017a, b; Di Fabio & Blustein, 2016), it is critical that workers feel not only hedonic well-being but above all eudaimonic well-being, identifying personal meanings and authentic aspects of the self in order to reach full self-realization at work. Meaning is the intrinsic motivational energy rooted in personal values that nurtures real sustainability for workers (Di Fabio, 2017a). In a prevention framework, the psychology of sustainability and sustainable development thus emphasizes the importance of early interventions to enhance both the hedonic and eudaimonic well-being of workers at many levels—individual, group, organization, and inter-organization—to foster healthy organizations.

In the psychology of sustainability and sustainable development perspective for healthy organizations, regarding meaning at work, it is also possible to underline the importance of relationships (Blustein, 2011). Relationships arise and are incrementally constructed through meanings that shape the organizational life. Positive work relationships are essential to promote organizational well-being (Di Fabio, 2017a; Di Fabio et al., 2020). In the framework of healthy organizations (Di Fabio, 2017a; Di Fabio et al. 2020; Tetrick & Peiró, 2012), relationships deserve particular attention when considering leadership. The psychology of sustainability and sustainable development offers insights into relationships and leadership, addressing the importance of management and leadership styles that respect the fundamental influence of relationships on worker well-being. Today's organizations need an aware managerial sensibility that ensures sustainable organizational development; therefore, a new construct of human capital sustainability leadership was introduced (Di Fabio & Peiró, 2018). This construct is centered on "healthy people as flourishing and resilient workers, on healthy organizations as thriving and successful environments characterized by the positive circle of long-term well-being and performance" (Di Fabio & Peiró, 2018, p. 3). One factor that emerged as a critical ingredient was a managerial style that was aware of meaningful details, conscious of the importance of relationships, and engaged in the construction of positive narratives. This kind of leadership will help an organization mobilize energies and meet challenges while growing in light of sustainable development. Managers and leaders of today's organizations must be able to use a resilient and nurturing perspective regarding workers and organizational environments, sustaining organizational well-being by constructing and maintaining meanings and positive narratives about work and relationships (Di Fabio, 2017b).

In the framework of healthy organizations rooted in positive organizational health psychology, performance is a basic condition for healthy business and healthy workers (Di Fabio, 2017a). The psychology of sustainability and sustainable development gives important suggestions. Recent developments underline the importance of work characteristics that affect job performance, and also the mediating role of individual dispositions on the work design and performance (Peiró et al., 2020). In this study, the importance of work characteristics for job performance emerged as well as the relevance of individual values in affecting the association between work characteristics and job performance. To facilitate a sustainable and thriving performance, it is essential to consider both the environmental attributes of the work and the preferences of the workers. These aspects are key factors in sustainable resources management processes, particularly in terms of selection and work redesign.

In the framework of healthy organizations rooted in positive organizational health psychology, work-related stress is another critical issue. In relation to work-related stress, the psychology of sustainability and sustainable development helps in considering not only the dark side of stress—characterized by negative outcomes for both workers and organizations—but also the positive side of stress, characterized by positive effects in terms of well-being, work satisfaction, organizational commitment, and engagement (Di Fabio et al., 2018). In this regard, there is value in the Rodríguez et al. (2013) Valencia Eustress-Distress Appraisal Scale (VEDAS), which simultaneously detects if the individual perceives a stressful situation as a challenge or opportunity (eustress) and as a source of pressure (distress). This scale allows for more than a merely negative view of stimuli, allowing subjective evaluation of demanding work situations in terms of challenges and opportunities for growth, as well as pressures and threats. Allowing workers to focus not only on the negative aspects of stress but also on its positive side—linked to the positive subjective evaluation by workers of the demanding situation associated with productive activation and vital energy—opens up new perspectives of evaluation while supporting the tailoring and precision of resources. Focusing on the positive side of demanding situations can critically influence challenging situations and organizational practices. On this basis, the VEDAS could be placed inside the frame of the psychology of sustainability and sustainable development.

1.3 Harmonization and Decent Work

Starting from an analysis of harmony from both eastern and western sources, Di Fabio and Tsuda (2018) reflect on the fact that harmony recalls a kind of static balance. The authors propose therefore the value of harmonization in terms of a balancing process. The complex construction of harmonization from a psychological understanding emphasizes the concept of relationality. The concept of harmonization considers different levels: within the individual, between individuals, and between human beings and the natural world/universe. Di Fabio and Tsuda (2018) describe the levels of harmonization in the following way:

> Within the individual, harmony refers to the person and the process of harmonizing various parts of the body, mind-heart, and different purposes in life in a well-functioning organic whole. Between individuals, harmony focuses on individual/individuals, as harmony can be realized between individuals at different levels: Family, community, nation, and the world. Between human beings and the natural world/universe refers to a special relationship including the harmony of human societies and the natural world with the ultimate goal of reaching grand harmony throughout the cosmos. (p. 4)

The psychology of harmonization represents a pillar of the psychology of sustainability and sustainable development. It considers harmonization "in geographical and temporal perspectives, including meaningful construction processes from the past, to the present, and into the future using reflexivity processes at the individual, group, community, social, and national levels" (Di Fabio & Tsuda, 2018, p. 2).

The psychology of sustainability and sustainable development and the psychology of harmonization call for a primary prevention perspective as well as strength-based prevention perspectives, addressing critical psychological issues in order to advance sustainability and sustainable development while promoting well-being in every environment. The critical focus of the research in this area is on psychological strengths amenable to training associated with both hedonic and eudaimonic well-being (Di Fabio, 2017a, b; Di Fabio & Rosen, 2018).

Well-being in organizations also involves the issue of decent work. Current challenges that workers face include a decrease in the amount and worth of accessible jobs; continuous precariousness and transitions are accelerating at an exponential rate (Blustein et al., 2019), generating anxiety over finding decent work for a decent life (Di Fabio & Blustein, 2016). Researchers have tried to improve the concept of decent work (Blustein et al., 2016; Di Fabio & Kenny, 2016; Di Fabio & Maree, 2016; Ribeiro et al., 2016) by including not only social and economic factors but also psychological factors.

The psychology of working theory (PWT) (Duffy et al., 2016), which is based on the previously developed psychology of working framework (PWF) (Blustein, 2006), has highlighted the importance of a vocational psychology that can assume inclusiveness and multidisciplinarity, fostering decent work for people with less opportunity and volition due to economic barriers and marginalization. The PWF (Blustein, 2006) and the PWT (Duffy et al., 2016) underline decent work as relevant for satisfying survival needs, social connection, and self-determination. With the aim to study decent work in depth, Duffy et al. (2016, 2017) emphasized the necessity to operationalize this construct, realizing a specific measure that overcomes the evaluation of macrolevel economic indices. They defined the decent work construct (Duffy et al., 2016, 2017) as involving (1) physically and interpersonally safe working conditions (e.g., absent of physical, mental, and emotional abuse), (2) work hours that allow for free time and adequate rest, (3) organizational values that complement family and social values, (4) adequate compensation, and (5) access to adequate healthcare. Duffy et al. (2017) developed the Decent Work Scale (DWS), which considers how people experience their working life. The DWS has five dimensions: (1) physically and interpersonally safe working conditions, (2) access to healthcare, (3) adequate compensation, (4) work hours that allow for free time and rest, and (5) organizational values that complement family and social values.

A study by Di Fabio and Kenny (2019) aimed at a deeper comprehension of decent work in Italy. The validity of the Italian version of the DWS was examined through both quantitative and qualitative approaches. The quantitative approach supported the reliability and validity of the Italian version of the DWS for evaluating decent work in Italy. With respect to the original version, only the three items for the *access to healthcare* dimension were adapted to Italy, which has a strong public healthcare system. Qualitative analysis added to the comprehension of decent work in Italy by both confirming and enlarging the perspective that emerged from the quantitative analysis. Some dimensions of decent work found by qualitative analysis corresponded to DWS factors, or subscales. In particular, the dimensions *remuneration, safety, work organization,* and *protection* and *free time and rest*, which emerged from qualitative analysis, correspond to the DWS factors *adequate compensation, physically and interpersonally safe working conditions*, and *free time and rest* (Duffy et al., 2017). Qualitative analysis showed additional significant dimensions, such as *relationships* and *meaning*. The dimension *relationships* (good relations with colleagues, leaders, collaborators, users—including collaboration, possibility of dialogue, support, and the creation of a good working climate) is explained following the second pillar of the PWF and the PWT, which considers work as an important social opportunity. The dimension *meaning* can be understood in the light of the third pillar of the PWF and the PWT, which considers work as an opportunity for self-determination. *Meaning* includes those aspects of work related to authentic and consistent values and beliefs. Following the PWF and PWT, meaning could be included in the area of self-determination. A sense of self-determination sees workers meaningfully engaged in tasks illuminated by intrinsic or extrinsic motivations (Duffy et al., 2016). Meaning thus appears fundamental in defining the concept of decent work.

At this point, it is essential to consider the shift from the motivation paradigm to the meaning paradigm (Di Fabio & Blustein, 2016), stressing the relevance of meaning in recognizing how workers construct meaningful lives and meaningful work experiences. The meaning paradigm is a key concept for sustainability and healthy organizations (Di Fabio, 2017a, b).

In the psychology of sustainability and sustainable development perspective, meaning is vital for a sustainable construction, which includes coherence, direction, significance, and belonging (Di Fabio, 2017b, Schnell et al., 2013; Bernaud, 2015). The psychology of sustainability and sustainable development calls for the construction of authentic meanings for individuals. It underlines the importance of meaning, purpose, and connections (Di Fabio, 2017b; Di Fabio & Blustein, 2016; Kaufman, 2018; Maslow, 1971). The psychology of sustainability and sustainable development contributes to the sustainable development of every individual, allowing his or her interpersonal and intrapersonal talents to flourish. In this framework, self-attunement in fostering talents is introduced. Self-attunement describes objective talents and potential (*what I am able to do*) and also subjective talents and potential (*what energizes and motivates me*) in the construction of a self that is full of concreate meaning for each individual (Di Fabio, 2014). Self-attunement permits to "go between the concepts" (Guichard, 2005), dialoguing between objectivity and subjectivity and facilitating identity awareness (Di Fabio, 2014) to improve worker performance through meaningful goals. Purposeful identitarian awareness (Di Fabio, 2014) represents the contact between the most authentic aspects of one's self and one's professional life and deepest purposes as an aid in the construction of meaning.

The psychology of sustainability and sustainable development, with its pillar of the psychology of harmonization, contributes to the concepts of authentic meaning and decent work—harmonizing different aspects of individuals, environments, and individuals in environments, considering both temporal and spatial perspectives.

1.4 Conclusion

The psychology of sustainability and sustainable development represents a critical research and intervention area that contributes to the understanding of sustainability through a psychological framework within the transdisciplinary domain of sustainability science. The foundation of a specific psychological research area within the domain of sustainability that goes beyond existing limited psychological contributions (in most cases, such contributions published in books are of only a theoretical nature) permits the valuable use of the psychological approach for sustainability and sustainable development. This psychological approach could also contribute toward the United Nations' (2020) Sustainable Development Goals, which are included in the 2030 agenda as essential aims for the future of our planet and all human beings. This psychological perspective could also contribute to the well-being of humans, environments, and humans in their environments, overcoming the traditional eco-socio-economic perspective. Its aim is to support an effective and sustainable well-being for the natural environment, as well as for the many different environments (e.g., personal, social, and organizational) and individuals. Healthy organizations, healthy business, healthy workers, harmonization, and decent work—all with a focus on well-being—are at the basis of this perspective.

Regarding healthy organizations and healthy business—from a primary prevention perspective, as well as from strength-based prevention perspectives—the psychology of sustainability and sustainable development calls for a new awareness regarding preventive resources to promote well-being in organizations. It is also important to construct specific interventions to increase these preventive resources, thus fostering the sustainable well-being of workers and organizations. Specific trainings could also be conducted early on with young people as future workers, to enhance preventive resources associated with both hedonic and eudaimonic well-being. Furthermore, these resources could be particularly useful for new managers to promote suitable leadership for twenty-first century organizations (Di Fabio & Peiró, 2018).

Challenges are inherently and first of all opportunities for growth. The psychology of sustainability and sustainable development offers a fundamental way to more adaptively construct opportunities not only to promote organizational well-being (Robertson & Cooper, 2010; Johnson et al., 2018) but also to cope in unpredictable modern workplaces. This psychological framework is a response to the call to foster healthy organizations and healthy business through healthy workers (Di Fabio, 2017a; Di Fabio et al., 2020).

The psychology of harmonization also highlights the value of introducing a psychological perspective into the process of balancing different levels (within individuals, between individuals, and between individuals and the natural world). This framework also engages a primary prevention perspective (Di Fabio & Kenny, 2015, 2016) and strength-based prevention perspectives (Di Fabio & Saklofske, 2021), underlining the relevance of increasing psychological strengths toward greater sustainable development and well-being.

The themes of well-being and sustainability are closely connected with decent work. Psychological processes are inherently embedded in decent work, particularly considering the importance of relational factors and meaning-making processes (Di Fabio & Blustein, 2016; Di Fabio & Kenny, 2019). In the present challenging and difficult period, exacerbated by the COVID-19 pandemic (Gori et al., 2020), meaning represents an essential resource for the construction of career and life paths that are anchored to the most authentic aspects of the self, thus representing a crucial point of reference in promoting well-being in sustainability and sustainable development.

References

Antoniou, A. S. G., & Cooper, C. L. (Eds.). (2005). *Research companion to organizational health psychology*. Edward Elgar Publishing.

Bernaud, J. L. (2015). *Psychologie de L'accompagnement: Concepts et Outils pour Développer le Sens de la vie et du Travail*. Dunod.

Blustein, D. L. (2006). *The psychology of working: A new perspective for career development, counseling, and public policy*. Routledge.

Blustein, D. L. (2011). A relational theory of working. *Journal of Vocational Behavior, 79*(1), 1–17. https://doi.org/10.1016/j.jvb.2010.10.004

Blustein, D. L., Olle, C., Connors-Kellgren, A., & Diamonti, A. J. (2016). Decent work: A psychological perspective. *Frontiers in Psychology, 7*, 407. https://doi.org/10.3389/fpsyg.2016.00407

Blustein, D. L., Kenny, M. E., Di Fabio, A., & Guichard, J. (2019). Expanding the impact of the psychology of working: Engaging psychology in the struggle for decent work and human rights. *Journal of Career Assessment, 27*, 3–28. https://doi.org/10.1177/1069072718774002

Brundtland Report. (1987). *Our common future*. Butterworth.

Cox, T., Baldursson, E. B., & Rial-González, E. (2000). Occupational health psychology. *Work and Stress, 14*(2), 101–104.

De Smet, A., Loch, M., & Schaninger, B. (2007). Anatomy of a healthy corporation. *Mckinsey Q., 2*, 64–73.

Di Fabio, A. (2014). The new purposeful identitarian awareness for the twenty-first century: Valorize themselves in the Life Construction from youth to adulthood and late adulthood. In A. Di Fabio & J.-L. Bernaud (Eds.), *The Construction of the Identity in 21st Century: A Festschrift for Jean Guichard* (pp. 157–168). Nova Science Publishers.

Di Fabio, A. (2017a). Positive Healthy Organizations: Promoting well-being, meaningfulness, and sustainability in organizations. *Frontiers in Psychology. Organizational Psychology, 8*, 1938. https://doi.org/10.3389/fpsyg.2017.01938

Di Fabio, A. (2017b). The psychology of sustainability and sustainable development for well-being in organizations. *Frontiers in Psychology. Organizational Psychology, 8*, 1534. https://doi.org/10.3389/fpsyg.2017.01534

Di Fabio, A., & Blustein, D. L. (2016). Editorial. In A. Di Fabio & D. L. Blustein (Eds.), Ebook research topic "from meaning of working to meaningful lives: The challenges of expanding decent work" in *Frontiers in Psychology. Organizational Psychology, 7*, 1119. https://doi.org/10.3389/fpsyg.2016.01119

Di Fabio, A., & Kenny, M. E. (2015). The contributions of emotional intelligence and social support for adaptive career progress among Italian youth. *Journal of Career Development, 42*, 48–59. https://doi.org/10.1177/0894845314533420

Di Fabio, A., & Kenny, E. M. (2016). From decent work to decent lives: Positive Self and Relational Management (PS&RM) in the twenty-first century. *Frontiers in Psychology. Section Organizational Psychology, 7*, 361. https://doi.org/10.3389/fpsyg.2016.00361

Di Fabio, A., & Kenny, M. E. (2019). Resources for enhancing employee and organizational well-being beyond personality traits: The promise of Emotional Intelligence and Positive Relational Management. *Personality and Individual Differences (Special Issue Personality, Individual Differences and Healthy Organizations), 151*. https://doi.org/10.1016/j.paid.2019.02.022

Di Fabio, A., & Maree, J. G. (2016). Using a transdisciplinary interpretive lens to broaden reflections on alleviating poverty and promoting decent work. *Frontiers in Psychology. Section Organizational Psychology, 7*, 503. https://doi.org/10.3389/fpsyg.2016.00503

Di Fabio, A., & Peiró, J. M. (2017, May). Positive Organizational Health Psychology. *Sustainability MDPI, 10*(7), 2413. https://doi.org/10.3390/su10072413. First International Cross-cultural Conference Healthier societies fostering healthy organizations: A cross-cultural perspective, University of Florence, Florence, Italy

Di Fabio, A., & Peiró, J. M. (2018). Human capital sustainability leadership to promote sustainable development and healthy organizations. *Sustainability MDPI, 10*(7), 2413. https://doi.org/10.3390/su10072413

Di Fabio, A., & Rosen, M. A. (2018). Opening the Black Box of psychological processes in the science of sustainable development: A new Frontier. *European Journal of Sustainable Development Research, 2*(4), 47. https://doi.org/10.20897/ejosdr/3933

Di Fabio, A., & Rosen, M. A. (2020). An exploratory study of a new psychological instrument for evaluating sustainability: The Sustainable Development Goals Psychological Inventory. *Sustainability, 12*, 7617. https://doi.org/10.3390/su12187617

Di Fabio, A., & Saklofske, D. H. (2021). The relationship of compassion and self-compassion with personality and emotional intelligence. *PAID 40th Anniversary Special Issue. Personality and Individual Differences, 157*. https://doi.org/10.1016/j.paid.2020.110109

Di Fabio, A., & Tsuda, A. (2018). The psychology of harmony and harmonization: Advancing the perspectives for the psychology of sustainability and sustainable development. *Sustainability, 10*(12), 4726. https://doi.org/10.3390/su10124726

Di Fabio, A., Peiró, J. M., Rodríguez, I., & Kozusznik, M. W. (2018). The Valencia Eustress-Distress Appraisal Scale (VEDAS): Validation of the Italian version. *Sustainability MDPI, 10*(11), 3903. https://doi.org/10.3390/su10113903

Di Fabio, A., Cheung, F., & Peiró, J.-M. (2020). Editorial Special Issue Personality and individual differences and healthy organizations. *Personality and Individual Differences, 166*. https://doi.org/10.1016/j.paid.2020.110196

Dincer, I., & Rosen, M. A. (2013). *Exergy energy, environment and sustainable development* (2nd ed.). Elsevier Science Publishers.

Duffy, R. D., Blustein, D. L., Diemer, M. A., & Autin, K. L. (2016). The psychology of working theory. *Journal of Counseling Psychology, 63*, 127–148. https://doi.org/10.1037/cou0000140

Duffy, R. D., Allan, B. A., England, J. W., Blustein, D. L., Autin, K. L., Douglass, R. P., et al. (2017). The development and initial validation of the Decent Work Scale. *Journal of Counseling Psychology, 64*(2), 206–221. https://doi.org/10.1037/cou0000191

Gori, A., Topino, E., & Di Fabio, A. (2020). The protective role of life satisfaction, coping strategies and defense mechanisms on perceived stress due to COVID-19 emergency: A chained mediation model. *PLOS ONE, 15*(11), e0242402. https://doi.org/10.1371/journal.pone.0242402

Grawitch, M. J., & Ballard, D. W. (2016). *The psychologically healthy workplace: Building a win-win environment for organizations and employees*. American Psychological Association.

Guichard, J. (2005). Life-long self-construction. *International Journal for Educational and Vocational Guidance, 5*(2), 111–124. https://doi.org/10.1007/s10775-005-8789-y

Hage, S. M., Romano, J. L., Conyne, R. K., Kenny, M., Matthews, C., Schwartz, J. P., & Waldo, M. (2007). Best practice guidelines on prevention practice, research, training, and social advocacy for psychologists. *The Counseling Psychologist, 35*(4), 493–566. https://doi.org/10.1177/0011000006291411

Harris, J. M. (2003). Sustainability and sustainable development. *International Society for Ecological Economics, 1*(1), 1–12.

Hart, P. M., & Cooper, C. L. (2001). Occupational stress: Towards a more integrated framework. In N. Anderson et al. (Eds.), *Handbook of work, industrial and organizational psychology: Volume 2*. Sage.

Henry, J. (2005). The healthy organization. In A. S. G. Antoniou & C. L. Cooper (Eds.), *Research companion to organizational health psychology* (pp. 382–391). Edward Elgar.

Johnson, S., Robertson, I., & Cooper, C. L. (2018). *Wellbeing: Productivity and happiness at work* (2nd ed.). Palgrave Macmillan.

Kaufman, S. B. (2018). Self-actualizing people in the 21st century: Integration with contemporary theory and research on personality and well-being. *Journal of Humanistic Psychology, 7*. https://doi.org/10.1177/0022167818809187

Maslow, A. H. (1971). *The Farther reaches of human nature*. Viking Press.

Peiró, J. M. (2008). Stress and coping at work: New research trends and their implications for practice. In K. Näswall, J. Hellgren, & M. Sverke (Eds.), *The individual in the changing working life* (pp. 284–310). Cambridge University Press.

Peiró, J. M., & Rodríguez, I. (2008). Work stress, leadership and organizational health. *Papeles del Psicólogo, 29*(1), 68–82.

Peiró, J. M., & Tetrick, L. (2011). Occupational health psychology. In P. R. Martin, F. M. Cheung, M. C. Knowles, M. Kyrios, L. Littlefield, J. B. Overmier, & J. M. Prieto (Eds.), *Wiley-Blackwell IAAP handbooks of applied psychology. IAAP handbook of applied psychology* (pp. 292–315). Wiley-Blackwell.

Peiró, J. M., Bayonab, J. A., Caballer, A., & Di Fabio, A. (2020). Importance of work characteristics affects job performance: The mediating role of individual dispositions on the work design-performance relationships. *PAID 40th Anniversary Special Issue. Personality and Individual Differences, 157*. https://doi.org/10.1016/j.paid.2019.109808

Raymond, J. S., Wood, D. W., & Patrick, W. K. (1990). Psychology doctoral training in work and health. *American Psychologist, 45*(10), 1159–1161. https://doi.org/10.1037/0003-066X.45.10.1159

Ribeiro, M. A., Silva, F. F., & Figueiredo, P. M. (2016). Discussing the notion of decent work: Senses of working for a group of Brazilian workers without college education. *Frontiers in Psychology, 7*, 207. https://doi.org/10.3389/fpsyg.2016.00207

Robertson, I., & Cooper, C. L. (2010). *Wellbeing: Productivity and happiness at work*. Palgrave Macmillan.

Rodríguez, I., Kozusznik, M. W., & Peiró, J. M. (2013). Development and validation of the Valencia Eustress-Distress Appraisal scale. *International Journal of Stress Management, 20*(4), 279–308. https://doi.org/10.1037/a0034330

Rosen, M. A. (2017). Sustainable development: A vital quest. *European Journal of Sustainable Development Research, 1*(1), 2. https://doi.org/10.20897/ejosdr.201702

Schaufeli, W. B. (2004). The future of occupational health psychology. *Applied Psychology, 53*(4), 502–517.

Schnell, T., Höge, T., & Pollet, E. (2013). Predicting meaning in work: Theory, data, implications. *The Journal of Positive Psychology, 8*(6), 543–554.

Tetrick, L. E., & Peiró, J. M. (2012). Occupational safety and health. In *The Oxford handbook of organizational psychology* (Vol. 2). Oxford University Press.

United Nations. (2020). *Sustainable development: The 17 goals*. https://sdgs.un.org/goals. Accessed 10 Nov 2020.

Chapter 2
Methodological Approaches for Sustainability Research: Modeling Beyond Time in International Adaptation

Lynette H. Bikos, Kaitlin Patton, Thomas Pankau, and Lauren Hirsch

Transitioning across national borders requires the sojourner to respond to affective, behavioral, and cognitive challenges in both interpersonal and environmental contexts (Bikos et al., 2021). In this chapter, we will use the term *international adaptation* to refer to this multidimensional process of sociocultural adaptation. Called by several names, but widely recognized as the "U-curve adjustment theory" (Black & Mendenhall, 1991, p. 225), many have speculated there to be a natural unfolding of adaptation that is a function of time. Lysgaard (1955) was the first to collect and analyze data that provided evidence of change over time (Black & Mendenhall, 1991). In fact, Black and Mendenhall catalogued 18 articles written between 1955 and 1982 when Torbiörn (1982) published his longitudinal study of expatriate managers. Black and Mendenhall were critical of the methods of many of the 12 studies that claimed support for the U-curve; 10 offered no statistical tests of their descriptive-level data, and only 2 were longitudinal.

In the 30 years following Black and Mendenhall's (1991) review, the desire to model international adaptation as a function of time remains high and calls for "more longitudinal research" abound (e.g., Richardson, 2017; Wan et al., 2017). Yet, multi-wave studies are rare (e.g., Bhaskar-Shrinivas et al., 2005); research designs and statistical analyses are frequently criticized; and strong theories suggest that predicting well-being in international or intercultural contexts is multileveled.

The psychology of sustainability and sustainable development (Di Fabio, 2017a, b; Di Fabio & Rosen, 2018) is useful as a theoretical framework to guide the research design and statistical analysis of international adaptation. The psychology of harmony and harmonization (Di Fabio & Tsuda, 2018) is a recent pillar within the psychology of sustainability. Harmony is focused on well-being in geographical

L. H. Bikos (✉) · K. Patton · T. Pankau · L. Hirsch
Seattle Pacific University, Seattle, WA, USA
e-mail: lhbikos@spu.edu

A. Di Fabio (ed.), *Cross-cultural Perspectives on Well-Being and Sustainability in Organizations*, https://doi.org/10.1007/978-3-030-86709-6_2

and temporal perspectives. As such, it concerns the investigation of well-being at interpersonal, intrapersonal, and contextual levels of analysis.

In "A Spectrum of Statistical Approaches" section of this chapter, we review what can (and cannot) be learned from an array of statistical procedures. Organized by the complexity of the statistic utilized, we sample the quantitative research literature with examples relevant to international adaptation. In "The Statistical Choice Matters" section, we address statistical theory at three levels: paradigmatic/epistemological (null hypothesis significance testing versus statistical modeling), computational algorithm (ordinary least squares versus maximum likelihood estimation), and statistical (repeated measures analysis of variance versus hierarchical linear modeling). Next, we offer recommendations for longitudinal researchers and provide three demonstrations of modeling time with multilevel modeling. Consistent with the recent conceptualizations of the psychology of sustainability (Di Fabio & Rosen, 2018; Di Fabio & Tsuda, 2018), well-being is an outcome for all three demonstrations, and the examples include intrapersonal, interpersonal, and contextual predictors in a global/cross-cultural environment.

2.1 A Spectrum of Statistical Approaches

We begin with an overview of the statistical approaches (arranged in order of complexity) used to analyze literature that purports to address change over time. Many of these articles included multiple analyses to investigate their research question; owing to space, we only review only the ones pertinent to the intended illustrations.

At the outset we note that our utilization of the term *longitudinal* is reserved to mean, "repeated observations on the same individuals… collected over three or more points in time" (O'Connell et al., 2013, p. 125). We call attention to two aspects of this definition. First, at least some of the measures are administered at all time points. Second, there are at least three points in the observation. Authors of some articles we reviewed have used "longitudinal" when (a) there are only two observations over time, (b) there is only one observation but it includes a retrospective perspective, and (c) different variables are collected at each wave. Throughout the chapter, we try to be as precise as possible regarding modeling time. Thus, terms we use include variations of change over time, repeated measures, within-subjects' analysis, growth trajectories, and growth curves.

Descriptives Providing the foundation for data analysis, descriptive statistics encompass the summarization and description of data for a given sample. Utilizing a cross-sectional design, Torbiörn (1982) plotted means of "satisfaction in the host country" scores for 1,094 Swedes as a function of clusters of time (e.g., 0–3 months, 5–7 years, >15 years). A decade later, de Cieri et al. (1991) plotted means to represent psychological adjustment across five points associated with pre-departure, expatriation, and repatriation. A curious aspect of de Cieri et al.'s cross-sectional design was the collection of *retrospective* estimates during the single interview. Neither Torbiorn nor de Cieri et al. conducted statistical comparison of the waves, so, while the plots look convincing, it is impossible to know if the mean differences are statistically significant.

Correlations Correlations have been used in longitudinal designs to investigate the strength and direction of relationship between two variables at several time points during expatriation. For example, Torbiörn (1982) correlated in-country satisfaction with degree of affiliation with fellow nationals and local nationals. Using the same time intervals as described above, he observed that the correlation between satisfaction and home country attitude was positive at the beginning of the expatriation but transitioned to becoming negative after the first year.

Correlations are useful when they analyze variables that, for ethical or logistical reasons, cannot be manipulated by researchers. Fisher and Shaw (1994) conducted a hierarchical multiple regression analysis but focused the majority of their interpretation on the bivariate relations of 23 attitudinal and contextual factors. For example, the number of moves prior to expatriation was not correlated with adjustment difficulty, yet a strong relationship between prior and current adjustment difficulties existed.

While researchers can generate hypotheses about the mechanisms, interventions, or future research directions based upon correlated relations, these analyses do not involve the manipulation of variables and, therefore, cannot claim causality or directionality of the effect. Especially because the results may appear clear and straightforward, critical thinking must always focus on unexamined variables that may influence the results of a correlation coefficient.

***t*-tests** Paired (or dependent) *t-tests* evaluate whether mean differences between paired observations are significantly different than zero. In a sample of 86 expatriate employees from Australia and New Zealand, Fee et al. (2013) reported a significant increase in cognitive complexity from pre-departure to 12 months later.

Independent *t*-tests compare the means of distinct groups. Both a grouping variable (e.g., expatriate or national) and a test variable (e.g., scores on adjustment) must be present for each point of data. Bikos et al. (2009) used normative comparisons, a specialized pairing of the traditional independent *t*-test in tandem with *clinical equivalency t*-tests to compare mental health scores of two groups of female expatriate spouses with each other and with norms from the test's manual. Comparisons were made for each of the five waves in the first year of expatriation. While mental health scores fluctuated over time and between these two groups, the spouse group on assignment with missionary or humanitarian aid organizations was significantly lower than other expatriates scores at all waves and significantly lower than the normative sample of chronically ill medical patients at 3 and 9 months.

Analysis of Variance (ANOVA) ANOVA is used when researchers assess the relationship of one or more factors with a dependent variable. Bikos et al. (2007) assessed a variety of adaptation indices (e.g., mental health, life role salience, substance use) in 32 female expatriate spouses across five waves within the first year of adaptation to Turkey. Results of the one-way repeated measures ANOVA indicated no significant changes across time.

An advantage of factorial ANOVA is the ability to isolate effects through the inclusion of moderators. Two such studies augmented the repeated measures ANOVA by including a non-traveling condition. Over a three-wave, 2-year investigation, Anderzén and Arnetz (1999) reported a decrease in mental health and increase in stress (including stress indicators) for an expatriate group of Swedish workers ($n = 47$; but not the non-moving control group [$n = 15$]). In a two-wave, 1-year investigation, Fee and Gray (2012) reported significant increases in cognitive flexibility and creative thinking for expatriates from Australian and New Zealand ($n = 54$; but not for non-expatriates [$n = 15$]).

In a rare instance of experimental design, Van Bakel et al. (2014) randomly assigned 65 expatriate employees in the Netherlands into groups that were or were not paired with a Dutch host. Social initiative decreased over 9 months for the group who had no host.

Multiple Linear Regression In multiple linear regression, a predicted dependent variable is formed from the linear combination of the multiple independent variables. The strength and significance of weights can be evaluated in the presence of each other.

Holopainen and Björkman (2005) examined the effects of pre-departure psychological traits (e.g., stress tolerance and communication ability assessed by trained psychologists) of 117 Finnish expatriate employees on self- and supervisor- evaluated work performance after a minimum of 12 months on assignment. Controlling for gender, cultural distance, previous international experience, and cross-cultural training, relational ability was negatively related to self-rated performance, whereas communication ability was positively related to both supervisor and self-rated performance.

van Erp et al., (2014), using a two-wave approach analyzed expatriate employee and expatriate spouse personality traits on adjustment. Time 1 (T1) participants were 105 couples (210 participants) who had been in the country for 5.5 years or less; T2 (1 year later) included 45 of the original participants. Controlled for T1 measures of adjustment, results suggested that specific intercultural personality dimensions influenced specific domains of adjustment somewhat differently. For example, emotional stability affected psychological adjustment for both partners at T1, but only the spouse's at T2.

A series of hierarchical linear regression analyses was used to analyze a three-wave longitudinal data from expatriate English teachers in Hong Kong (Fu et al., 2017). T1 ($n = 207$) data were collected at the beginning of the new school year, T2 ($n = 117$) 3 months later, and T3 ($n = 117$) end of school year. T3 also included 116 responses from the teacher's school principal. Data were evaluated in stages, with a first set of analyses separately predicting the outcome variables (e.g., T2 social integration) from control variables (e.g., age, cultural novelty) and T1 predictor variables (three organizational socialization tactics). In the second stage, T3 outcome variables (e.g., cultural adjustment) were predicted from the control variables and T2 predictor variables (e.g., adjustment variables, social integration). In the third stage, different T3 outcome variables (e.g., task or relationship performance) were predicted from the control variables, T2 predictors, and interaction effects. From this patched-together series of analyses, the authors concluded that organizational socialization facilitated social integration and learning speed, which, in turn, were related to adjustment.

Meta-analysis Meta-analyses combine multiple, similar studies to observe larger patterns. In their meta-analysis of expatriate adjustment, Bhaskar-Shrinivas et al. (2005) analyzed 66 studies that involved a total of 8,474 expatriate workers. Only 3 of the 66 studies in the meta-analysis included a panel or longitudinal design. Therefore, to evaluate change over time, they transformed the cultural adjustment means from 20 cross-sectional studies and used these as a dependent variable in a weighted regression analysis. The authors identified a quartic function, the "sideways S," that takes about 4 years to unfold. The authors cautioned that several points of the curve may have been unduly influenced by single studies.

Structural Equation Modeling (SEM) SEM estimates the relationship between latent variables and measured variables (the measurement model) and among both measured variables and latent variables themselves (the structural model; Cohen et al., 2003). Two SEM studies predicted supervisor-evaluated job performance at T3 from expatriate self-ratings of different variables at earlier ratings (Takeuchi et al., 2009; Wang & Takeuchi, 2007). Takeuchi et al.'s final model suggested that perceived organizational support (T1) in the current assignment and in off-the-job life interacted with each other to predict work and general adjustment (T2). Further, higher affective commitment (at T2) by the expatriate employee mediated the relations between work (T2) and general adjustment with job performance (T3). These designs were similar in that they did not follow one (or more) aspects of adjustment over time. Rather, different variables at earlier times predicted manager appraisal of performance at the final time. Further, the timing of assessments was not queued to entry or departure, making time a non-critical factor.

Gibbs et al. (2020) used SEM to model interpersonal, intrapersonal, and contextual factors on the psychological and sociocultural adjustment of international students in Turkey. Interpersonal connections with host nationals predicted greater sociocultural adaptation; interpersonal connections with co-nationals resulted in poorer psychological adjustment. Host country language proficiency led to better sociocultural adaptation and perceptions of greater cultural distance and negative effects on psychological and sociocultural adaptation. Thus, students who engaged in relations with host country nationals fared better.

Multilevel modeling In MLM we can ask two types of questions: How does each person change over time (within-subjects)? What predicts differences among people in their changes (between-subjects)? Firth et al. (2014) analyzed work adjustment over three waves of data for 70 expatriate employees during their first 4 months of international assignment. Work adjustment was characterized by a negative linear slope; those with the highest initial levels of work adjustment saw the sharpest declines. Controlling for between-participants levels of variables such as organizational tenure and language skills, significant positive relations existed between initial levels of work adjustment with both cross-cultural motivation and psychological empowerment.

Firth et al. (2014) utilized MLM to model the mediating role of initial work adjustment between cross-cultural motivation and psychological empowerment through work adjustment change. Results suggested that expatriate employees' success in maintaining higher adjustment was negatively related to initial levels of work adjustment but positively related to the presence of challenge stressors. Further, the impact of motivational states and the experiences of stressors on work adjustment was time-dependent.

2.2 The Statistical Choice Matters

Statistics contain assumptions about the form of research questions and the characteristics of data. Because a primary purpose of our chapter is to demonstrate the utility of MLM as an approach to analyzing longitudinal data that addresses limitations of other statistic approaches, we use this section to review statistical theory at three levels. First, we compare traditional null hypothesis significance testing (NHST) to statistical modeling. Second, we compare ordinary least squares (OLS) and maximum likelihood estimation (MLE) as approaches to regression. Third, we contrast the assumptions and data-specific requirements between repeated measures ANOVA and MLM.

NHST has been the primary epistemological system for the social sciences (Rodgers, 2010) and was prominent in the research we reviewed. At the core of NHST is evaluation of the null hypothesis – that a chance process generated the data. The decision to reject the null is dichotomous and based on the point estimate of the *p* value, with most null hypotheses being rejected if $p < .05$ (Cumming & Finch, 2005). The computational basis of NHST is analyzing variance and includes *t*-tests, correlations, ANOVA, regression, and their variants.

Resistance to NHST began with criticisms of the null. Cohen (1994) argued (a) any null could be rejected with a large enough sample size, (b) rejecting a null is not strong support for the researcher's alternative hypothesis, (c) the logic of evaluating the probability of the data given the hypothesis is backward, and (d) statistical significance does not address practical significance. Similarly, Rodgers (2010) described a "quiet methodological revolution" such that the "epistemological basis of statistics has moved away from being a set of procedures, applied mechanistically, and moved toward building and evaluating scientific and statistical models" (p. 1). Requirements to report exact *p* values, confidence intervals, and effect sizes;

examine Type I and II error; and conduct power analyses move researchers away from dichotomous decisions (i.e., $p < .05$) and instead encourage utilization of a dashboard of metrics to evaluate results (Cumming & Finch, 2005; Wilkinson, 1999).

As our chapter's review of the statistics transitioned from ANOVA and regression to SEM and MLM, the analytic paradigm shifted from NHST to *statistical modeling*. A *statistical model* has two characteristics: it aligns with the reality it describes but is simpler than that reality (Rodgers, 2010). Modelers begin by examining the relatively simpler relations between variables and assessing the viability of progressively more complex ones based on the results (e.g., McCoach, 2010; O'Connell et al., 2013). Models are evaluated with a host of criteria such as strength and significance of regression weights, fit statistics, residuals, model parsimony, and degrees of freedom.

The epistemological transition from NHST to statistical modeling parallels a significant shift in the computational algorithms. The generalized linear model is the foundation for NHST statistics, and regression is completed with OLS. Modeling approaches such as SEM and MLM utilize MLE. Each has unique characteristics, assumptions, and limitations.

OLS is based on the least squares criterion. The solution is an analytic one with known sets of equations, where the mean of errors is zero, and the errors have zero correlations with each predictor (Cohen et al., 2003). OLS regression assumes that observations are independent of one another. If data are repeated observations, standard OLS regression assumes that these observations are independent across time. Longitudinal researchers know that this is an unlikely scenario.

MLE differs from OLS because there is no analytic solution. Unlike OLS, regression coefficients are obtained through an iterative process until a convergence criterion is achieved. A *likelihood* is the probability that a score or set of scores could occur given the values of a set of parameters in a specified model; the *maximum likelihood* is the highest probability possible (Cohen et al., 2003). Singer and Willett (2003) noted that the asymptotic nature of ML estimates makes them desirable because they are consistent and efficient. However, this asymptotic advantage is predicated on sample sizes much larger than typically required in OLS.

If researchers were to analyze longitudinal data in a strictly NHST frame, they would likely choose repeated (or mixed design) measures ANOVA. While repeated measures ANOVA is designed to accommodate dependency across multiple measures, it assumes that there are no significant (unaccounted for) individual differences in systematic changes over time (such as linear slopes) and that there are equal correlations among all possible pairs of the values of the dependent variable measured at different times (Cohen et al., 2003). Further, the statistic requires evenly spaced (structured, time invarying) observations and requires the same number of observations per person. Together, the statistical assumptions and the characteristics of the data are nearly impossible to meet.

In contrast, MLM measures the relationships of predictors to outcome separately for each participant, including variance due to individual differences in the intercepts of the equation that characterize individual changes over time; the variance across individuals in the slopes of these equations; and the covariance between the individual slopes and intercepts across all occasions. These variance components are termed "random effects" (Cohen et al., 2003). Although MLM requires at least three observations per person (repeated measures ANOVA requires only two), the estimation processes associated with MLM can easily accommodate time-varying data and different numbers of observations per person.

Considering the information presented thus far, we urge researchers to become aware of the assumptions and requirements of data characteristics of the statistics utilized for their analysis and consider transitioning to a statistical modeling approach. The next section in our chapter provides three demonstrations of evaluating the role of time in understanding first-year adaptation for female, expatriate spouses. First, we identify some general implications.

1. If we wish to model the role of time in adjustment, then we must specify the structure of time in our statistical model.
   (a) This begins with the starting point (Hippler et al., 2015). The timing of T1 in longitudinal research is critical. Enrolling research participants for multiple waves irrespective of their arrival or repatriation date tells us nothing about the predictive utility about time. Further, if we suspect participant characteristics (i.e., previous international assignments) have bearing on the role of time, we must specify these in our model.
   (b) We must model the shape or effect of time. A minimum of three repeated measures observations are required to estimate a linear growth model. If curvilinearity is hypothesized, there must be at least one additional occasion of data collection than the polynomial being modeled (O'Connell et al., 2013). For example, to estimate a quadratic model, at least four occasions are needed; a cubic model requires at least five.
   (c) Linear and curvilinear modeling are not the only options. Mitchell and James (2001) identified five time variations: lag, durations, rate of change, dynamic relationships, and reciprocal causation.
   (d) Time alone is an insufficient predictor. We must begin to specify mechanisms for adaptation that expand beyond time and identify the nature of the relationship: covarying, mediating, and moderating.
2. A broad spectrum of theories contributes to research that is relevant to international adaptation. Most acknowledge that internal and external predictors of adaptation should be simultaneously considered (Hippler et al., 2015). Correspondingly, the assumptions and data requirements of the chosen statistic should accommodate this requirement.
3. Research designs rule out plausible rival hypotheses when they control for method bias. Longitudinal designs are strengthened when they include input beyond self-report (e.g., significant others, supervisor, biological samples [e.g., Anderzén & Arnetz, 1999]).
4. Attrition is a threat to longitudinal research. In fact, Fu et al. (2017) noted changes to their statistical approach from SEM to hierarchical regression because of the substantial loss of participants from T1 (207) to T3 (116). Researchers

who plan multi-wave studies must strategize for retention (e.g., robust incentives) while planning for attrition. As research design and operationalization matters, so do choice of statistics. For example, latent growth curve modeling in AMOS requires evenly spaced data with the same number of waves per participant (Byrne, 2016).

5. Longitudinal research in the global context is resource-intensive. Success likely involves collaborative relationships with practitioners, and statistical modeling is complex. Researchers who are accustomed to NHST approaches may need supplementary training in new statistical methods. Further, researchers must convey findings to practitioners in an accessible manner and work together to disentangle implications and applications.

2.3 Potential Contributions of Multilevel Modeling

MLM (also termed hierarchical linear, linear mixed effects, and random coefficients modeling) allows the simultaneous modeling of individual (level 1, L1) and contextual (level 2, L2) variables on the dependent variable of interest, as well as modeling cross-level interactions of the individual and contextual variables on the outcome (McCoach, 2010). We accompany our description of MLM with three demonstrations that incorporate the role of time as a predictor. These include (a) modeling time as varying/unstructured to investigate well-being in first-year, expatriate, spouses, (b) modeling discontinuous shifts in psychological distress resulting from the terrorist attacks on September 11, 2001, and (c) demonstrating the time-covarying effects of marital role importance and marital satisfaction on mental health with time as the context. We first describe the steps common to all three demonstrations. We acknowledge that MLM is complex. Our demonstration is written to facilitate a quick conceptual understanding. We recommend the interested researcher access the excellent resources cited.

Our demonstrations use a sample of 32 female, expatriate spouses in Ankara, Turkey (all US citizens). In this mixed methods design, each participant completed a packet of measures at five intervals beginning with just-arrived and ending around 12 months in country. All surveys took place between December 2000 and December 2002. A description of the research methods and participant characteristics is found in Bikos et al. (2007).

While an advantage of MLM is the capacity to accommodate variations in the timing of data collection across individuals as well as randomly missing data points (O'Connell et al., 2013), MLM uses MLE, which functions better in large sample sizes. Thus, sample size is a critical and complex issue in multilevel models. Power in the longitudinal multilevel model is a function of the number of participants and the number of repeated measures per person, the intraclass correlation, and the effect size. McCoach's (2010) summary of the literature suggests that while "a bare minimum of 10 clusters [participants]" (p. 129) could be sufficient, at least 30 participants are required to produce unbiased estimates of variance components and at least 100 are necessary to have reasonable estimates of the standard errors of the level-two variance components. Additionally, the number of repeated measures and degree of missingness in the longitudinal design can be problematic. Our sample of 32 participants is represented with 143 repeated measures observations ($M = 4.46$ observations per person), with 2 participants having only 2 observations.

Demonstration 1: Mental health growth trajectories as a function of unstructured time. In our first demonstration, we ask, "Does mental health change over the first year of expatriation to an international location?" Our dependent variable is an index of mental health functioning (MHI32, 1 to 6 scaling) where higher scores represent stronger mental health.

Model building begins by exploring the raw, unconditioned data. Researchers are encouraged to visually scan the data to identify possible patterns. Figure 2.1 plots MHI32 raw scores from all participants as a function of months since arrival. Readers may notice that the asterisks (representing the time when a survey is taken) are variably spaced. Our impressions at the initial scanning were (a) there is variability over time, (b) for many (but not all) there are decreases after arrival that turn upward in later months, and (c) variability seems greatest for those with the lowest, overall, mental health.

Formal statistical modeling begins with an unconditional (also termed null or empty) model without predictors. The purpose of the empty model is to obtain estimates of the L1 (within-persons, repeated measures) and L2 (between persons) variance components. Together, these estimate an intraclass correlation. Further, comparison of the initial variance components to those from later, more parameterized models allows for evaluation of adequacy of the model.

In our empty model where the MHI is the dependent variable, the estimated overall mean MHI score across all participants and all waves is $\beta_{00} = 4.92$, ($p < .001$). Random error between participants on the overall intercept is presented with the variance component r_{0i}; e_{ti} represents random error within students from their own mean score. Although this model does not describe growth, it is a useful starting point because it allows for partitioning of between (r_{0i}) and total ($r_{0i} + e_{ti}$) variance. The resultant intraclass correlation (ICC) for MHI suggests that 49% of the variance lies within participants across occasions; 51% is due to variation between participants. This encourages adding L1 and L2 predictors.

Because our interest is to understand the mental health trajectory during the first year of expatriation, our next task is to fit a growth model. Singer and Willett (2003) have suggested that forcing naturally occurring variably spaced models into time-structured formats such as waves or months introduces error into the model because there is more unexplained variation in initial status and growth rates. A variably spaced approach provides more precise information and likely produces a better-fitting growth trajectory. Thus, we represent time in months (with fractional portions retained). Each time interval is calculated by subtracting each person's host-country entrance date from the survey date. For example, our first participant arrived 12/03/2000 and completed the first survey on 12/05/2000. The value for MONTHS at T1 is .0657. To evaluate a quadratic growth curve, the MONTHS variable is squared; to evaluate a cubic growth curve, the MONTHS variable is cubed.

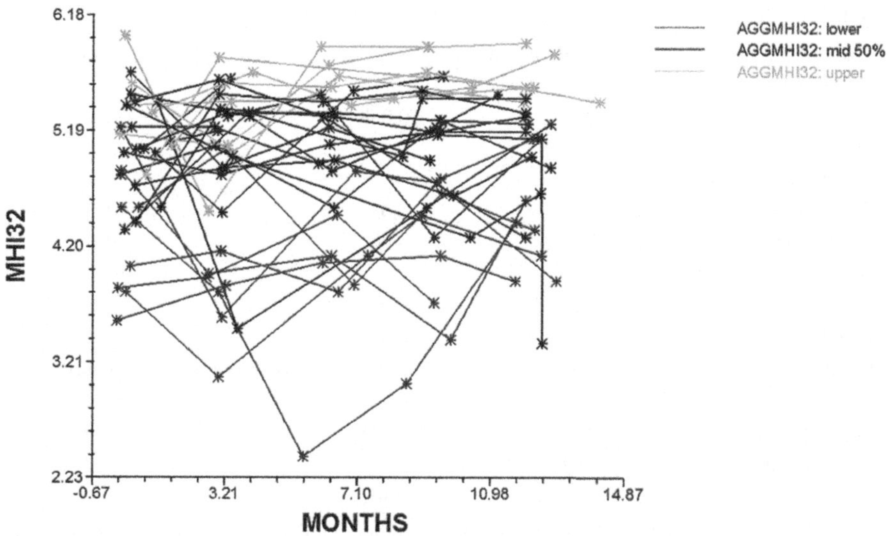

Fig. 2.1 Plot for visual inspection of raw data MHI32 (scaling 1 to 6) raw scores for all participants ($N = 32$) as a function of time in country. Top cluster (light grey) are 25% with the highest aggregated MHI32 scores across all waves; middle cluster (black) are those in the middle 50%; lowest cluster (dark grey) are those in the bottom 25%

Following the exploratory and sequential approach outlined by O'Connell et al. (2013), we evaluate the fit of linear, quadratic, and cubic models by adding MONTHS, MONTHSQ, and MONTHSCB as within-subjects (L1) predictors. No trajectory fits the data well; time alone does not result in linear, quadratic, or cubic change.

In combination with the models of time, we next assess the contribution of the L2 predictor, overall mental health functioning (AGGMHI32). Our decision to include this predictor is to account for both within- and between-persons variance. A subjective sense of psychological well-being has been hypothesized not only to remain relatively stable between subjects but also to be highly variable within persons as influenced by such factors as genetic set point and life events (Diener et al., 2006). AGGMHI32 is created by calculating the mean of each MHI score for each individual. As is common in MLM, we investigate a variety of models that include variations in centering our variables (see Enders & Tofighi, 2007, for an excellent review) and allowing intercepts (i.e., where people start; in our case at the just-arrived wave) and slopes (i.e., how people change during the first year of expatriation) to be fixed to the same value or free to vary. We determine that the best representation is a quadratic growth curve with AGGMHI32 (centered at the grand mean) added as predictors to the intercept and curvature (instantaneous rate of change at baseline and rate of change) functions. Further, these effects are free to vary. Freeing a random intercept to vary means we hypothesize that the participants started at different levels of mental health; freeing random slopes means that we hypothesize that individuals' mental health change at different rates. In Table 2.1,

Table 2.1 Model predicting mental health from overall mental health and time

Fixed effect	Coefficient	*SE*	*p*-value
For INTRCPT1, π_0			
INTRCPT2, β_{00}	4.874	0.088	<0.001
AGGMHI32, β_{01}	0.701	0.157	<0.001
For MONTHS slope, π_1			
INTRCPT2, β_{10}	0.000	0.037	0.985
AGGMHI32, β_{11}	0.145	0.067	0.039
For MONTHSSQ slope, π_2			
INTRCPT2, β_{20}	0.001	0.003	0.767
AGGMHI32, β_{21}	−0.011	0.005	0.052
Random effect	***SD***	**Variance component**	***p*-value**
INTRCPT1, r_0	0.278	0.077	0.043
MONTHS slope, r_1	0.133	0.018	0.009
MONTHSSQ slope, r_2	0.010	0.000	0.017
level-1, *e*	0.367	0.135	

Note. AGGMHI32 is aggregate mental health. Together, MONTHS and MONTHS (squared) assess change over time as a quadratic function. Dependent variable is MHI32, mean score of Mental Health Inventory (1 to 6 scaling)

$\beta_{00} = 4.87$ ($p < 0.001$) is the just-arrived MHI score for someone whose overall mental health is at the mean. For every 1-point increase in AGGMHI32, the just-arrived MHI is 0.70 ($\beta_{01;}$ $p < 0.001$) points higher. Time alone does not result in change to the MHI trajectory; however, a cross-level interaction suggests that overall mental health impacts change over time. The values for the components associated with the hypothesized quadratic function – the instantaneous rate of change at baseline ($\beta_{11} = 0.145, p = 0.039$) and rate of change ($\beta_{21} = -0.011, p = 0.052$) – suggest that those with the lowest overall mental health functioning experience a slight decrease in mental health functioning (lowest around 6 months) and return to baseline by 12 months, whereas those with greater overall mental health experience a slight increase and those with average overall mental health remain relatively flat (see Fig. 2.2).

MLM offers more to analyze and interpret than space in this chapter (and most journal articles) permits. Of critical importance is the variance components. In this example, our empty model produces variance components estimating variance between subjects (r_0) and within subject (e). We use the parallel estimates in the final model to learn that, compared to the empty model, the final model has reduced within-persons variance by 41% and between-persons variance by 68%. Further, we examine the p values associated with the intercept (r_0) and slopes (r_1, r_0) to see if significant variance remains to be explained. Because these values are statistically significant, we could add variables and/or reconsider the variables we have chosen.

Demonstration 2: The effect of 911 – modeling discontinuous time. External events and changes in life status may introduce discontinuity in growth trajectories that could be otherwise plotted as continuous and represented with linear or polynomial functions. If a researcher knows when and why a discontinuous shift may occur, it is possible to model it (Singer & Willett, 2003). The occurrence of the terrorist attacks in the USA on September 11, 2001 occurred approximately 10 months into the 2 years of our data collection. Thus, all participants were scheduled to have at least one wave post-911, and two participants had all waves post-911. This allows us to ask, "Did the level or trajectory of psychological distress increase after 911?" Our dependent variable is an index of psychological distress (DISS22; 1 to 6 scaling) where higher scores represent greater distress.

Discontinuity may lead to four possible combinations of change/no change to the intercept slope. These are specified differently in the L1 individual growth model. We followed Singer and Willett's (2003) recommendations to specify and statistically compare all four specifications (often in combination with other key variables) before selecting the best representation. For efficiency, we present the model that specifies discontinuities in both elevation and slope.

As before, modeling begins with estimation of the intraclass correlation in the empty model where only the time changing-dependent variable (DISS22) is specified. With between- and within-subjects variance components of .21 and .22, respectively, our intraclass correlation indicates that 48% of the psychological distress variance is between persons and 52% within persons.

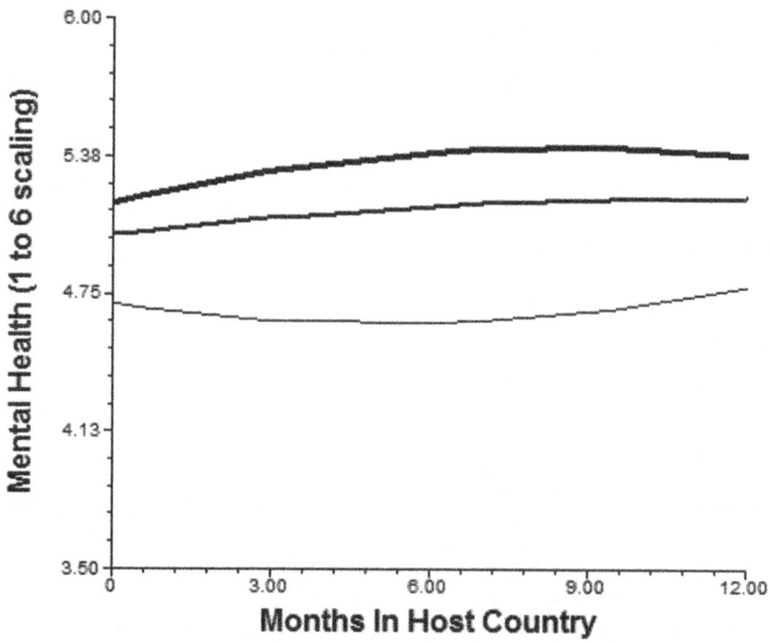

Fig. 2.2 Modeling a quadratic growth trajectory with unstructured time MHI32 score (1 to 6 scaling with higher scores indicating stronger mental health functioning) as a function of months in country. The top (darkest) curve represents a participant at the 75th percentile, the middle (medium) at the 50th percentile, and the lowest (lightest) at the 25th percentile of overall (aggregate) mental health functioning

Modeling discontinuity requires (a) creating a set of time-identifying variables in the data file and (b) entering their proper combination into the L1 equation. To fit a model specifying discontinuities in both elevation and slope, three variables were entered. The first, MONTHS (clocking time since entry in a variably spaced fashion) was described earlier. The second, PRE911_0 codes all pre-911 waves as 0 and post-911 waves as 1. The third MNTHx911 is an interaction term created by multiplying the MONTHS and PRE911_0 variables. This combination creates four equations (one intercept, three slopes) in the L2 model. To each of these, we add AGGDIS22 – the aggregated variable representing psychological distress across all waves.

As in the prior analysis, overall distress clarified the effect of 911 within the first year of expatriation. In Table 2.2, DISS22(β_{00}) is 1.88 ($p < 0.001$) at just-arrived, when it preceded 911, for those with average aggregated psychological distress (AGGDIS22). For each point higher in AGGDIS22, the just-arrived distress (β_{01}) increases by .65 ($p < 0.001$). Fitting this model reveals a discontinuity in intercept, but not slope. The values for the PRE911_0 slope suggest an increase in distress of .23 (β_{20}, $p = 0.067$) following 911. Further, for each point higher in AGGDIS22, DISS22 increases by.76 (β_{21}, $p = 0.002$). Values on the MONTHS and MONTHSX911 slope indicate neither a main nor interaction effect for time. Examining Fig. 2.3 facilitates interpretation. The bottom two lines (more solid) represent individuals

Table 2.2 Model predicting change in psychological distress following 911

Fixed effect	Coefficient	*SE*	*p*-value
For INTRCPT1, π_0			
INTRCPT2, β_{00}	1.878	0.082	<0.001
AGGDIS22, β_{01}	0.650	0.173	<0.001
For MONTHS slope, π_1			
INTRCPT2, β_{10}	0.009	0.035	0.802
AGGDIS22, β_{11}	−0.013	0.075	0.863
For PRE911_0 slope, π_2			
INTRCPT2, β_{20}	0.232	0.125	0.067
AGGDIS22, β_{21}	0.763	0.236	0.002
For MNTHX911 slope, π_3			
INTRCPT2, β_{30}	−0.032	0.037	0.381
AGGDIS22, β_{31}	−0.027	0.078	0.733
Random effect	***SD***	**Variance component**	***p*-value**
INTRCPT1, r_0	0.01025	0.00011	>0.500
MONTHS slope, r_1	0.00168	0.00000	0.417
level-1, *e*	0.38733	0.15002	

Note. AGGDIS22 is aggregate psychological distress. Together, MONTHS, PRE911_0, and MNTHX911 assess for discontinuity, as a function of 911, following expatriation. Dependent variable is DISS22, mean score of psychological distress subscale of Mental Health Inventory (1 to 6 scaling)

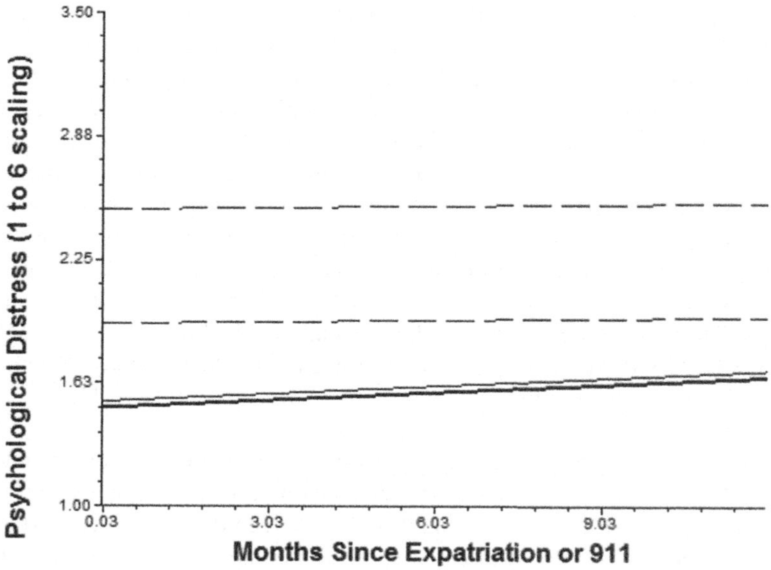

Fig. 2.3 Modeling discontinuity in time
DIS22 (1 to 6 scaling with higher scores indicating greater psychological distress) as a function of months following expatriation and 911. In this modeling of discontinuous data, the top two lines (dashed) represent individuals at the 75th percentile of aggregated psychological distress. The lower of the top two lines (2nd from top) represents their trajectory following their arrival to the host country. The upper of the top two lines represents their trajectory following 911. The bottom two lines represent individuals at the 25th percentile of aggregated psychological distress. An individual trajectory could be estimated by "jumping" from the lower to the upper line

with lower levels of psychological distress; the upper two lines (dashed) represent individuals with higher levels of psychological distress. In each pair, the lower line represents the DISS22 trajectory beginning with just-arrived; the upper line represents the DISS22 trajectory beginning immediately following 911. If a person experienced 911 at 3 months after arrival, they would "jump" from their place on the lower of the pair of lines to the line above and continue along that trajectory. Those functioning with the highest overall levels of psychological distress clearly experienced greater increases in distress than those at lower levels of distress. Evaluating the variance components suggests that there is no further variance to explain in the intercept ($r_0 = 0.0001$, $p > .500$) nor slope ($r_1 = .0000$, $p = .417$). Further, compared to the empty model, this model has reduced within-persons variance by 32% and between-persons variance by 99%.

If this example were more elaborated in a research manuscript, we would have provided much more detail about the process. Further, we would have presented the more parsimonious model where we only predict discontinuity in elevation, but not slope. Given only five waves in the study, our data is limited in its ability to determine the two trajectories (linear or polynomial) required in this discontinuity.

Demonstration 3: The moderating effect of marital role salience on the relationship between marital satisfaction and mental health – time-varying covariates. Time-varying covariates are those whose values differ over time. These values can change naturally (i.e., as a function of context or person) or by research design (Singer & Willett, 2003). While the values of time-covarying covariates change across time, the parameters estimating the effects of time-covarying covariates are assumed to be constant across time (McCoach & Kaniskan, 2010). In this example, we model the effects of marital role salience (MARITAL) and marital satisfaction (MARSATIS) on the dependent variable, mental health (MHI32). These (MHI32, MARITAL, MARSATIS) were assessed at each wave of the study. Because we continue to be interested in change over time, we include MONTHS as an L1 predictor.

We use the compositional effects approach to center our variables (Enders & Tofighi, 2007). A strength of this approach is that it perfectly partitions within- and between-subjects variance. At L1, the variables are centered within context (group mean centered); at L2, the aggregated form of the variable is used. Thus, the L2 variable indicates an individual's average level on that variable across all waves while the L1 variable represents an individual's fluctuation around that mean at any given assessment point. Our compositional effects model predicts an individual's MHI32 from MARITAL, MARSAT, and MONTHS entered at L1. The aggregates of marital role salience and marital satisfaction (AGGMAR and AGGMSAT) are entered in the L2 equations.

The values in our empty model are the same as in our first demonstration. In our full model (Table 2.3), 4.84 (β_{00}, $p < 0.001$) is the MHI32 value at just-arrived for expatriates at the mean of AGGMAR and AGGMSAT. MHI32 values start at 0.37 (β_{01}, $p < .010$) higher for each additional point in marital satisfaction and 0.35 (β_{02}, $p < .021$) points lower for each additional point in overall marital role salience.

In the presence of the other predictors, time in country is not useful in predicting MHI32 scores ($\beta_{30} = 0.01$, $p = 0.137$). Graphed as a function of time, Fig. 2.4 demonstrates that MHI32 scores are highest for those with the combination of high marital satisfaction and low marital role salience. The lowest MHI32 scores are experienced by those with the reverse (e.g., low marital satisfaction, high marital role salience). Were we modeling this for an empirical paper, we would likely favor parsimony and drop MONTHS.

Given that time in country is not a useful predictor, graphing the predictors differently (Fig. 2.5) explains the cross-level interactions. We note significant effects of (a) MARSATIS ($\beta_{10} = 0.22$, $p = 0.059$) on MHI32 and (b) AGGMSAT ($\beta_{21} = 0.64$, $p = 0.047$) on the relationship between MARITAL and MHI32. Marital satisfaction at any given wave and overall is a strong predictor of MHI32. That is, fluctuations

Table 2.3 Model predicting mental health over time from the time-varying covariates marital satisfaction and marital role salience

Fixed effect	Coefficient	SE	p-value
For INTRCPT1, π_0			
INTRCPT2, β_{00}	4.836	0.089	<0.001
AGGMSAT, β_{01}	0.368	0.134	0.010
AGGMAR, β_{02}	0.351	0.144	0.021
For MARSATIS slope, π_1			
INTRCPT2, β_{10}	0.223	0.114	0.059
AGGMSAT, β_{11}	0.0244	0.166	0.884
AGGMAR, β_{12}	0.084	0.121	0.493
For MARITAL slope, π_2			
INTRCPT2, β_{20}	−0.104	0.213	0.629
AGGMSAT, β_{21}	0.636	0.306	0.047
AGGMAR, β_{22}	−0.193	0.330	0.564
For MONTHS slope, π_3			
INTRCPT2, β_{30}	0.014	0.009	0.137
Random effect	***SD***	**Variance component**	***p*-value**
INTRCPT1, r_0	0.353	0.124	0.071
MARSATIS slope, r_1	0.252	0.063	0.008
MARITAL slope, r_2	0.681	0.464	0.404
MONTHS slope, r_3	0.016	0.000	0.102
level-1, e	0.407	0.166	

Note. MARSATIS and MARITAL are marital satisfaction and marital role salience, respectively, assessed at each wave. AGGMSAT and AGGMAR are the aggregate forms of the same variables. MONTHS is the metric for assessing time in country. The dependent variable is mental health assessed by the Mental Health Inventory (1–6 scaling)

in marital satisfaction correspond with fluctuations in mental health. Marital role salience has a curious moderating role. At any given wave, those with the highest MHI32 scores (a) have low marital role salience (wave) but high marital satisfaction (wave; irrespective of aggregate marital satisfaction) or (b) high marital role salience (wave) and high marital satisfaction (aggregate). Discrepancies in marital role salience and satisfaction produce lower MHI32.

While there is additional variance that could be explained in the intercept (r_0 = 0.12, p = 0.071) and MARSATIS slope (r_1 = 0.06, p = 0.008), the MARITAL (r_2 = 0.46, p = 0.404) and MONTHS (r_3 = 0.00, p = 0.102) slopes have no significant variance remaining. Compared to the empty model, this model reduces within-persons variance by 27% and between-persons variance by 48%.

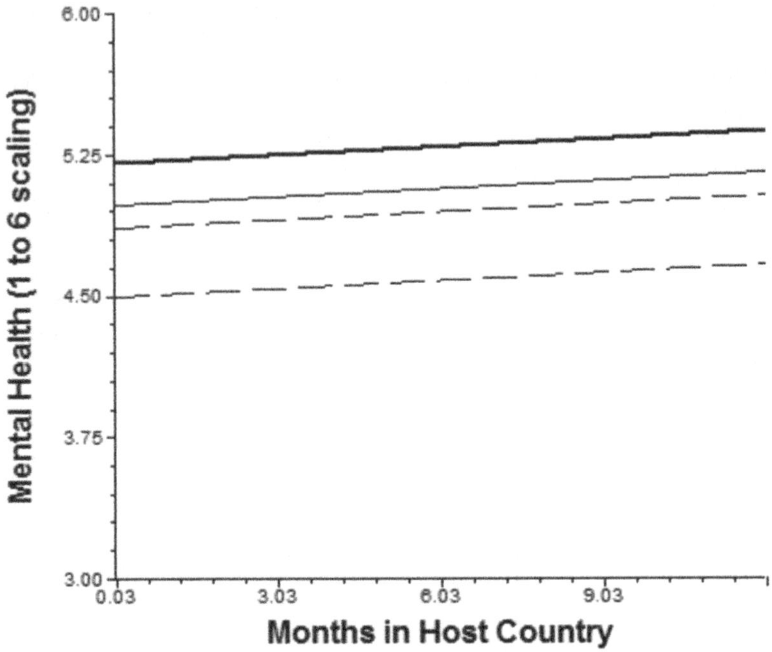

Fig. 2.4 Modeling time-covarying variables as a function of time
MHI32 (1 to 6 scaling with higher scores indicating stronger mental health) as a function of months in host country. The marital satisfaction and role salience variables are time-covarying (not aggregate). The top line (dark, solid) represents an individual at the 75th percentile of marital satisfaction at the 25th percentile of marital role salience. The second line (lighter, solid) represents someone at the 75thth percentile of both marital satisfaction and role salience. The third line (dashed) line represents someone at the 25th percentiles of marital role satisfaction and role salience. The bottom line (also dashed) represents someone at the 25th percentile of marital satisfaction but the 75th percentile of marital role salience. Although the slopes appear to trend upward, there is no significant change over time

2.4 Conclusion

As researchers, particularly those committed to a psychology of sustainability, continue to explore well-being in international adaptation, the validity and utility of their work are strengthened or limited by the statistics used to analyze the data. In our three demonstrations of modeling time, we encourage the utilization of MLM as a powerful yet complex resource that addresses the limitations of NHST and many of the statistics associated with it.

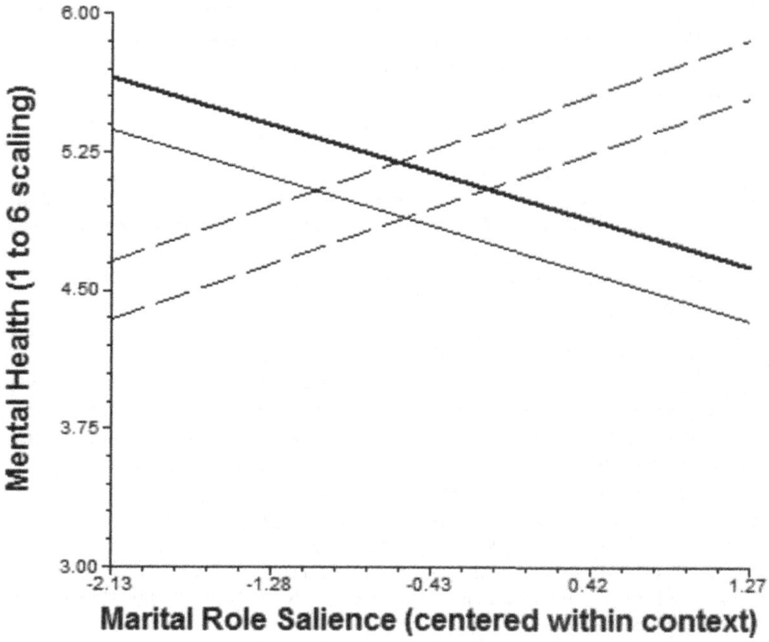

Fig. 2.5 Modeling time-covarying variables separate from time
MHI32 (1 to 6 scaling with higher scores indicating stronger mental health) as a function of marital satisfaction and role salience. The two solid lines (descending toward the right) represent individuals at the 25th percentile of overall (aggregated) marital satisfaction. The upper of the two lines (darker) represents individuals at the 75th percentile (the lower of the two represents individuals at the 25th percentile) of marital satisfaction at any given point in time. The two dashed lines (ascending toward the right) represent individuals at the 75th percentile of overall (aggregated) marital satisfaction. The upper of the two lines represents individuals at the 75th percentile marital satisfaction (and the lower of the two represents someone at the 25th percentile) at any given point in time

References

Anderzén, I., & Arnetz, B. B. (1999). Psychophysiological reactions to international adjustment: Results from a controlled, longitudinal study. *Psychotherapy and Psychosomatics, 68*(2), 67–75. https://doi.org/10.1159/000012315

Bhaskar-Shrinivas, P., Harrison, D. A., Shaffer, M., & Luk, D. M. (2005). Input-based and time-based models of international adjustment: Meta-analytic evidence and theoretical extensions. *Academy of Management Journal, 48*(2), 257–281. https://doi.org/10.5465/AMJ.2005.16928400

Bikos, L. H., Çiftçi, A., Güneri, O. Y., Demir, C. E., Sümer, Z. H., Danielson, S., DeVries, S., & Bilgen, W. A. (2007). A repeated measures investigation of the first-year adaptation experiences of the female expatriate spouse living in Turkey. *Journal of Career Development, 34*(1), 5–27. https://doi.org/10.1177/0894845307304063

Bikos, L. H., Forman, R., & Patton, K. M. (2021). The Self-Efficacy for Sociocultural Adaptation Scale (SESCAS): Development and initial psychometric evaluation. *The Counseling Psychologist, 49*(1), 33–79. https://doi.org/10.1177/0011000020951861

Bikos, L. H., Klemens, M. J., Randa, L. A., Barry, A., Bore, T., Gibbs, R., & Kocheleva, J. (2009). First-year adaptation of female, expatriate religious and humanitarian aid workers: A mixed methods analysis. *Mental Health, Religion & Culture, 12*(7), 639–661. https://doi.org/10.1080/13674670903167573

Black, J. S., & Mendenhall, M. (1991). The U-curve adjustment hypothesis revisited: A review and theoretical framework. *Journal of International Business Studies, 22*(2), 225–247. https://doi.org/10.1057/palgrave.jibs.8490301

Byrne, B. M. (2016). *Structural equation modeling with AMOS: Basic concepts, applications, and programming* (3rd ed.). Routledge. http://ebookcentral.proquest.com/lib/spu/detail.action?docID=4556523

Cohen, J. (1994). The earth is round (p < .05). *American Psychologist, 49*(12), 997–1003. https://doi.org/10.1037/0003-066X.49.12.997

Cohen, J., Cohen, P., West, S. G., & Aiken, L. S. (2003). *Applied multiple regression/correlation analysis for the behavioral sciences* (3rd ed.). LErlbaum Associates.

Cumming, G., & Finch, S. (2005). Inference by eye: Confidence intervals and how to read pictures of data. *American Psychologist, 60*(2), 170–180. https://doi.org/10.1037/0003-066X.60.2.170

de Cieri, H., Dowling, P. J., & Taylor, K. F. (1991). The psychological impact of expatriate relocation on partners. *International Journal of Human Resource Management, 2*(3), 377–414. https://doi.org/10.1080/09585199100000074

Di Fabio, A. (2017a). The psychology of sustainability and sustainable development for well-being in organizations. *Frontiers in Psychology, 8*. https://doi.org/10.3389/fpsyg.2017.01534

Di Fabio, A. (2017b). Positive healthy organizations: Promoting well-being, meaningfulness, and sustainability in organizations. *Frontiers in Psychology, 8*. https://doi.org/10.3389/fpsyg.2017.01938

Di Fabio, A., & Rosen, M. A. (2018). Opening the black box of psychological processes in the science of sustainable development: A new frontier. *European Journal of Sustainable Development Research, 2*(4), 47. https://doi.org/10.20897/ejosdr/3933

Di Fabio, A., & Tsuda, A. (2018). The psychology of harmony and harmonization: Advancing the perspectives for the psychology of sustainability and sustainable development. *Sustainability, 10*(12), 4726. https://doi.org/10.3390/su10124726

Diener, E., Lucas, R. E., & Scollon, C. N. (2006). Beyond the hedonic treadmill: Revising the adaptation theory of well-being. *American Psychologist, 61*(4), 305–314. https://doi.org/10.1037/0003-066X.61.4.305

Enders, C. K., & Tofighi, D. (2007). Centering predictor variables in cross-sectional multilevel models: A new look at an old issue. *Psychological Methods, 12*(2), 121–138. https://doi.org/10.1037/1082-989X.12.2.121

Fee, A., & Gray, S. J. (2012). The expatriate-creativity hypothesis: A longitudinal field test. *Human Relations, 65*(12), 1515–1538. https://doi.org/10.1177/0018726712454900

Fee, A., Gray, S. J., & Lu, S. (2013). Developing cognitive complexity from the expatriate experience: Evidence from a longitudinal field study. *International Journal of Cross Cultural Management, 13*(3), 299–318. https://doi.org/10.1177/1470595813484310

Firth, B. M., Chen, G., Kirkman, B. L., & Kim, K. (2014). Newcomers abroad: Expatriate adaptation during early phases of international assignments. *Academy of Management Journal, 57*(1), 280–300. https://doi.org/10.5465/amj.2011.0574

Fisher, C. D., & Shaw, J. B. (1994). Relocation attitudes and adjustment: A longitudinal study. *Journal of Organizational Behavior, 15*(3), 209–224. https://doi.org/10.1002/job.4030150304

Fu, C., Hsu, Y.-S., Shaffer, M. A., & Ren, H. (2017). A longitudinal investigation of self-initiated expatriate organizational socialization. *Personnel Review, 46*(2), 182–204. https://doi.org/10.1108/PR-05-2015-0149

Gibbs, R., Guneri, O. Y., Pankau, T., & Bikos, L. H. (2020). Birds of a feather fare less well together: Modeling predictors of international student adaptation. *Sustainability, 12*(6), 2317. https://doi.org/10.3390/su12062317

Hippler, T., Brewster, C., & Haslberger, A. (2015). The elephant in the room: The role of time in expatriate adjustment. *The International Journal of Human Resource Management, 26*(15), 1920–1935. https://doi.org/10.1080/09585192.2015.1041762

Holopainen, J., & Björkman, I. (2005). The personal characteristics of the successful expatriate: A critical review of the literature and an empirical investigation. *Personnel Review, 34*(1), 37–50. https://doi.org/10.1108/00483480510578476

Lysgaard, S. (1955). Adjustment in a foreign society: Norwegian Fulbright grantees visiting the United States. *International Social Science Bulletin, 7*, 45–51.

McCoach, D. B. (2010). Hierarchical linear modeling. In G. R. Hancock, R. O'Mueller, & L. M. Stapleton (Eds.), *The reviewer's guide to quantitative methods in the social sciences* (pp. 123–140). Routledge.

McCoach, D. B., & Kaniskan, B. (2010). Using time-varying covariates in multilevel growth models. *Frontiers in Psychology, 1*(17), 1–12.

Mitchell, T. R., & James, L. R. (2001). Building better theory: Time and the specification of when things happen. *Academy of Management Review, 26*(4), 530–547. https://doi.org/10.5465/AMR.2001.5393889

O'Connell, A. A., Logan, J. A. R., Pentimonti, J. M., & McCoach, D. B. (2013). Linear and quadratic growth models for continuous and dichotomous outcomes. In Y. Petscher, C. Schatschneider, & D. L. Compton (Eds.), *Applied quantitative analysis in education and the social sciences*. Taylor & Francis Group. http://ebookcentral.proquest.com/lib/spu/detail.action?docID=1143716

Richardson, J. (2017). Case study research on expatriates. In Y. McNulty & J. Selmer (Eds.), *Research handbook of expatriates* (pp. 434–447). Edward Elgar Publishing. https://doi.org/10.4337/9781784718183

Rodgers, J. L. (2010). The epistemology of mathematical and statistical modeling: A quiet methodological revolution. *American Psychologist, 65*(1), 1–12. https://doi.org/10.1037/a0018326

Singer, J. D., & Willett, J. B. (2003). *Applied longitudinal data analysis: Modeling change and event occurence*. Oxford University Press. http://public.ebookcentral.proquest.com/choice/publicfullrecord.aspx?p=3054153

Takeuchi, R., Wang, M., Marinova, S. V., & Yao, X. (2009). Role of domain-specific facets of perceived organizational support during expatriation and implications for performance. *Organization Science, 20*(3), 621–634. https://doi.org/10.1287/orsc.1080.0403

Torbiörn, I. (1982). *Living abroad: Personal adjustment and personnel policy in the overseas setting*. Wiley.

Van Bakel, M., Gerritsen, M., & Van Oudenhoven, J. P. (2014). Impact of a local host on the intercultural competence of expatriates. *The International Journal of Human Resource Management, 25*(14), 2050–2067. https://doi.org/10.1080/09585192.2013.870292

van Erp, K. J. P. M., van der Zee, K. I., Giebels, E., & van Duijn, M. A. J. (2014). Lean on me: The importance of one's own and partner's intercultural personality for expatriate's and expatriate spouse's successful adjustment abroad. *European Journal of Work and Organizational Psychology, 23*(5), 706–728. https://doi.org/10.1080/1359432X.2013.816088

Wan, M., Singh, R., & Shaffer, R. (2017). Global families. In Y. McNulty & J. Selmer (Eds.), *Research handbook of expatriates* (pp. 468–489). Edward Elgar Publishing. https://doi.org/10.4337/9781784718183

Wang, M., & Takeuchi, R. (2007). The role of goal orientation during expatriation: A cross-sectional and longitudinal investigation. *Journal of Applied Psychology, 92*(5), 1437–1445. https://doi.org/10.1037/0021-9010.92.5.1437

Wilkinson, L. (1999). Statistical methods in psychology journals: Guidelines and explanations. *American Psychologist, 54*(8), 594–604. https://doi.org/10.1037/0003-066X.54.8.594

Chapter 3
Casting and Advocacy Model of Training Helpers: Principle-Based Supervision in a Training Clinic to Strengthen Health and Well-Being

Mickey Stein, Larry E. Beutler, Satoko Kimpara, Nancy A. Haug, Hannah Brunet, Kathleen Someah, Christopher J. Edwards, and Sandra Macias

3.1 Introduction

Supervision of psychotherapy trainees has, from the beginning, been the core of the psychotherapy training process (Watkins, 1997). There is both evidence that the theoretical orientation of these supervisors influences the allegiances of their supervisees (Guest & Beutler, 1988) and that "…research has not shown a strong link between level and type of professional training of providers, and psychotherapy outcome" (Lambert, 2004; p. 5). In the midst of this confusion, it is no wonder that among the many brands of psychotherapy that tout their superiority over others, any advantage that falls to any of them is exceedingly difficult to discern (Alavi & Sanderson, 2015). The widespread conclusion that all legitimate brands of psychotherapy produce equivalent outcomes on the average (the so-called Dodo bird verdict) has persisted in the face of nearly four decades of research designed to prove it wrong (c.f., Alavi & Sanderson, 2015; Norcross & Lambert, 2011).

The Dodo bird verdict (Luborsky et al., 1975) posits that each therapeutic modality generally produces equivalent outcomes (Wampold et al., 2002). In other words, when each therapeutic modality of psychotherapy is compared to one another, they generally produce equivalent outcomes (Wampold et al., 1997, 2002). This finding is so persuasive that it has been proposed as a benchmark standard against which any claims of superiority must be tested (Minami et al., 2008) and is a critical aspect.

More specifically and despite the ever-growing types of interventions or modalities that are available to clinicians, the unique qualities that characterize the different approaches only account for a small amount of variation, at most. This poses

M. Stein (✉) · L. E. Beutler · S. Kimpara · N. A. Haug · H. Brunet · K. Someah · C. J. Edwards · S. Macias
Department of Psychology, Palo Alto University, Palo Alto, CA, USA
e-mail: mstein@paloaltou.edu

A. Di Fabio (ed.), *Cross-cultural Perspectives on Well-Being and Sustainability in Organizations*, https://doi.org/10.1007/978-3-030-86709-6_3

various challenges for clinicians in that it leaves mental health professionals with the nebulous assignment of investigating what factors produce the relatively large therapeutic changes that they see in practice. Therefore, further study of the components that inform divergent treatment outcomes not only benefits clinical practice but also better informs the field of psychology as a whole. Research shows, however, that despite what are often large differences in how the approaches are thought to operate, when utilizing evidence-based treatments for a specific disorder, these different modalities tend to produce similar results, therefore leaving one to wonder what factors underlie effective psychotherapy (Luborsky et al., 1975).

The pervasiveness of the Dodo bird verdict (e.g., Minami et al., 2008) argues for the hypothesis that common factors, rather than unique or distinctive one's, account for most of the change attributable to psychotherapy. An alternative view asserts that while the principles that underlie psychotherapy are common to effective therapies, some of the principles themselves predict differential outcomes that are moderated by quite distinctive therapy procedures. Thus, STS guides helpers to select most appropriate treatment for each patient, giving a relevant contribution in the process of strengthening health and well-being in the perspective of the research area of the psychology of sustainability and sustainable development (Di Fabio, 2017a, b; Di Fabio & Rosen, 2018; Di Fabio & Tsuda, 2018). Accordingly, the fit of the treatment to the unique patient, rather than just its context, is given credit for changes over and above those that can be attributed to a common relationship (e.g., Castonguay & Beutler, 2006).

Beutler and colleagues (e.g., Beutler & Clarkin, 1990; Beutler et al., 2000) were among the first to pick up Garfield's challenge to identify common principles. Beutler, Clarkin, and Bongar identified a preliminary list of 18 principles emphasizing the roles of patient X treatment fit that are associated with enhanced improvement. These early efforts were followed by empirical studies to validate the impact of the principles (e.g., Beutler, 2009; Beutler et al., 2012). From these studies, a standard set of criteria was applied to recurrent findings to transform them into empirically established "principles." A principle constitutes a generalized restatement of a recurrent finding that establishes a connection between a treatment factor, a patient factor, and a treatment outcome (Beutler et al., 2000; Castonguay & Beutler, 2006).

Under the designation, *Systematic Treatment Selection* (STS), a manual (e.g., Beutler & Harwood, 2000), was created and tested to ensure the systematic application of a coherent set of principles. Controlled trial research confirmed the interworking of clusters of principles (Beutler et al., 2003). In 2016, the Substance Abuse and Mental Health Services Administration (SAMHSA) (personal communication), followed by Division 12 of the American Psychological Association (personal communication), accorded STS the status of an empirically supported treatment, not in small part because of the effectiveness study that was the precursor to that contained in this paper.

The current study was the second of two converging studies addressing the thesis that supervised psychotherapy trainees who are provided with STS-assisted supervision (SAS) will produce better results among their patients than trainees receiving supervision as usual (SAU). The first study (Holt et al., 2015) extracted and pilot tested a list of eight interacting principles from the longer lists, focusing on those that could be taught and applied over a relatively small number of sessions. A program of supervisor training was piloted on 25 supervisors and then applied to 9 patients/supervisor/therapist pairings. Eight weeks of treatment produced an effect size improvement of r = .96 over and above a benchmark value derived from extant research on therapist feedback (e.g., Lambert & Shimokawa, 2011).

The current study extended the patient, trainee, and supervisor sample piloted in the earlier study via a comparison of patients of therapists who received training in the use of therapeutic principles (the SAS group) and a sample of therapist-patient pairs who were extracted from a research pool and time yoked to the SAS therapists and who receive the usual type of supervision (the SAU group). The yoking procedure ensured that SAU and SAS therapist-patient pairs were assigned to treatment within 2 weeks of one another. This procedure also ensured that the SAU therapist pairs were blind to the research procedures. In this study, supervisors were guided by a manual and, in the case of SAS supervision, by the STS/innerlife cloud-generated treatment report (www.innerlife.com).

3.2 Method

The treatment setting was a community behavioral health clinic that served the training needs of and was owned and operated by Palo Alto University, a free-standing university that places emphasis on graduate training in Clinical Psychology. The university training clinic provided training to all second and several third year PhD and PsyD students. In accord with the stated policies of Palo Alto University, patient assignment patterns favored second year students who were designated to provide the primary service in the clinic. More advanced third year students are also allowed to provide service during an elective third, fourth, and fifth years, but these were relatively rare and frequently emphasized administrative roles and responsibilities rather than individual psychotherapy. Only one third year student was included in this study and assigned to the SAS condition.

Specific information about which year in the program each SAU student therapist was enrolled in was disallowed by the IRB committee. The heavy emphasis on the assignment of patients to second year therapy trainees before any others were assigned, combined with influence of early quarter assignment patterns, predicted that no more than one student therapist was likely to have filled a SAU slot in the study. Thus, both SAS and SAU student therapist groups can be considered to be comprised, overwhelmingly, by individuals who were seeing their first patients.

IRB Considerations

Prior to this study, the training clinic used in this study had little history of clinic wide research participation. To accommodate the several small dissertation projects that were occasionally supported, a general informed consent procedure (general IRB) was instituted that allowed data to be stored and used as archival data but did not permit any individual data (e.g., demographics, experience, background) to be gathered on patients or trainees for research purposes. The general IRB covered the use of the clinical interview and the periodic administration of an outcome measure for subjects in both groups. It also limited the data to be derived from SAU participants.

For this study, a special IRB proposal was submitted through the university, and it permitted personal data to be gathered only on participants in the SAS group. A research director was appointed by the university to monitor the interface of the SAU and SAS participants and to ensure compliance among SAU participants. The administrative directors of clinical training setting added several restrictions to those imposed by the IRB, some of which exemplify the difficulties of inserting research into ongoing clinical practice. Because of the limitations imposed, the project is considered to be an effectiveness or "generalization" study, with a blind comparison group. Trainees and procedures were modified to accommodate the limitations of the clinic and IRB imposed restrictions.

Setting

The training clinic was situated in a low- to middle-class neighborhood and served as a resource to the local and bay area communities. The patient population was largely comprised of working-class and middle-class individuals who received treatment on a sliding scale. The staff was comprised of three administrative psychologists who were full-time faculty members (non-tenure track) of the university and a varying number of practitioners (usually around 20–25) who provided supervision to second and third year practicum-level psychology trainees in one of two APA-approved programs (PhD and PsyD). It provided individual, group, and family interventions as well as provided consultation to other agencies. Medication and other medical needs were used sparsely and when needed were referred to one of several affiliated psychiatrists for medical treatment.

Procedure

Collectively, three restrictions required that the methodology used departed from that of a randomized trial. First, the university IRB that approved the project denied investigator access to the SAU therapists who were drawn from the archive pool. Second, the IRB disallowed the random assignment both of trainees and supervisors to SAS and SAU groups. Finally, the clinic administrators posed a restriction on the number of cases (n = 1) a SAS therapist and supervisor could be assigned at any one time, reducing the number of trainee-patient pairs that could be included within the time constraint imposed varying schedules. This was not considered as exclusion criteria because it complied with standard clinic procedure and imposed the same guidelines on the two groups.

The restrictions reduced the number of therapist and patient participants that could be recruited over the course of time. We could not compensate for the first stricture that is against extracting demographic and personal data on therapists and patients in the SAU condition. Partial compensation for the stricture against random assignment occurred because clinic procedure required rotating assignment of patients and therapists. This partially compensated for the stricture against randomization. The remaining issues were addressed by using the double-yoked control group extracted from recent therapy assignments as described in the following paragraphs.

Participants

The total number of participants in this study consisted of three groups drawn from those providing and receiving service in the clinic: supervisors (n = 18), trainees (n = 46), and patients (n = 51).

Trainees and Patients

Trainee therapists in this study included second and third year students in the PhD or PsyD program at Palo Alto University. All SAU trainees were yoked, post hoc, to a particular SAS patient assigned on rotation and a supervisor who was pre-assigned as per clinic policy. The recruiting period for trainee therapists extended from January 2014 to June 2014 and December 2014 to June 2015. Volunteers were sought among all of the students, and an invitation was tendered to attend SAS supervision group training. Those supervisors who volunteered and attended one or more group trainings were included in the SAS group. Those who did not attend the initial training were entered into the pool and assigned as SAU volunteers. SAU participants were not required to attend any training beyond their regularly scheduled supervision sessions.

SAS trainee volunteers were assigned to a participating SAS supervisor. Supervisors identified the order in which each of their designated SAS trainees would be assigned a training case. The yoking procedure was then invoked to ensure similarity of patients in the two groups. It should be noted that no patients participated in both formats (SAS and SAU). The final SAS trainee sample consisted of 15 second and 1 third year trainee volunteers who were assigned to supervisors as per clinic procedures, prior to receiving their first, second, or third training case. The second year students were assigned to supervisors prior to receiving one of their first three training cases. The third year student was assigned prior to receiving their fifth training case.

While IRB restrictions disallowed the investigators knowing how many prior training cases each therapist trainee in the SAU condition had treated, both the rotating assignment procedure and the priority given to second year students provided some assurance of sample comparability of experience. Collectively, these procedures, along with a nonsignificant chi-squared test of the demographic features available, supported a conclusion that the SAS and SAU groups were similarly inexperienced and naïve about psychopathology and psychotherapy.

Supervisors

The 18 supervisors taking part in this study were licensed clinical psychologists (PhD or PsyD) who had volunteered as part-time supervisors in the clinic. They were assigned a cadre of supervisees on a rotating schedule. As per clinic procedure, most supervision was provided within a group of six to eight trainees and involved individual discussion and direction based on therapists' particular orientations. Both SAS and SAU supervisors initially were selected from the same cadre of onboard supervisors and part-time faculty, and all were fully informed about the nature of their participation. Following the IRB mandates, among patients seen by SAU trainees, only grouped demographic data on patients and outcome data from the patient outcome measure were given permission to be used in this study (Table 3.1).

Procedure

Assignment

The clinic routinely followed a general informed consent procedure. All patients, students, and supervisors signed the general consent form specifying that grouped data describing incoming patients may be used for research purposes (unspecified). All patients completed an OQ-45.2 (Lambert et al., 1996), which was repeated at the end of treatment. They also completed a brief problem checklist before the intake session.

Table 3.1 Demographic data for patients

Demographic	*Total* n M(SD)	*SAS* n M(SD)	*SAU* n M(SD)
Patients	51	17	34
Gender			
Males	17	4	12
Females	31	11	21
Transgender	3	2	1
Age	37.02(14.83)	34.82(10.68)	36.44(16.65)
19–29	21	6	15
30–39	15	5	10
40–49	8	4	4
50 and older	7	2	5
Ethnicity			
Caucasian	27	11	16
Hispanic	11	2	9
Asian	4	2	2
African-American	2	0	2
Pacific Islander	1	1	0
Mixed	5	1	4
Declined to state	1	0	1

Note. *SAS* STS-assisted supervision, *SAU* supervision as usual

In order to compensate for not being able randomly to assign therapists to conditions, double-yoked, blind (contemporary, archival) comparison (SAU) group was employed. Specifically, after each SAS trainee assignment to a patient, both the immediately previous and next scheduled trainees were designated to the archival SAU condition. This meant both that the SAU group was twice as large as the SAS group and that some of the supervisors (n = 8) supervised students in both conditions. This within-subjects overlap served to homogenize the two supervisor groups and reduced the threat of unequal supervisor groups. This procedure was designed to (1) increase the total N, (2) reduce the unwanted effect of introducing new procedures (i.e., the Hawthorne effect), and (3) ensure temporal equivalence of the patient samples (McBride, 2013). It is notable that it also increased the likelihood of a Type II error.

Ultimately, the procedure resulted in having only the one third year student therapist in the SAS condition being assigned to two SAS patients, because she participated in both academic years in which the study was carried out.

SAS Training

SAS supervisor training proceeded in a three-step process: (1) All clinic supervisors were invited to a 2-h CE training on supervision (N = 31). Of these, 18 prospective supervisors accepted the invitation. (2) Supervisors who completed the initial general training were then invited to participate in the study as SAS or SAU supervisors. The total number of SAS and SAU supervisors was n = 10 and n = 16, respectively. Between the two conditions, eight supervisors overlapped and supervised at least one trainee-patient pair in each condition. (3) Extended SAS supervisor training took place over a period of 11 weeks.

SAS supervisors met six times, for 2 h each for training. The training was presented via live lectures with questions and answers along with PowerPoint presentations to present the format of supervision and the principles that would be addressed during the course of supervision. A training manual was also provided that outlined a week-by-week program of instruction and practice.

No special training was initiated for those supervisors, assigned to the SAU group. In the interests of maintaining blinds, they were told only a minimal amount about the study. Those supervisors, who supervised trainee therapists in both SAS and SAU conditions, were encouraged to use their usual methods when they were following SAU protocols but to shift to the SAS format if their student was assigned a patient designated for the SAS group.

Because of the limit imposed by the IRB on how many SAS patients could be seen by a student therapist at any one time, the study was extended over a second academic year. In this second period, the therapy was extended from 8 to 16 sessions to provide a test of therapy dose effects. Otherwise, the procedures were the same.

Measures

Two types of measures were required in this study: (1) a common, pre-posttreatment measure of therapeutic outcome and (2) specific SAS measures of therapist compliance with the training in psychotherapy.

Outcomes Questionnaire-45.2 (OQ-45.2; Lambert et al., 1996). The OQ-45.2 was used as the primary outcome measure to test the hypothesis that SAS facilitates greater therapeutic change than SAU. It is a 45-item self-report questionnaire that was administered to all patients upon entering treatment and at the beginning of each session. The OQ-45.2 monitors changes in distress over time as it relates to three subdomains: symptom distress (depression and anxiety), interpersonal functioning, and social role. Each item is scored on a five-point Likert scale ranging from 0 to 4 (0 = "Never"; 1 = "Rarely"; 2 = "Sometimes"; 3 = "Frequently"; 4 = "Almost Always"), and positive items are reverse scored. The questions included in this measure are structured to monitor a patient's feelings over a 1-week period. Total scores range from 0 to 180, in which a score of 63 or higher indicates symptoms of clinical significance (Lambert et al., 2004). The OQ-45.2 has high internal consistency of .93 and test-retest reliability of .84.

STS/Innerlife (Beutler et al., 2008). STS is a cloud-based psychological assessment (*www.innerlife.com)* designed to help a therapist select optimal types of intervention for each patient. It consists of 173 items and produces both an intake narrative that recommends specific classes of intervention and graphic information on 6 global symptom measures (depression, anxiety, somatic complaints, chemical abuse, thought disorder, and risk of self-harm) and up to 16 more specific domain scales (Beutler et al., 2000; Beutler & Harwood, 2000). These scales and the accompanying narrative report are used to construct tailored interventions for each participant. Interventions may be adjusted based on 11 questions addressing patients' self-help preferences, 4 questions tapping readiness for change, and 8 demographic questions. Scores on 9 of these scales were used directly in fitting the treatment to the eight core principles used in the therapy.

STS/innerlife scales earn high to moderate reliabilities (mean of rs = .85) and require from 15 to 35 minutes to complete on an iPad, iPhone, or computer. For the purposes of this study, only moderators were used in the assessment that relates to the eight principles comprising the training program for supervisors described by Holt et al. (2015) (e.g., functional impairment (FI), internalizing and externalizing coping styles, resistance, social isolation/support, symptom severity, chronicity, and readiness for change). Other scales and constructs associated with treatment planning were still under review and validation at the time of this study.

Supervision

Each trainee (n = 46) was part of a supervision group led by one of the faculty supervisors in either the SAS or SAU condition. Each group consisted of six to eight supervisees, and one half of the SAS supervisors saw trainees who were assigned to both conditions. This ensured that all SAS participants and at least half of the SAU participants received informational material about the STS principles but did not practice them. SAS supervisors (n = 10) "randomly" picked one of their students at a time to participate in SAS training, excluding those who had declined to enter the study.

SAS training was built around eight core principles of therapeutic change that were extracted from empirical literature (Beutler & Clarkin, 1990; Beutler et al., 2000; Castonguay & Beutler, 2006). Each session followed a common format which included (1) a review of patient status and feedback on trainee progress since the previous session, (2) the presentation (verbally and with DVD) of one of the eight principles to the trainee, (3) discussion and review using the STS/innerlife narrative and treatment recommendation, (4) review of graphs and scores as pertaining to the principle being presented, and (5) feedback and planning for the subsequent therapy session. A DVD of a treatment session with the patient-trainee pair was reviewed and scored for compliance every 3 or 4 weeks, and supervisor feedback about compliance was provided weekly or biweekly. The eight core principles that underlay STS supervision (Holt et al, 2014) began with the assessment of risk (impairment) and followed with the others in sequence either weekly (year #1) or semiweekly (year #2):

1. For all patients with moderate to severe impairment, the therapist should identify social service or medical care needs and arrange for attention to these needs. Those with low social support systems, in particular, need assistance from the therapist to develop social support and support services.
2. Therapy is likely to be beneficial if a strong working alliance is established and maintained during the course of treatment.
3. The qualities of a good working alliance are likely to be facilitated if the therapist relates to clients in an empathic way and adopts an attitude of caring, warmth, and acceptance and an attitude of congruence or authenticity.
4. Therapists are likely to resolve alliance ruptures when addressing such ruptures in an empathic and flexible way.
5. In dealing with the resistant client, the therapist's use of directive therapeutic interventions should be planned to inversely correspond with the patient's manifest level of resistant traits and states.
6. Clients whose personalities are characterized by relatively high "externalizing" styles (e.g., impulsivity, social gregariousness, emotional liability, and external blame for problems) benefit more from direct behavioral change and symptom reduction efforts, including building new skills and managing impulses, than they do from procedures that are designed to facilitate insight and self-awareness.
7. Clients whose personalities are characterized by relatively high "internalizing" styles (e.g., low levels of impulsivity, indecisiveness, self-inspection, and over-control) tend to benefit more from procedures that foster self-inspection, self-understanding, insight, interpersonal attachments, and self-esteem than they do from procedures that aim at directly altering symptoms and building new social skills.
8. Clients who are in more advanced stages of readiness for change (e.g., preparation, action, maintenance) are more likely to improve in psychotherapy than those at lower stages of readiness (pre-contemplation, contemplation).

These principles likely would be widely recognized by clinicians and researchers alike. The unique contributions of STS were to state the principles in a common (non-theoretical) language and to determine their relative importance to outcome. Foundational studies are summarized in the following references: Beutler, 2009; Beutler et al., 2000; Castonguay and Beutler, 2006; and www.innerlife.com.

Treatment

SAS and SAU followed similar formats of combined group and individual supervision. They also both used DVD recordings to monitor treatment processes and to measure treatment gains. The distinction between SAS and SAU was that the former groups of supervisors were trained in the eight principles and were monitored to ensure that they followed the supervision protocol having to do with the trainees' applications of the principles. The monitoring of supervision fidelity among SAS supervisors consisted of a weekly review of sessions, rating the degree of compliance using the Therapist Process Rating Scale (TPRS). This rating ensured that SAS trainees and supervisors had effectively learned and complied with the principles.

During the 2013/2014 academic year, the SAS patient group was administered eight treatment sessions, which was considered to be the shortest possible time necessary to learn and implement the eight principles. Teaching eight principles in eight supervision sessions proved to be unduly demanding on supervisors and trainees. Moreover, supervisors raised the concern that eight sessions were insufficient to accurately represent the changes taking place. Thus, for the 2014/2015 academic year, the treatment time was expanded to 16 sessions in order to (1) allow for more time to implement the eight principles and (2) to allow sufficient time for stable changes to occur (Howard et al., 1993).

For the 2013/2014 academic year, the OQ-45.2 was administered and scored at intake, as well as at the fourth and eight sessions. For the 2014/2015 academic year, the OQ-45.2 was administered and scored at intake, as well as at the 6th, 12th, and 16th sessions. For the patients in the SAU supervision group, the OQ-45.2 data were gathered from trainees at each of these same time points to ensure comparability.

3.3 Results

A preliminary analysis to ensure demographic comparability of the samples was conducted. Gender was defined as male, female, or transgender; ethnicity was defined as White, non-White, or mixed; and age was categorized into three groups. Specifically, patients between the ages of 18 and 29 were in group 1, ages 30 to 39 in group 2, and ages 40 and up in group 3. A series of independent sample chi-squared tests were performed to analyze these data check on group comparability. No significant relationships were found among either gender assigned (X^2 [2, N = 51] = 2.30, $p > .10$), ethnicity (X^2 [3 N = 51] = 1.82, $p > .10$), or age (X^2 ([, N = 51] = .514, $p > .10$) and the treatment assigned.

In order to test the major hypothesis, which predicted for differential outcomes between the two treatment groups, a two (SAS vs. SAU) by two (pre vs. post scores) repeated measure ANOVA was conducted using the OQ-45.2 as the dependent variable. As predicted, the main effect of time was significant, $F(1,49) = 51.38, p < .01$, as was the group by time interaction effect $F(1,49) = 4.83, p < .001$. There was a large effect size ($d = .62$, 95% CI [.05, 1.18]) indicating that the two groups differed in amount of change observed in the OQ-45.2 over time. A post hoc t-test was calculated to determine whether patients in the SAS group had significantly lower scores than those in the SAU group at the end of treatment. The results confirmed that the SAS group scored significantly lower on the OQ.45-2 (M = 57.65) at treatment end than those in the SAU group (69.09), $t(49) = 1.62, p = 0.1$.

A two-way ANOVA was conducted that examined the effect of treatment group (SAS vs. SAU) and different lengths of treatment (8 vs. 16 weeks). A statistically significant effect size was not noted between the effects of treatment groups and treatment length on treatment outcomes ($F[1,47] = .75$, $p > 0.5$). A main effect difference between treatments was determined ($F[1,47] = 5.60$, $p < .05$).

Reliable change (RCI) was calculated in order to assess the degree to which the change obtained placed the patient within the non-clinical range. Changes in patient scores were compared to the RCI non-clinical mean of the OQ-45.2, so as to determine the frequency of clinical improvement that returned patients to non-clinical levels, assessed between both groups. The RCI of the OQ-45.2 has been fixed at a score of 14. Twelve out of 17 patients (70.59%) in the SAS condition returned to normal limits within the brief period of the treatment studied. By comparison, only 16 of the 34 (47.06%) patients in SAU demonstrated a significant improvement over this same time period. The earned SE means was 8.38 and, at the .05 level of confidence, established the clinically reliable change criteria at 13.24. The earned difference of 22.94 indicates strong likelihood that the changes evoked by SAS were meaningful.

To calculate retention, patient's total number of sessions completed were counted and converted into a percentage value and then divided by a full treatment dose (i.e., 8 sessions in 2013/2014 academic year and 16 sessions in the 2014/2015 academic year). The results of the t-test with respect to retention indicated that there were no significant differences between treatment groups (SAS and SAU) in regard to prematurely terminating treatment, $t(49) = -0.86$, $p > .10$ (mean SAS = .88; mean SAU = .81). An additional chi-squared test was performed to determine whether or not there was a relationship between completers and non-completers and treatment groups. The results were nonsignificant (X^2 [1, N = 51] = 0.00, $p = .62$).

Finally, an analysis was conducted just on the supervisors in the SAU group, comparing outcomes on the supervisors who were trained in STS and those who were not trained. Again, there was not a significant effect of STS trained supervisors on OQ-45.2 outcomes, $F(1,34) = 0.02$, $p = 0.90$.[1]

3.4 Discussion

The overarching aim of this study was to determine if those patients whose trainees administer and receive supervision based on empirically derived principles of therapeutic change are more effective than those whose supervision was based on supervisor's theoretical preferences. As expected, patients whose therapists were trained

[1] A foreseen limitation of this study was that there were a limited number of supervisors available at the clinic. Therefore eight supervisors were reutilized in both the SAS and SAU conditions. Therefore, patient outcomes were compared among therapists who had been trained in STS when they were conducting SAU supervision and those who were trained in and practiced only one modality. An ANOVA was conducted, comparing STS- and non-STS-trained supervisors on composite change scores. There was not a statistically significant difference between supervisors who carried student therapists in both the SAS and SAU groups and those who carried students in only one modality of supervision, as determined by a one-way ANOVA ($F[1,49) = 1.85$, $p = .18$).

using STS principles (SAS) showed significantly greater symptom reduction than those whose therapists were supervised in the usual manner (SAU). Furthermore, neither the length of treatment (8 vs. 16 weeks) nor the year that patients entered into the study (2013/2014 vs. 2014/2015) had a significant impact on treatment outcomes.

The second hypothesis anticipated that dropout rates would be lower among those patients in the SAS group when compared to those in the SAU group. The means were in the expected direction, but the statistical results failed to demonstrate significance. It must be noted, however, that low dropout rates ($n = 4$) made the analysis of this hypothesis very weak. Little can be concluded on the basis of these small numbers of dropouts.

When analyzing the differences in outcomes obtained among supervisors who practiced SAS and those who used SAU, we noted that merely learning the structure of SAS was insufficient to provoke a treatment enhancement. Supervisors who learned by did not practice SAS, and those who participated in both SAS and SAU earned results suggesting that the actual practice of SAS is necessary to optimize cognitive knowledge of the procedures.

The findings add to a body of studies (Beutler & Clarkin, 1990; Beutler et al., 2000) that have demonstrated through various methods that when a therapist adheres to empirically derived principles of change, the effectiveness of their interventions is enhanced when compared to those that use abstract clinical theories. But these findings and others like them may easily be forgotten or discarded in the conceptual and practical discordance that exists among those whose mission is to preserve patient choice and those whose mission is to ensure scientific inquiry.

When transitioning research to natural clinical settings or expanding from isolated research projects to system-wide clinical programs, one must anticipate that the normal procedures of the clinical setting will limit the purity and generalizability of one's research. In this study, the IRB imposed limits on how participants are assigned, denied the general use of various measures, and limited what archival data could be used for comparing samples and treatments. While such strictures are often onerous in this type of research, one can, nonetheless, extract strong and reasonable findings using naturalistic and quasi-experimental methods, supplemented by statistical procedures that model more controlled methods (Beutler, 2009; Beutler & Forrester, 2014; Machado & Beutler, 2017). But doing so effectively also requires that both scientific journals and extant research guidelines offered by scientific organizations must become more tolerant of non-randomized designs.

A legitimate concern with this study is the fact that supervisors were asked (after the introductory training session) to "self-select" student therapists to receive training in STS and participate as SAS supervisors. Such self-selection may have inflated the results, since the supervisors who did not elect to undergo the additional training for SAS participation did not face the same choice. Their students were automatically placed in the SAU group. Mitigating the uniqueness of this concern and its contaminating effects on the findings are two facts. First, self-selection is present in RCT research almost universally and cannot be avoided. In most RCT research comparing two or more models of treatment, to our knowledge, therapists are recruited from a larger and poorly defined pool, they volunteer, and then assigned to the treatment that best fits their preferences and past training, they volunteer again. This procedure mirrors the self-selection used for supervisors in this study.

Second, while self-selection may impose a contaminating effect, it is likely to be small since non-SAS supervisors also were self-selected through a two-stage process. All supervisors were given the explicit choice to participate or not, and then those who selected participation were all asked if they would like to or be willing to undertake the additional training to then employ the SAS procedure. The selection was overt for all, and there is no real reason to think that selecting SAS imposed any more self-selection bias than selecting SAU.

Thus, in selecting the procedure used that while self-selection bias may exert a contaminating influence recruits, it was both balanced across groups and was consistent with extant RCT research on psychotherapy. Those who conduct efficacy studies and publish them are implicitly moved to accept the inevitability of non-randomness in assignments of therapist, patient, and supervisor roles.

The implicit nature of acceding to the presence of non-random factors in the use of controlled research designs underlines the need to develop alternative statistical and methodological paradigms to attenuate the effects of non-randomization. This study provides one example of methodological adaptation to explicit compromises to randomization, the use of yoked controls. Statistical methods (such as modeling and block entry regression analyses) also need to be expanded to help with these problems as they arise. But, perhaps most of all, there needs to be some general acceptance, within the scientific community, of the facts that there is no purely randomized model for conducting effectiveness research on psychotherapy and that there is knowledge to be gained via non-randomized research designs.

Finally, there are some questions that cannot easily be resolved in the current study, largely because options to conventional randomization designs are/were not available. The possibility that supervisors and supervisees in the SAS condition spent more time within supervision and/or psychotherapy with the respective patients is one such concern and can only be addressed if one is aware that different treatments must deal with time differently. The IRB forced this study to accommodate to the clinical settings use of group supervision, but many questions still arise about the equivalence of this type of training versus the usual one-on-one supervision. In order to gain a better understanding of therapy change, future research might benefit from comparing group and individual supervision models. Nonetheless, this research suggests that those who are training in the SAS model emphasize individual differences and its matching interventions in supporting health and well-being, offering a critical contribution in the research area of the psychology of sustainability and sustainable development (Di Fabio, 2017a, b; Di Fabio & Rosen, 2018; Di Fabio & Tsuda, 2018).

References

Alavi, S. B., & Sanderson, W. C. (2015). Mechanisms of action in psychotherapy. In *The encyclopedia of clinical psychology*.

Beutler, L. E. (2009). Making science matter in clinical practice: Redefining psychotherapy. *Clinical Psychology: Science And Practice, 16*(3), 301–317. https://doi.org/10.1111/j.1468-2850.2009.01168.x

Beutler, L. E., & Clarkin, J. F. (1990). Systematic treatment selection: Toward targeted therapeutic interventions. New York, NY: Brunner/Mazel.

Beutler, L. E., Clarkin, J. F., & Bongar, B. (2000). *Guidelines for the systematic treatment of the depressed patient*. Oxford University Press. https://doi.org/10.1093/acprof:oso/9780195105308.001.0001

Beutler, L. E., & Forrester, B. (2014). What needs to change: Moving from 'research informed' practice to 'empirically effective' practice. *Journal Of Psychotherapy Integration, 24*(3), 168–177. https://doi.org/10.1037/a0037587

Beutler, L. E., Forrester, B., Gallagher-Thompson, D., Thompson, L., & Tomlins, J. B. (2012). Common, specific, and treatment fit variables in psychotherapy outcome. *Journal Of Psychotherapy Integration, 22*(3), 255–281. https://doi.org/10.1037/a0029695

Beutler, L. E., & Harwood, T. M. (2000). *Prescriptive psychotherapy: A practical guide to systematic treatment selection*. Oxford University Press.

Beutler, L. E., Moleiro, C., Malik, M., Harwood, T. M., Romanelli, R., Gallagher-Thompson, D., & Thompson, L. (2003). A comparison of the dodo, EST, and ATI factors among comorbid stimulant-dependent, depressed patients. *Clinical Psychology & Psychotherapy, 10*(2), 69–85. https://doi.org/10.1002/cpp.354

Castonguay, L. G., & Beutler, L. E. (Eds.). (2006). *Principles of Therapeutic Change that work: Integrating relationship, treatment, client, and therapist factors, Vol. I*. Oxford University Press.

Di Fabio, A. (2017a). The psychology of sustainability and sustainable development for well-being in organizations. In G. Arcangeli, G. Giorgi, N. Mucci, J.-L. Bernaud, & A. Di Fabio (Eds.), Emerging and re-emerging organizational features, work transitions and occupational risk factors: The good, the bad, the right. An interdisciplinary perspective. Research Topic in *Frontiers in Psychology. Organizational Psychology, 8, 1534*. https://doi.org/10.3389/fpsyg.2017.01534

Di Fabio, A. (2017b). Positive Healthy Organizations: Promoting well-being, meaningfulness, and sustainability in organizations. In G. Arcangeli, G. Giorgi, N. Mucci, J.-L. Bernaud, & A. Di Fabio (Eds.), Emerging and re-emerging organizational features, work transitions and occupational risk factors: The good, the bad, the right. An interdisciplinary perspective. Research Topic in *Frontiers in Psychology. Organizational Psychology, 8, 1938*. https://doi.org/10.3389/fpsyg.2017.01938

Di Fabio, A., & Rosen, M. A. (2018). Opening the black box of psychological processes in the science of sustainable development: a new frontier. *European Journal of Sustainable Development Research, 2(4), 47*. https://doi.org/10.20897/ejosdr/3933

Di Fabio, A. & Tsuda, A. (2018). The psychology of harmony and harmonization: Advancing the perspectives for the psychology of sustainability and sustainable development. In A. Di Fabio (Ed.), Psychology of Sustainability and Sustainable Development. Special Issue in *Sustainability MDPI 10*(12), 4726. https://doi.org/10.3390/su10124726

Guest, P. D., & Beutler, L. E. (1988). Impact of psychotherapy supervision on therapist orientation and values. *Journal Of Consulting and Clinical Psychology, 56*(5), 653–658. https://doi.org/10.1037/0022-006X.56.5.653

Holt, H., Beutler, L. E., Kimpara, S., Macias, S., Haug, N. A., Shiloff, N., Goldblum, P., Temkin, R. S., & Stein, M. (2015). Evidence-based supervision: Tracking outcome and teaching principles of change in clinical supervision to bring science to integrative practice. *Psychotherapy, 52*(2), 185–189. https://doi.org/10.1037/a0038732

Howard, K. I., Lueger, R. J., Maling, M. S., & Martinovich, Z. (1993). A phase model of psychotherapy outcome: Causal mediation of change. *Journal of Consulting and Clinical Psychology, 61*(4), 678–685. https://doi.org/10.1037/0022-006X.61.4.678

Lambert, M. J. (2004). The efficacy and effectiveness of psychotheraphy. In M. J. Lambert (Ed.), Handbook of psychotheraphy and behavior change. Hoboken, NJ: Wiley.

Lambert, M. J., Burlingame, G. M., Umphress, V., Hansen, N. B., Vermeersch, D. A., Clouse, G. C., & Yanchar, S. C. (1996). The reliability and validity of the Outcome Questionnaire. *Clinical Psychology & Psychotherapy, 3*(4), 249–258.

Lambert, M. J., Vermeersh, D., Brown, G. C., & Burlingame, G. (2004). Administration and Scoring Manual for the OQ-30.2. OQ Measures. Available at: http://www.oqmeasures.com/measures/adult-measures/oq-30/

Lambert, M. J., & Shimokawa, K. (2011). Collecting client feedback. In J. C. Norcross (Ed.), *Psychotherapy relationships that work* (2nd ed., pp. 203–223). Oxford University Press.

Luborsky, L., Singer, B., & Luborsky, L. (1975). Comparative studies of psychotherapies: Is it true that "everyone has won and all must have prizes"? *Archives of general psychiatry, 32*(8), 995.

Machado, P. P. P., & Beutler, L. E. (2017). Methods of psychotherapy research. In A. J. Consoli, L. E. Beutler, & B. Bongar (Eds.), *Comprehensive textbook of psychotherapy: Theory and practice* (2nd ed., pp. 445–461). Oxford University Press.

McBride, D. M. (2013). *The process of research in psychology*. Sage Publications.

Minami, T., Wampold, B. E., Serlin, R. C., Hamilton, E. G., Brown, G., & Kircher, J. C. (2008). Benchmarking the effectiveness of psychotherapy treatment for adult depression in a managed care environment: A preliminary study. *Journal Of Consulting And Clinical Psychology, 76*(1), 116–124. https://doi.org/10.1037/0022-006X.76.1.116

Norcross, J. C., & Lambert, M. J. (2011). Psychotherapy relationships that work II. *Psychotherapy, 48*(1), 4–8. https://doi.org/10.1037/a0022180

Wampold, B. E., Minami, T., Baskin, T. W., & Callen Tierney, S. (2002). A meta-(re) analysis of the effects of cognitive therapy versus 'other therapies' for depression. *Journal of Affective Disorders, 68*(2), 159–165.

Wampold, B. E., Mondin, G. W., Moody, M., Stich, F., Benson, K., & Ahn, H. N. (1997). A meta-analysis of outcome studies comparing bona fide psychotherapies: Empirically, "all must have prizes.". *Psychological Bulletin, 122*(3), 203.

Watkins, C. E. (Ed.). (1997). Handbook of psychotherapy supervision. New York: Wiley.

Chapter 4
Psychobiological Responses with Work-Related Stress in Japanese Female Workers

Hisayoshi Okamura, Jumpei Yajima, Kengo Mihara, and Akira Tsuda

4.1 Introduction

The Psychology of Sustainability and Sustainable Development (Di Fabio, 2017a, b; Di Fabio & Rosen, 2018) is focused on health and well-being of individuals in different environments. In this framework, Psychology of Harmonization (Di Fabio & Tsuda, 2018) represents a new theoretical point of reference. It highlights the worth of recomposing different levels and aspects of internal and external environments for enhancing health and well-being. Harmonization of natural and cultural resources (Di Fabio & Tsuda, 2018) for well-being, happiness, and sustainable development constitutes a cross-cultural challenge. The present study is placed in this framework with the aim of better understanding work-related stress in Japanese female workers.

Recent research has focused on a stress model that evaluates the integration of multiple biological systems and the long-term effects of stress. The body adapts to stress through fluctuations in the autonomic nervous, endocrine, and immune systems (referred to as allostasis). When these adaptations are chronic, the body's regulatory functions become irregular, exhausted, or disrupted—a phenomenon known as allostatic load (McEwen, 2006). Much research focusing on cortisol as an

H. Okamura (✉)
Cognitive and Molecular Research Institute of Brain Diseases, Kurume University, Fukuoka, Japan
e-mail: okamura_hisayoshi@med.kurume-u.ac.jp

J. Yajima
Department of Human Studies, Beppu University, Beppu, Oita, Japan

K. Mihara
Department of Human Community, Kyushu Ryukoku Junior College, Tosu, Saga, Japan

A. Tsuda
Graduate School of Medical Sciences, Teikyo University of Science, Senju, Tokyo, Japan

A. Di Fabio (ed.), *Cross-cultural Perspectives on Well-Being and Sustainability in Organizations*, https://doi.org/10.1007/978-3-030-86709-6_4

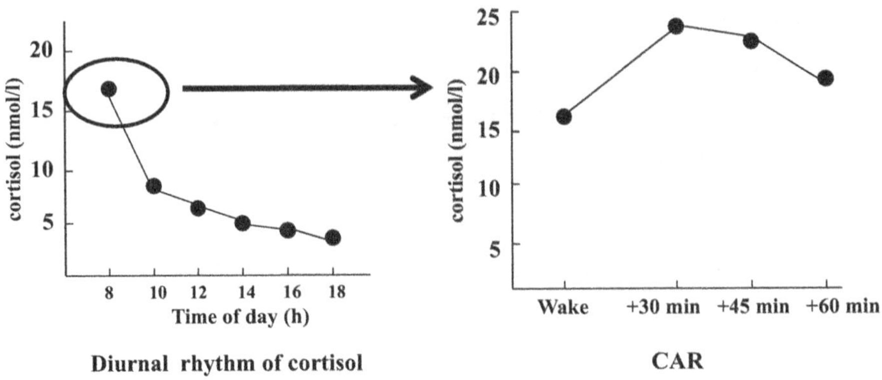

Fig. 4.1 Diurnal rhythm of cortisol and cortisol awakening response (CAR)

indicator of allostatic load has been conducted alongside the development of allostasis theory. These studies have placed a particular emphasis on the cortisol awakening response (CAR)—the elevation in cortisol 20–40 min after awakening—as a precise indicator of stress and health (Fig. 4.1). CAR also reflects a switch from a state of rest to that of activity and is thus thought to be a precise indicator of hypothalamic–pituitary–adrenal (HPA) axis activation (Stalder et al., 2016).

Studies on CAR and stress have found that CAR is promoted in individuals with high levels of everyday stress, who work long hours, or who have many worries about their job (Gustafsson et al., 2008; Schlotz et al., 2004; Steptoe et al., 2004a). CAR in workers has also been shown to be higher on weekdays (workdays) than on weekends (Kunz-Ebrecht et al., 2004). In contrast, CAR is reduced in states of chronic stress and chronic fatigue (Miller et al., 2007; Lindeberg et al., 2008; O'Connor et al., 2009) and is lower in individuals with high positive affect, including subjective happiness (Okamura et al., 2010; Stalder et al., 2010). A recent meta-analysis exploring the relation between deterioration of psychosocial function and CAR demonstrated that everyday stress, negative affect (depression, anxiety), and neurotic personality are factors that elevate CAR, whereas posttraumatic stress disorder (PTSD) and physical fatigue are factors that inhibit CAR (Boggero et al., 2017). These findings suggest that CAR cannot be interpreted unilaterally but rather is a dynamic response involving multiple variables. Thus, studies exploring the relation between psychosocial factors and CAR must be comprehensive.

According to the Ministry of Health, Labour and Welfare (2019), women comprise an increasing percentage of the Japanese workforce every year, currently reported at 44.1%. As the number of Japanese working women has increased, so too has the number of women complaining of stress, fatigue, and insomnia (IIDA & Ogino, 2019). Further, Japanese men spend little time on childcare and housework compared with other countries; indeed, Japanese husbands participate minimally in such tasks regardless of whether their wife is also working (Statistical Bureau of Japan, 2016). The subjective burden of housework and childcare on women has been particularly high in recent years with Japan's culture of nuclear households.

Many women simultaneously take on both a career and housework/education and are faced with the same work-related health problems as men. As such, work-related stress and fatigue among women workers have been receiving attention as serious concerns (Miyauchi et al., 2010).

Despite the need for updated health care for women workers, few studies have considered the differences in the effects that subjective work-related stress and everyday affective states have on mental and physical health and stress response between weekdays (workdays) and weekends. There are even fewer studies simultaneously exploring biomarkers, such as cortisol; positive affective states, such as subjective happiness; and psychosocial factors.

McEwen (2001) considered the temporary increase in cortisol secretion in response to short-term stress (acute stress response) to be a mechanism for adapting to stress but indicated that chronic elevation not only leads to a depressive state but also plays a role in causing stress-related illnesses, including hypertension, high cholesterol, and stomach ulcers. These findings suggest the critical importance of resting, both psychologically and physiologically, on weekends—which are seen as a chance to escape from work-related stress—to stay healthy and avoid stress-related diseases. Objective research using biomarkers is thought to be of great significance in Japan, where the most common methodology for health psychology studies is survey research using questionnaires or psychological tests.

Therefore, in this study, we explored the relation between psychosocial factors and CAR in Japanese women workers through three separate studies focusing on subjective happiness (study 1), fatigue (study 2), and loneliness (study 3). We considered how these three psychosocial factors and CAR on weekdays and weekends differ while also incorporating comparison by subjective work-related stress.

4.2 Study 1 Subjective Happiness and CAR on Weekdays/Weekends (Okamura et al., 2010)

The Japanese Subjective Happiness Scale (JSHS) was developed by Shimai et al. (2004) to reveal the characteristics of people who are consistently happy under a variety of circumstances; they found that those with a high JSHS score have high life satisfaction, have high self-esteem, and are healthy. Further, the JSHS score has been shown to have a negative association with the objective stress indicators of cortisol response and peripheral inflammatory cytokine concentration. Thus, the JSHS score is useful as a standardized indicator of subjective happiness.

In this study, we explored how the dependent variable CAR differed in groups with high and low subjective happiness as evaluated by the JSHS. Next, we explored how subjective work-related stress, measured by the Japanese version of the Effort–Reward Imbalance Questionnaire (ERIQ) (Tsutsumi et al., 2001), and day of the week (weekday vs. weekend) impacted CAR as mediating variables. Subjective work-related stress was evaluated as exposure to stressful working conditions and individual attitude toward work and behavioral patterns.

Method

Research Participants

Participants were recruited over the phone and by email. The JSHS was administered to 58 healthy women university staff members (20–50 years old) who consented to study participation. High (n = 15; mean + 1/2 SD) and low (n = 12; mean – 1/2 SD) subjective happiness groups were extracted. We confirmed in advance that the participants had no history of serious illness and were not taking any medications.

Study Design

The participants followed the same procedure. They collected saliva samples on two separate days: once on a weekday (a normal workday) and once on a weekend. Saliva samples were collected six times on each of these days: before sleeping, upon awakening, 30 min after awakening, at 10:00, at 12:00, and at 15:00. They were also instructed to complete a mood scale when they collected each saliva sample. The participants were asked to visit the lab for a brief explanation of the experiment procedure and saliva sample collection method and to provide basic data on their attributes. After this, the participants were provided with the ERIQ and Center for Epidemiologic Studies Depression (CES-D) scale, and then they were requested to complete them by the end of the experiment. Saliva samples were collected using Salisoft (SARSTEDT, Numbrecht, Germany) for a period of 2 min (without moving the jaw or stimulating saliva production in any way).

Assessment of Psychosocial Variables

Happiness was assessed using the Japanese version of the Subjective Happiness Scale (SHS) (Shimai et al., 2004). The SHS is a questionnaire measuring subjective happiness comprising four items self-evaluated on a seven-point scale. For example, the first item asks participants to characterize themselves using absolute ratings from “not a very happy person” (1) to “a very happy person” (7). The scores from each item were averaged into single JSHS composite score, with higher score demonstrates a higher level of subjective happiness. The test–retest reliability, convergent validity, and discriminant validity of the SHS were confirmed through a series of studies to determine the characteristics of people who continue to be happy regardless of their circumstances.

Work-related stress was measured using the Japanese version of the ERIQ (Tsutsumi et al., 2001). A Japanese version of the ERIQ has three main scales: extrinsic effort, reward, and overcommitment. Extrinsic effort refers to the demanding aspects of the work environment (subjective evaluation of the workload).

Reward refers to three subscales: financial remuneration, esteem, and career opportunities, which includes job security. The mean scores on each component were computed (range 0–3), with higher scores reflecting greater effort and greater rewards. Overcommitment indicates a state of exhaustive coping that reflects frustrated but continued efforts and associated negative feelings. The ratings were totaled so that higher scores indicated greater overcommitment (rating 0–15).

Depression was assessed using the Japanese version of the Center for Epidemiologic Studies Depression (CES-D) scale, a 20-item self-report instrument that is a valid and reliable measure for assessing the level of depressive symptoms (Shima, 1998). The CES-D score is a measure of the chronically depressed state that contrasts with the laboratory ratings of depression, which is a reflection of transient, acute mood state. Based on the conventional cutoff score of 16, participants were placed into either high depression (CES-D > 16) or low depression (CES-D < 16) symptom groups.

Measurement of Cortisol

Saliva samples were frozen at −80 °C until assayed. Cortisol concentrations were measured using an enzyme-linked immunosorbent assay (ELISA Kit; DRG International, Mountainside, New Jersey, USA). The limit of detection of this assay was 0.53 ng/mL, with intra- and inter-assay coefficients of variance (CV) of 2.61% and 3.63%, respectively. The assay plates were read at 450 nm for cortisol, using a Multiskan Ascent microplate photometer (Thermo Labsystems Oy, Helsinki, Finland).

Ethical Considerations

This study was carried out with the approval of the Ethics Committee of Kurume University. Participants provided consent after both an oral and written explanation that participant safety would be prioritized, that research results data anonymized and used only for research purposes, that privacy would be strictly observed, that there would be no disadvantages to declining to participate, and that subjects could revoke their participation freely at any time even after participating.

Results and Discussion

Concerning the relation between subjective happiness and work-related stress, individuals with high subjective happiness were found to have significantly lower self-evaluations of external effort ($t(27) = -2.18$, $p < 0.05$), effort–reward ratio (representing imbalance between effort and reward) ($t(27) = -2.11$, $p < 0.05$), and overcommitment toward their jobs ($t(27) = -3.46$, $p < 0.01$), as measured by the

Table 4.1 Characteristics of participants (Study 1)

Characteristics	High happiness group (n=15)	Low happiness group (n=15)	
Age	35.0 ± 11.1	31.5±9.1	
BMI	20.1 ± 1.4	21.9 ± 2.0	
Marital status			
Married (%)	2 (13.3)	1 (6.3)	
SHS	5.8 ± 0.4	3.1 ± 0.8	**
CES-D	13.6 ± 4.8	22.3 ± 9.6	**
ERIQ			
Extrinsic effort	12.3 ± 5.6	17.3 ± 6.1	*
Reward	42.2 ± 5.8	40.9 ± 6.6	
Effort/reward ratio	0.5 ± 0.3	0.8 ± 0.3	*
Over-commitment	12.6 ± 2.6	18.1 ± 3.5	**
Time of waking (h:min)			
Work days	6:06 (1:17)	6:25 (1:01)	
Weekends	7:23 (1:08)	7:57 (0:57)	

*p<.05, **p<.01

ERIQ, compared with individuals with low subjective happiness (Table 4.1). This result suggests that higher subjective happiness is associated with a sense of satisfaction, in which the individual feels that they are suited to their job. This in turn leads to less pressure from the work environment and fewer anxious thoughts about the job (Connnolly & Viswesvaran, 2000; Tait et al., 1989; Weiss et al., 1999).

A three-way analysis of variance was used to explore differences in CAR and cortisol secretion changes on weekdays and weekends between the groups with high and low JSHS scores. The results confirmed interactions for subjective happiness, day of the week (weekday/weekend), and time of the day ($F(5, 235) = 3.18$, $p < 0.01$). The results of multiple comparisons revealed that individuals with high subjective happiness showed CAR on weekdays ($p < 0.05$) but not on weekends. In contrast, weekends and weekdays showed no differences for those with low subjective happiness. Further, although neither group showed CAR on weekends, the high subjective happiness group had significantly lower cortisol secretion 30 min after awakening and at 10:00 compared with the low subjective happiness group ($p < 0.05$, respectively) (Figs. 4.2 and 4.3).

Studies have shown that CAR in workers is lower on weekends than on weekdays (workdays) (Kunz-Ebrecht et al., 2004) and that individuals with higher positive affect have lower CAR and cortisol secretion throughout the day (Stepote & Wardle, 2005). These findings suggest that higher happiness maintains the HPA axis's ability to adapt to work-related stress through CAR and cortisol secretion allostasis. Meanwhile, those with low subjective happiness might experience high levels of work-related stress and negative affect constantly and may thus be unable to appropriately switch over, both mentally and physically (affect and HPA axis function), from workday to weekend.

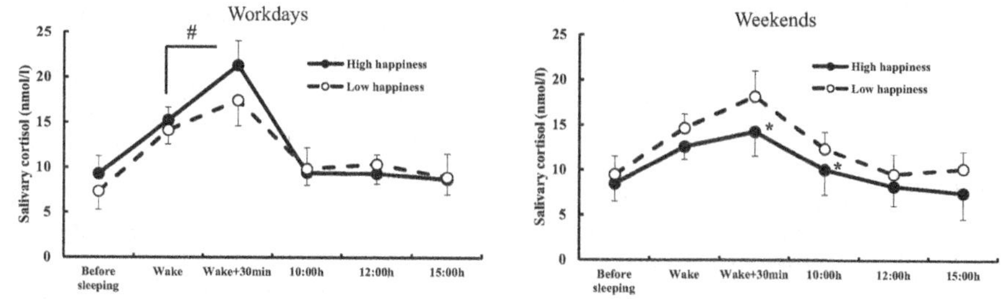

Fig. 4.2 Comparison of the cortisol awakening response in the high- and low-happiness groups on workdays and weekends. #$p < .05$ compared with on awakening, *$p < .05$ compared with the low happiness group

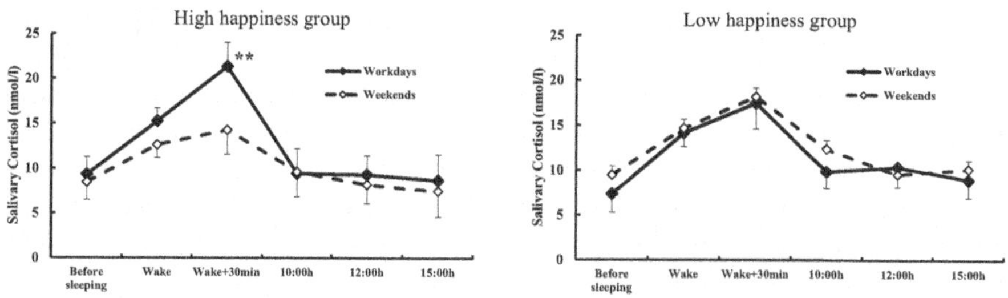

Fig. 4.3 Comparison of the cortisol awakening response on workdays and weekends in the high- and low-happiness groups. **$p < .05$ compared with weekends

4.3 Study 2 Fatigue and CAR on Weekdays/Weekends (Okamura et al., 2011a)

Fatigue acts as a trigger for a number of diseases. In particular, fatigue from work-related stress speeds the progression of cardiovascular diseases and is a factor in sudden cardiac death (Uehata, 1991). Fatigue also reduces attention, concentration, and task performance (Van der Linden et al., 2005). Thus, fatigue is not simply an individual health issue but a serious concern of modern society. We defined subjective fatigue as subjective feelings of mental and physical tiredness or weariness arising in one's daily life. Stone and Shiffman (2002) found that most studies dealing with positive and negative affect conduct only one mood evaluation and that participants' responses are strongly impacted by their affective state from shortly before the survey (a few minutes), even when these evaluations asked about the past several weeks or months. Addressing these issues, Steptoe and Wardle (2005) found that a more reliable evaluation of affect could be achieved by assessing mood several times throughout the day and averaging the results. Therefore, with reference to Steptoe et al. (2005), we asked our participants to rate their subjective fatigue on a five-point scale (1 = not at all to 5 = very much) each time they collected a saliva sample. Differences in CAR on weekends and weekdays between groups with high and low subjective fatigue were then evaluated while also incorporating comparison by subjective work-related stress.

Method

Research Participants

Research participants were recruited over the phone and by email. Eighty healthy women university staff members (20–50 years old) who consented to study participation were asked to collect saliva samples and rate their subjective fatigue at the same time. High (n = 15; average score> 3 on weekdays and weekends) and low (n = 20; average score <3 on both weekdays and weekends) fatigue groups were extracted. We confirmed in advance that the participants had no history of serious illness and were not taking any medications. Further, there were no major differences in the participants' working hours or job duties: managers were excluded from the study, and all participants were university office staff who worked 5 days per week.

The experiment procedure, saliva collection method, cortisol measurement, and ethical considerations were identical to those in study 1. However, here we only show cortisol levels at awakening and +30 min.

Results and Discussion

Self-evaluated fatigue was significantly higher on both weekdays and weekends for the high subjective fatigue group compared with the low subjective fatigue group ($t = -8.72$, $df = 33$, $p < 0.01$; $t = -6.98$, $df = 33$, $p < 0.01$, respectively). There were no differences in the mean age, body mass index, wake times on weekdays and weekends, and percentage of married persons between the two groups. External effort and overcommitment scores were significantly higher in the high subjective fatigue group compared with the low subjective fatigue group ($t = -2.57$, $df = 33$, $p < 0.05$; $t = -2.09$, $df = 33$, $p < 0.05$, respectively) (Table 4.2).

Concerning differences in CAR on weekdays and weekends between the high and low subjective fatigue groups, cortisol increased significantly from the time of awakening to 30 min after awakening in both groups ($p < 0.05$, respectively), and although it was only a significant trend, cortisol secretion 30 min after awakening was higher in the high subjective fatigue group than in the low subjective fatigue group. On weekends, the high subjective fatigue group showed CAR ($p < 0.05$), whereas the low subjective fatigue group did not. Cortisol secretion was significantly higher at awakening and 30 min after awakening in the high subjective fatigue group than in the low subjective fatigue group ($p < 0.05$, respectively). Concerning differences in CAR between weekdays and weekends, the low subjective fatigue group had significantly lower cortisol 30 min after awakening on weekends compared with weekdays ($p < 0.05$) (Fig. 4.4).

Table 4.2 Characteristics of participants (Study 2)

	High fatigue group (n=15)	Low fatigue group (n=20)	
Age	33.8 ± 11.1	35.9 ± 10.1	
BMI	22.3 ± 1.4	22.8 ± 2.0	
Marital status			
Married (%)	5 (33.3)	7 (35.0)	
Subjective fatigue			
Work days	3.6 ± 0.4	2.3 ± 0.5	**
Weekends	3.2 ± 0.3	2.2 ± 0.6	**
SHS	16.1 ± 4.7	19.6 ± 3.1	**
CES-D	22.6 ± 8.1	14.2 ± 3.3	**
ERIQ			
Extrinsic effort	16.13 ± 5.6	11.95 ± 5.6	*
Reward	39.87 ± 5.8	38.76 ± 6.6	
Over-commitment	15.93 ± 2.6	13.24 ± 3.5	*
Time of waking (h:min)			
Work days	6:12 (1:17)	6:25 (1:01)	
Weekends	7:41 (1:08)	7:16 (0:57)	

*p<.05, **p<.01

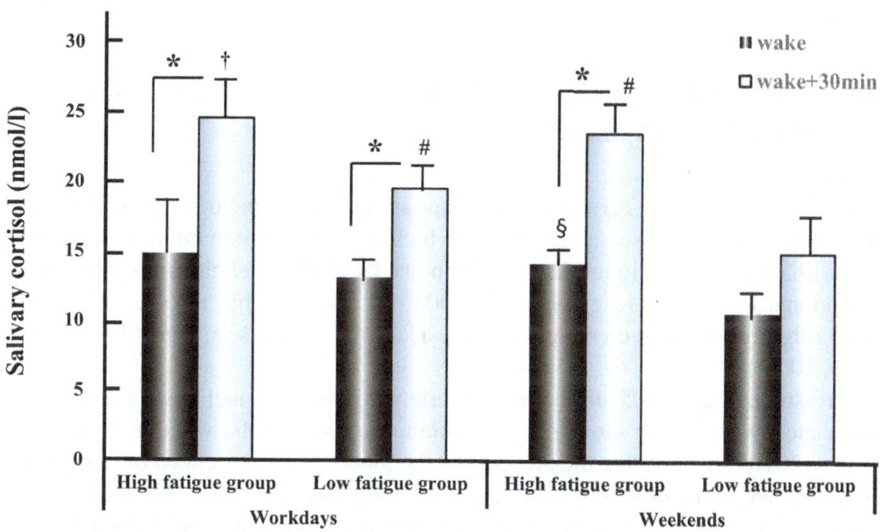

Fig. 4.4 Relationship between fatigue and CAR on workdays and weekends, †$p < .10$ compared with wake +30 min of low fatigue group on workdays, *$p < .05$ compared with on awakening, §$p < .05$ compared with wake of low fatigue group on weekends, # $p < .05$ compared with wake +30 min of low fatigue group on weekends

The results of this study reaffirm the previous finding that CAR is markedly high in those with many concerns about their job (Gustafsson et al., 2008; Schlotz et al., 2004). However, that there were no differences in CAR between weekdays and weekends for individuals with high subjective fatigue in the present study is inconsistent with past findings that CAR in workers is higher on weekdays (workdays) than on weekends (Kunz-Ebrecht et al., 2004) and that those with higher subjective work-related stress have marked CAR on weekdays (workdays) (Maina et al., 2009). One potential reason for this difference is that chronic fatigue is associated with HPA axis hyperactivity (Korte et al., 2005). Namely, our finding may reflect the fact that fatigue results from the inability to adapt successfully to stress and the environment. In other words, this result suggests that Japanese women workers with high subjective fatigue have intense anxious thoughts about their jobs even on weekends, when they are supposed to be free from work-related stress, resulting in HPA axis hyperactivity. A second reason may be that our research studied only women workers. Many past studies have shown sex differences in CAR. Women workers have more marked CAR on workdays compared with men workers, and men show more striking differences in CAR between workdays and weekends (Lundberg & Frankenhaeuser, 1999; Wright & Steptoe, 2005). This difference is likely because women often must care not only for themselves but for their children and the tasks around the house. As the present study included married participants, these factors may be reflected in our results.

4.4 Study 3 Loneliness and CAR on Weekdays/Weekends (Okamura et al., 2011b)

Loneliness arises when certain relationships are lacking. It includes the feelings that arise when one lacks a social network in which they can share in interests and activities with others (social loneliness) or when one lacks close relationships with others (emotional loneliness) (Steptoe et al., 2004b). Loneliness has also been shown to have a significant effect on mental and physical stress response and health (Cacioppo et al., 2003).

Pressman et al. (2005) found that experiencing intense loneliness in daily life is associated with higher morning and evening cortisol levels, whereas Doane and Adam (2010) revealed that the intensity of the loneliness experienced on a given day is associated with elevated CAR the following day. However, no studies have explored the differences in the association between loneliness and CAR on weekdays (workdays) and weekends, which would demonstrate that loneliness has a significant influence on both mental and physical health and work-related stress.

In this study, we explored the correlation between high and low loneliness and CAR in women workers with consideration for subjective work-related stress (ERIQ) and day of the week (weekday vs. weekend) as mediating variables, to consider the HPA axis allostatic load.

Method

Research Participants

Participants were recruited over the phone and by email. The Revised UCLA Loneliness Scale was administered to 90 healthy women university staff members (20–50 years old) who consented to study participation. High (n = 20; mean + 1 SD) and low (n = 20; mean −1 SD) loneliness groups were extracted. We confirmed in advance that the participants had no history of serious illness and were not taking any medications. Further, there were no major differences in working hours or job duties, as managers were excluded from the study and all participants were university office staff who worked 5 days per week.

The experiment procedure, saliva collection method, cortisol measurement, and ethical considerations were identical to those in study 1. However, here we only show cortisol levels at awakening and +30 min.

Results and Discussion

The high loneliness group had significantly higher scores for the UCLA Loneliness Scale ($t(38) = -12.38$, $p < .01$) and CES-D ($t(38) = -5.02$, $p<.01$) and significantly lower JSHS scores ($t(38) = -12.38$, $p<.01$) compared with the low loneliness group. These results support the report of past studies that subjective loneliness evaluated using the UCLA Loneliness Scale is associated with higher negative affect and lower positive affect (e.g., Pressman et al., 2005). The high loneliness group had significantly higher scores for external effort and overcommitment and significantly lower score for external reward compared with the low loneliness group ($t(38) = -2.22$, $p<.05$; $t(38) = -4.76$, $p<.01$; $t(38) = 2.52$, $p<.05$, respectively). There were no differences in the mean age, body mass index, wake times on weekdays and weekends, and percentage of married persons between the two groups. Loneliness is associated with a small social network and low emotional support (Steptoe et al., 2004b). Further, social support in the workplace serves to alleviate subjective stress and regulate the relation between stressors and stress (Viswesvaran et al., 1999). These findings suggested that individuals experiencing intense loneliness are constantly feeling severe work-related stress because they are dissatisfied with their work environment and do not receive sufficient social and emotional support from their coworkers. Thus, loneliness is thought to be a vital factor in maintaining the well-being of workers (Table 4.3).

A three-way analysis of variance was used to explore differences in mean CAR on weekdays and weekends between the groups with high and low subjective loneliness. The analyses of the CAR on workdays and weekends revealed that there were significant effects of the day (weekday/weekend) (F (1,76) = 6.41, $p < .05$), time (upon awakening/30 min after awakening) (F (1,76) =12.15, $p < .01$), and a

Table 4.3 Characteristics of participants (Study 3)

Characteristics	High loneliness group (*n*=20)	Low loneliness group (*n*=20)	
Age	38.2 ± 4.8	39.9 ± 1.1	
BMI	21.12 ± 2.4	22.02 ± 0.4	
Marital status			
Married (%)	7 (35%)	8 (40%)	
Loneliness scale	47.6 ± 6.9	27.1 ± 2.8	**
SHS	16.6 ± 4.7	20.5 ± 3.8	**
CES-D	22.5 ± 8.7	12.0 ± 4.9	**
ERIQ			
Extrinsic effort	16.2 ± 5.2	12.1 ± 5.1	*
Reward	39.6 ± 6.4	43.8 ± 4.6	**
Over-commitment	17.4 ± 3.5	12.9 ± 2.6	*
Time of waking (h:min)			
Work days	6:26 (0:52)	6:07 (0:39)	
Weekends	7:43 (0:56)	7:15 (1:02)	

*p<.05, **p<.01

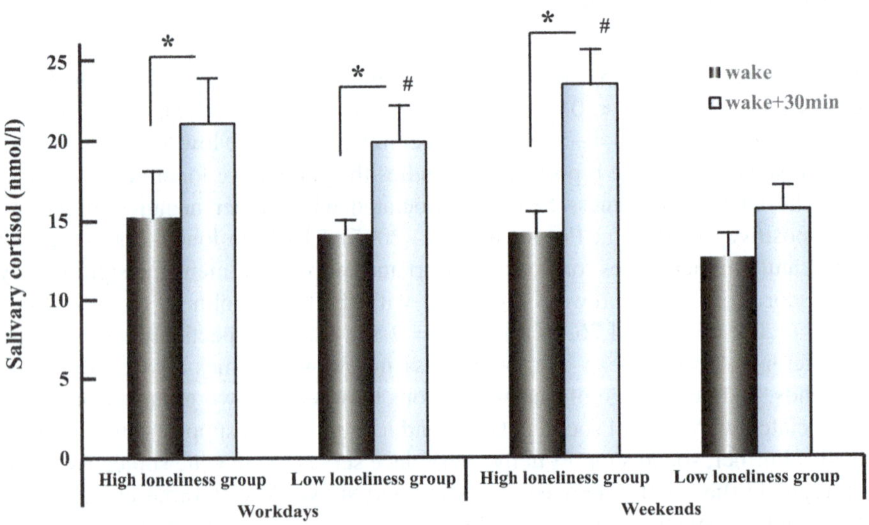

Fig. 4.5 Relationship between loneliness and CAR on workdays and weekends.*p < .05 compared with on awakening, # p < .05 compared with wake +30 min of low loneliness group on weekends

significant three-way interaction between the day, time, and group (F (1,76) = 5.19, p < .05). Participants with high subjective loneliness had consistently high CAR, with no differences between weekdays and weekends. Meanwhile, those with low subjective loneliness had lower weekend CAR than weekday CAR (p < 0.05). Those with high loneliness had significantly higher weekend CAR compared with those with low loneliness (p < 0.05). We found no difference in weekday CAR between the high and low loneliness groups (Fig. 4.5). These results confirm that loneliness is a psychosocial factor that promotes CAR. Further, among those with severe subjective loneliness, the loneliness felt on weekends seems to become a cause of stress in itself, leading to a stressed state of HPA axis hyperactivity even on weekends, which should be restful. In other words, this promotion of CAR on weekends likely reflects a psychological state in which the individual is unable to enjoy weekends because they lack ties to a social network. Considering that the ability to switch, mentally and physically, between weekdays (workdays) and weekends is essential to maintaining health (Okamura et al., 2012), these findings suggest that loneliness plays a role in the prevalence of diseases.

4.5 General Discussion

We considered how happiness (study 1), fatigue (study 2), and loneliness (study 3) are associated with weekday and weekend CAR in female workers.

The results suggested that higher happiness maintains the HPA axis's ability to adapt to work-related stress through CAR and cortisol secretion allostasis. Happiness was thus shown to reduce CAR through the normalization of the HPA axis function that accompanies stress relief. In contrast, we found no difference in CAR between weekdays and weekends in individuals with high fatigue and loneliness, suggesting that the HPA axis was functioning abnormally even on weekends—days that should be restful. Thus, fatigue and loneliness were shown to be factors promoting CAR. That weekend and weekday CAR differed between individuals with high and low self-evaluations of happiness, fatigue, and loneliness is objective evidence demonstrating the importance of evaluating psychological and physiological stress response not only on workdays but on weekends as well, when considering the impact of work-related stress and everyday stress on the mind and body.

As the studies reported in this article utilized small-scale data, they could not offer sufficient evidence for the relation between CAR and psychosocial factors, considered to mediate subjective stress response and HPA axis function. In future, a large-scale study with consideration for factors such as stress management practice, presence of social support, and socioeconomic status will be necessary to clarify further the impact of psychosocial factors on subjective stress response and CAR. Further, cortisol was measured for only 1 day during the week and 1 day during the weekend in the present study. As CAR is influenced by situational factors, collecting data multiple times is recommended to clarify an individual's unique responses (Hellhammer et al., 2007). Future research collecting data multiple times from equivalent subjects is necessary both to increase CAR reliability and to reveal whether the results of the present research can be generalized to the larger group of Japanese workers.

In the framework of the Psychology of Sustainability and Sustainable Development (Di Fabio, 2017a, b; Di Fabio & Rosen, 2018) and of the Psychology of Harmonization (Di Fabio & Tsuda, 2018), this study offers results in relation to work-related stress in the Japanese context opening future opportunities for research and intervention in a cross-cultural perspective for health and well-being of people and organizations (Di Fabio et al., 2020).

References

Boggero, I. A., Hostinar, C. E., Haak, E. A., Murphy, M. L. M., & Segerstrom, S. C. (2017). Psychosocial functioning and the cortisol awakening response: Meta-analysis, P-curve analysis, and evaluation of the evidential value in existing studies. *Biological Psychology, 129*, 207–230.

Cacioppo, J., Hawkley, L., & Berntson, G. (2003). The anatomy of loneliness. *Current Directions in Psychological Science, 12*(3), 71–74.

Connnolly, J. J., & Viswesvaran, C. (2000). The role of affectivity in job satisfaction: A meta-analysis. *Personality and Individual Differences, 29*(2), 265–281.

Di Fabio, A. (2017a). Positive Healthy Organizations: Promoting well-being, meaningfulness, and sustainability in organizations. *Frontiers in Psychology, 8*, 1938. https://doi.org/10.3389/fpsyg.2017.01938

Di Fabio, A. (2017b). The psychology of sustainability and sustainable development for well-being in organizations. *Frontiers in Psychology, 8*, 1534. https://doi.org/10.3389/fpsyg.2017.01534

Di Fabio, A., Cheung, F., & Peiró, J.-M. (2020). Editorial special issue personality and individual differences and healthy organizations. *Personality and Individual Differences, 166*. https://doi.org/10.1016/j.paid.2020.110196

Di Fabio, A., & Rosen, M. A. (2018). Opening the Black Box of psychological processes in the science of sustainable development: A new frontier. *European Journal of Sustainable Development Research, 2*(4), 47. https://doi.org/10.20897/ejosdr/3933

Di Fabio, A., & Tsuda, A. (2018). The psychology of harmony and harmonization: Advancing the perspectives for the psychology of sustainability and sustainable development. *Sustainability MDPI, 10*(12), 4726. https://doi.org/10.3390/su10124726

Doane, L. D., & Adam, E. K. (2010). Loneliness and cortisol: Momentary, day-to-day, and trait associations. *Psychoneuroendocrinology, 35*(3), 430–441.

Gustafsson, K., Lindfors, P., Aronsson, G., & Lundberg, U. (2008). Relationship between self-rating of recovery from work and morning salivary cortisol. *Journal of Occupational Health, 50*(1), 24–30.

Hellhammer, J., Fries, E., Schweisthal, O. W., Schlotz, W., Stone, A. A., & Hagemann, D. (2007). Several daily measurements are necessary to reliably assess the cortisol rise after awakening state and trait components. *Psychoneuroendocrinology, 32*(1), 80–86.

Iida, N., & Ogino, T. (2019). The relationship between the use of daily time, childcare support and parenting stress in mothers. *Bulletin of Faculty of Health and Welfare Science, Okayama Prefectural University, 26*, 11–19. (in Japanese).

Korte, S. M., Koolhaas, J. M., Wingfield, J. C., McEwen, B., & S. (2005). The Darwinian concept of stress: Benefits of allostasis and costs of allostatic load. *Neuroscience and Biobehavioral Reviews, 29*(1), 3–38.

Kunz-Ebrecht, S. R., Kirschbaum, C., & Steptoe, A. (2004). Differences in cortisol awakening response on workdays and weekends in women and men from the Whitehall II cohort. *Psychoneuroendocrinology, 29*(4), 516–528.

Lindeberg, S. I., Eek, F., Lindbladh, E., Ostergren, P. O., Hansen, A. M., & Karlson, B. (2008). Exhaustion measured by the SF-36 vitality scale is associated with a flattened diurnal cortisol profile. *Psychoneuroendocrinology, 33*(4), 471–477.

Lundberg, U., & Frankenhaeuser, M. (1999). Stress and workload of men and women in high-ranking positions. *Journal of Occupational Health Psychology, 4*(2), 142–151.

Maina, G., Palmas, A., Bovenzi, M., & Filon, F. L. (2009). Salivary cortisol and psychosocial hazards at work. *American Journal of Industrial Medcine, 52*(3), 251–260.

McEwen, B. S. (2001). Plasticity of the hippocampus: Adaptation to chronic stress and allostatic load. *Annals of the New York Academy of Sciences, 933*(1), 265–277.

McEwen, B. S. (2006). Protective and damaging effects of stress mediators. *Dialogues in Clinical Neuroscience, 8*(4), 171–179.

Miller, G. E., Chen, E., & Zhou, E. S. (2007). If it goes up, must it come down? Chronic stress and the hypothalamic-pituitary-adrenocortical axis in humans. *Psychological Bulletin, 133*(1), 25–45.

Ministry of Health, Labor and Welfare. (2019). *The actual situation of working women* (in Japanese). https://www.mhlw.go.jp/bunya/koyoukintou/josei-jitsujo/dl/18-01.pdf. Accessed on 4 Dec 2020.

Miyauchi, K., Mochizuki, Y., Ishida, S., & Sato, C. (2010). Effects of an educational effort with a leaflet on depression for middle-aged female workers. *Maternal Health, 50*(4), 646–655. (in Japanese).

O'Connor, D. B., Hendrickx, H., Dadd, T., Elliman, T. D., Willis, T. A., Talbot, D., et al. (2009). Cortisol awakening rise in middle-aged women in relation to psychological stress. *Psychoneuroendocrinology, 34*(10), 1486–1494.

Okamura, H., Tsuda, A., & Matsuishi, T. (2010). Perceived happiness and cortisol awakening response on workdays and weekends. *The Japanese Journal of Health Psychology, 23*(2), 11–21. (in Japanese).

Okamura, H., Urata, H., Ueda, Y., & Yajima, J. (2011a). The relationship between perceived fatigue and salivary cortisol responses on workdays and weekends in women. *Behavioral Science Research, 49*(2), 105–112. (in Japanese).

Okamura, H., Tsuda, A., & Matsuishi, T. (2011b). The relationship between perceived loneliness and cortisol awakening responses on workdays and weekends. *Japanese Psychological Research, 53*(2), 113–120.

Okamura, H., Tsuda, A., & Yajima, J. (2012). Cortisol awakening response (CAR) and psychosocial factors: Application of objective indicator for assessing the effectiveness of stress management. *Stress Management Research, 9*(1), 3–17. (in Japanese).

Pressman, S. D., Cohen, S., Miller, G. E., Barkin, A., Rabin, B. S., & Treanor, J. J. (2005). Loneliness, social network size, and immune response to influenza vaccination in college freshmen. *Health Psychology, 24*(3), 297–306.

Schlotz, W., Hellhammer, J., Schulz, P., & Stone, A. A. (2004). Perceived work overload and chronic worrying predict weekend-weekday differences in the cortisol awakening response. *Psychosomatic Medicine, 66*(2), 207–214.

Shima, S. (1998). *CES-D scale*. Chiba Test Center. (In Japanese, translated by the author of this article.).

Shimai, S., Otake, K., Utsuki, N., Ikemi, A., & Lyubomirsky, S. (2004). Development of a Japanese version of the Subjective Happiness Scale (SHS), and examination of its validity and reliability. *Japanese journal of public health, 51*(10), 845–853. (In Japanese).

Statistical Bureau of Japan. (2016). *Survey on time use and leisure activities* (in Japanese). https://www.stat.go.jp/data/shakai/2016/pdf/youyaku2.pdf. Accessed on 4 Dec 2020.

Stalder, T., Evans, P., Hucklebridge, F., & Clow, A. (2010). Associations between psychosocial state variables and the cortisol awakening response in a single case study. *Psychoneuroendocrinology, 35*(2), 209–214.

Stalder, T., Kirschbaum, C., Kudielka, B. M., Adam, E. K., Pruessner, J. C., Wüst, S., et al. (2016). Assessment of the cortisol awakening response: Expert consensus guidelines. *Psychoneuroendocrinology, 63*, 414–432.

Steptoe, A., Siegrist, J., Kirschbaum, C., & Marmot, M. (2004a). Effort-reward imbalance, overcommitment, and measure of cortisol and blood pressure over the working day. *Psychosomatic Medicine, 66*(3), 323–329.

Steptoe, A., Owen, N., Kunz-Ebrecht, S. R., & Brydon, L. (2004b). Loneliness and neuroendocrine, cardiovascular, and inflammatory stress responses in middle-aged men and women. *Psychoneuroendocrinology, 29*(5), 593–611.

Steptoe, A., & Wardle, J. (2005). Positive affect and biological function in everyday life. *Neurobiology of Aging, 1*, 108–112.

Steptoe, A., Wardle, J., & Marmot, M. (2005). Positive affect and health-related neuroendocrine, cardiovascular, and inflammatory processes. *Proceedings of the National Academy of Sciences of the United States of America, 102*(18), 6508–6512.

Stone, A. A., & Shiffman, S. (2002). Capturing momentary, self-report data: A proposal for reporting guideline. *Annals Behavioral Medicine, 24*(3), 236–243.

Tait, M., Padgett, M. Y., & Baldwin, T. (1989). Job and life satisfaction: A reevaluation of the strength of the relationship and gender effects as a function of the data of the study. *Journal of Applied Psychology, 74*(3), 502–507.

Tsutsumi, A., Ishitake, T., Peter, R., Siegrist, J., & Matoba, T. (2001). The Japanese version of the Effort-Reward Imbalance Questionnaire: A study in dental technicians. *Work and Stress, 15*(1), 86–96.

Uehata, T. (1991). Long working hours and occupational stress-related cardiovascular attacks among middle-aged workers in Japan. *Journal of Human Ergology, 20*(2), 147–153.

Van der Linden, D., Keijsers, G. P. J., Eling, P., & Van Schaijk, R. (2005). Work stress and attentional difficulties: An initial study on burnout and cognitive failures. *Work & Stress, 19*, 23–36.

Viswesvaran, C., Sanchez, J. I., & Fisher, J. (1999). The role of social support in the process of work stress: A meta-analysis. *Journal of Vocational Behavior, 54*, 314–334.

Weiss, H. M., Nicholas, J. P., & Daus, C. S. (1999). An examination of the joint effects of affective experiences and job beliefs on job satisfaction and variations in affective experiences over time. *Organizational Behavior and Human Decision Processes, 78*(1), 1–24.

Wright, C. E., & Steptoe, A. (2005). Subjective socioeconomic position, gender, and cortisol responses to waking in an elderly population. *Psychoneuroendocrinology, 30*(6), 582–590.

Chapter 5
Trait Emotional Intelligence and the Six Rings of Positive Self-Capital for Optimal Performance and Sustainability

José-Antonio Gutiérrez-Carrasco, Gabriela Topa, and Juan-Carlos Pérez-González

5.1 Introduction

To a large extent, as Furnham (2001) has stressed, many of the concepts, ideas, and approaches in theory, research, and practice in organizational behavior have not been proposed by non-pragmatic thinkers but have actually emerged from the response or demand of employers and organizations to the new social, economic, and political conditions they have encountered. Thus, we can see the emergence and survival in the literature of some concepts that were new a decade or two ago but are now established globally, such as the concepts of "emotionally intelligent organizations," "positive organizational psychology," and "psychological capital," among others. In this line of evolution in accordance with the demand of sociohistorical changes, we also find, more recently, revolutionary concepts and ideas such as "positive self-capital" (Borgogni & Miraglia, 2020), "the Psychology of Sustainability and Sustainable Development" (Di Fabio, 2017a, b; Di Fabio & Rosen, 2018), and "the Psychology of Harmonization" (Di Fabio & Tsuda, 2018).

Over time, the different branches of psychology have progressively separated, and this has caused, for example, that few theorists of personality and individual differences have paid attention to the role that individual differences play in productivity, performance, and job satisfaction, or motivation in organizations. Likewise, a

J.-A. Gutiérrez-Carrasco · G. Topa
Faculty of Psychology, Universidad Nacional de Educación a Distancia (UNED), Madrid, Spain

J.-C. Pérez-González (✉)
EDUEMO Lab, Faculty of Education, Universidad Nacional de Educación a Distancia (UNED), Madrid, Spain

Department of Research Methods and Assessment in Education (MIDE II), Faculty of Education, UNED, Madrid, Spain
e-mail: jcperez@edu.uned.es; http://eduemo.com

A. Di Fabio (ed.), *Cross-cultural Perspectives on Well-Being and Sustainability in Organizations*, https://doi.org/10.1007/978-3-030-86709-6_5

few researchers of organizational behavior have done more than recognize the importance of personality traits in occupational behavior (Furnham, 1992). Overcoming the person-situation debate during the 1970s and 1980s, nowadays this debate "has now brought about co-operation and collaboration and more fruitful ways of understanding the interaction between persons and situations" (Matthews et al., 2003, p. 49). Nonetheless, the characteristics of situations are, empirically, poorly understood, since we do not have tested which aspects of a situation are the important ones in determining behavior change (for a novel approach, see Rauthmann et al., 2014). Contrary, while we have enough evidence that the prediction of behavior from personality traits for an isolated situation is unreliable, in contrast we can make reliable predictions across many situations from certain combinations of individual differences (Matthews et al., 2003). A clear example of this evidence-based fact is the high cross-cultural criterion and incremental validity of trait emotional intelligence (see Siegling et al., 2015; Petrides et al., 2016).

This chapter is built on three premises. In tune with our times, the first premise is our conviction that it is imperative to collectively pursue the third sustainable development goal proposed by the 2030 agenda of the United Nations (i.e., promoting good health and well-being). In particular, the COVID-19 pandemic has cruelly underlined the necessary to join forces in the search for this specific sustainable goal inside and outside the organizations. It should be noted that the year 2020 has revalidated the emotional intelligence (EI) construct as a buffer against maladjustment, as a personal resource to deal with uncertainty and change, and, finally, to cope with the stress caused by the COVID-19 pandemic (Moroń & Biolik-Moroń, 2020; Zysberg & Zisberg, 2020). So, the promotion of emotionally intelligent citizens, workers, teams, and organizations may be a powerful way forward in the desired direction.

Our second premise is at present indisputable that EI is one of the most demanded competences by the labor market worldwide to professionals with or without a university degree. This idea has been previously expressed by the World Economic Forum's (2016, 2020) reports on the Future of Jobs, where EI was in the top 10 skills demanded for 2020 and the top 15 skills demanded for 2025. Likewise, despite some controversial issues, experts on Work and Organizational Psychology have supported it beyond the theoretical reasons why EI is so attractive to business (e.g., Ashkanassy & Daus, 2020; Furnham, 2009; Petrides, 2010). An example of this widespread conviction in the business world about the importance of EI for work and career success is the EI series books edited by the Harvard Business Review. Indeed, in our liquid society (in the words of Zygmunt Bauman), EI have been many times suggested as essential objective of both higher education and business schools.

The third premise is that, specifically, EI operationalized as *Trait EI* constitutes a robust promoter of well-being (both hedonic and eudaimonic) and career success (Di Fabio & Saklofske, 2014; Di Fabio & Kenny, 2016; Pérez-González et al., 2020; Petrides, 2010, 2011; Petrides et al., 2016). Because of this, its outstanding practical utility in career counseling and human resources may be further understood extending its nomological network through the study of its empirical associations with other comprehensive and novel constructs related to human capital for optimal performance and sustainability, such as psychological strengths, positive psychological capital, high potentials, soft skills or noncognitive factors, career adaptability, and covitality.

In short, we see trait EI as the outstanding component of the so-called positive self-capital that should be promoted in organizations for the sake of well-being and the sustainable development.

5.2 Emotional Intelligence as a Hot Topic and Trait EI as the Gem

The interest in emotional intelligence in the organizational field has been constant since the mid-1990s, proving to be a very attractive construct to business.

Through the results of a quick systematic review in the Web of Science (WoS) and Scopus databases, it can be seen that the number of scientific articles published with the expression "emotional intelligence" captured in either the publications' title, abstract, or keywords has not stopped increasing progressively over the years in an overwhelming way. The results show that, in particular, the number of scientific publications on EI seems to double every 5-year period.

In 2008, Stough et al. (2009) conducted the same search in Scopus and found approximately 952 published papers, of which 337 (35.4%) were focused on the workplace. This same search carried out 13 years later, in January 2021, has shown us a total of 10,362 works recorded in Scopus and 9834 in WoS.

Later, limiting our search to the WoS database and the appearance of the expression "emotional intelligence" exclusively in the title of journal articles, 3703 articles appeared. An area-based analysis (WoS categories) showed that 579 of these articles (15.6%) were from management (392) and business (187) publications. Finally, a narrower search in WoS on only articles containing both "emotional intelligence" and "work" in title, abstract, or keywords showed a total of 2020 records, representing 54.5% of the 3703 articles with "emotional intelligence" in title, abstract, or keywords, a significant increase from the 35.4% identified by Stough et al. Consequently, it seems that currently more than half of the literature on IE connects in some way this concept with the workplace.

The EI literature has progressively led to two primary research streams, according to which EI can be conceptualized and measured in two distinct, albeit, complementary constructs of EI (Pérez-González et al., 2020): (1) as a *typical* performance-tested personality trait (called Trait EI), representing behavioral dispositions and perceptions of emotion-related competencies (Petrides et al., 2016), or (2) as a *maximum* performance construct, comprising a set of intellectual abilities for reasoning about emotion (called ability EI; e.g., Fiori et al., 2014).

Taken trait EI and ability EI together, as highlighted by Pérez-González et al. (2020), these two approaches to EI correspond conceptually and operationally with most of the components of emotional competence. Since trait EI is conceived as a personality trait and ability EI as kind of intelligence, Alba-Juez and Pérez-González (2019) suggested that the former may be reinterpreted as the noncognitive/automatic (temperamental) component of emotional competence while the latter may be conceptualized as the cognitive/conscious component.

On the one hand, objective assessment of the ability EI is still elusive, as it presents numerous theoretical and psychometric limitations without easy resolution that have been widely criticized and that escape the remit of this chapter (e.g., Fiori et al., 2014). On the other hand, trait EI has been suggested as the most useful indicator of both emotional competence or EI because of its superiority in terms of criterion and incremental validity (Andrei et al., 2016; Petrides, 2010, 2011; Siegling et al., 2015) as well as in terms of its theoretical and empirical connection with the scientific tradition of personality and individual difference research (e.g., Pérez-González et al., 2020; Van der Linden et al., 2018).

5.3 What Do the Meta-analyses of Emotional Intelligence Tell Us?

Regarding the criterion validity of trait EI, we can say, in a nutshell, that trait EI has been related to the most relevant areas of interest for prosperity and happiness across the life span, such as well-being (both hedonic and eudaimonic), health, satisfactory romantic and social relationships, leadership, psychosocial adjustment, adaptive emotional regulation strategies, higher academic performance, lower burnout, higher engagement, higher job satisfaction and job performance (mainly organizational citizenship), negotiation outcomes and conflict management, and work attitudes (e.g., Petrides et al., 2016). These relationships have been confirmed not only by independent studies but also through meta-analytical research.

A summary of some key meta-analyses on EI is depicted in Table 5.1. A relevant finding is that the average correlation with those life outcomes is usually stronger for trait EI ($r = 0.37$) than for ability EI ($r = 0.20$), with even much higher validity coefficients ($r = 0.55$) for trait EI when it is specifically assessed through the Trait Emotional Intelligence Questionnaire or TEIQue (for further information on this gold pattern of trait EI, see O'Connor et al., 2019; see also Siegling et al., 2015). The exception for this rule is found for the relationship with academic performance. These meta-analytic findings are pretty robust indicating that the practical utility of trait EI versus ability EI is approximately twice as high in predictive power across outcomes and populations.

From a critical perspective, Zeidner et al. (2012) have argued that the superiority of trait EI versus ability EI in predicting success is the outcome of one of the components of trait EI known as "well-being." Particularly, Zeidner et al. (2012) argue that the problem with that factor as a predictor is it shares overlap with the criterion/outcome measures. While that might be partially right, Mavroveli et al. (2007) have argued the well-being component of trait EI relates to a person's confidence that

Table 5.1 Summary of criterion validity of both ability EI and trait EI according to meta-analyses

Criterion variables	Ability EI	Trait EI	Meta-analysis' authors
Health			Martins et al. (2010)
Mental	.17	**.36**	
		.53▲	
Physical		.27	
		.52▲	
Psychosomatic		**.33**	
		.50▲	
Well-being	.22	**.35**	Sánchez-Álvarez et al. (2016)
		.55▲	
Leadership	.20	**.58**	Harms and Credé (2010)
	.07	**.39**	Miao et al. (2018a)
Job performance	.18	**.47**	Joseph and Newman (2010)
	.24	.28	O'Boyle et al. (2011)
		.46	Miao et al. (2018b)
Job satisfaction	.06	**.31**	Miao et al. (2017)
Academic performance		.20	Perera and DiGiacomo (2013)
	.24	.19	MacCann et al. (2020)

Adapted from Pérez-González and Qualter (2018)
Note: Validity coefficients as average correlations >0.30 are in bold; ▲validity coefficient when trait EI was assessed via TEIQue

they can regulate emotions in a way to promote well-being rather than being an actual state of well-being. Likewise, Siegling et al. (2015) explained that the trait EI label "well-being" is misleading and might be better termed "trait positivity," given it is not a marker of a person's health or happiness but relates, instead, to a person's current state of mind about how they may use emotions under an optimist look to enable better coping in everyday life. However, it should be noted that the incremental validity of trait EI over and above general personality and mood variables has been consistently proven (e.g., Di Fabio & Kenny, 2016; Petrides et al., 2007a).

In the same critical vein, Joseph et al. (2015), proponents of the ability EI model, having found a strong relationship between trait EI (they speak of "mixed EI") and job performance (Joseph & Newman 2010), performed a new critical meta-analysis eliminating the shared variance of trait EI with a collection of seven factors (i.e., ability EI, cognitive ability, emotional stability, conscientiousness, extraversion, general self-efficacy, and self-rated job performance). This statistical trick resulted in a zero relationship between the remaining trait EI and job performance, which was interpreted by Joseph et al. (2015) as proof that trait EI (mixed EI) is "guilty" of being a mix, since they think that trait EI measures reflect a heterogeneous combination of traits that have long been known to predict job performance, not as an authentic "new" construct. Attempting to discredit the trait EI construct because its variance can be explained by a combination of seven cognitive and noncognitive constructs is fruitless, as the variance of many psychological constructs can be explained by a combination of cognitive ability and broad personality traits such as the Big Five. In all cases, "variance that can be easily explained within the trait EI framework (e.g., high trait EI individuals are more sensitive to affect-laden stimuli) requires convoluted combinations of the Big Five to be predicted, but not really explained (e.g., high E, low N, above average C and O people tend to be more sensitive to affect-laden stimuli than those who are low E, high N, below average C and O)" (Petrides et al., 2007b, p. 286). So even in the hypothetical case that trait EI could be reduced to the combination of seven or more general constructs, the trait EI construct would still be useful, as it would always be more pragmatic to use a single construct to explain and predict a person's behavior than to conjugate a panoply of them.

These authors, despite their reticence, finally acknowledge the predictive utility of trait EI when they concluded that their meta-analytic results suggest that practitioners could use a single trait EI measure "as part of a selection system because they tap into a diffuse, compound construct of personality and self-perceptions that exhibits reasonable criterion-related validity" (Joseph et al., 2015, p. 317), instead of using a complex battery of a constellation of seven constructs of "knowledge, skills, abilities, and other characteristics." In short, the theoretical conclusion reached by Joseph et al. (2015) is that trait EI is a good predictor of job performance because it is "a new blend of old good wines." Since this could be partially correct, again, it has been argued that this recurrent view may be overly pessimistic and reductionist (Pérez et al., 2005), given that there is compelling evidence "in support of the discriminant and incremental validity of trait EI, including the isolation of an oblique trait EI factor in Eysenckian as well as Big Five factor space and mounting data showing that several of the measures used to operationalize the construct are able to predict criteria in the presence of the basic personality traits" (Pérez et al., 2005, p. 198; see also Pérez-González & Sanchez-Ruiz, 2014). Additionally, maybe the conclusions by Joseph et al. (2015) were not simply anchored under the right framework, since they omitted any allusion to the trait EI theory (e.g., Petrides, 2010) to interpret their results. However, at the end of the day, they explicitly recognized the high value of trait EI for practice.

5.4 Definition and Sampling Domain of Trait Emotional Intelligence

Trait EI theory was introduced by Petrides in 2001, and trait EI "describes our perceptions of our emotional world: what our emotional dispositions are and how good we believe we are in terms of perceiving, understanding, managing, and utilizing our own and other people's emotions" (Petrides et al., 2018, p. 50). Alternatively, following Alba-Juez and Pérez-González (2019), trait EI refers to a general noncognitive characteristic comprising a set of generally advantageous personal dispositions for how to procure, maintain, and optimize psychological well-being, how to regulate emotions (one's own and others'), and how to communicate emotions and socialize.

Table 5.2 The sampling domain of trait emotional intelligence in adults

Global trait EI	High scorers perceive themselves as…
Well-being or trait positivity factor (W)	
Self-esteem	…successful and self-confident
Trait happiness	…cheerful and satisfied with their lives
Trait optimism	…confident and likely to "look on the bright side" of life
Emotionality factor (E)	
Emotion perception (self and others)	…clear about their own and other people's feelings
Emotion expression	…capable of communicating their feelings to others
Relationships	…capable of having fulfilling personal relationships
Trait empathy	…capable of taking someone else's perspective
Sociability factor (S)	
Social awareness	…accomplished networkers with excellent social skills
Emotion management (others)	…capable of influencing other people's feelings
Assertiveness	…forthright, frank, and willing to stand up for their rights
Self-control factor (C)	
Emotion control	…capable of controlling their emotions
Stress management	…capable of withstanding pressure and regulating stress
Impulse control	…reflective and less likely to give into their urges
Adaptability[a]	…flexible and willing to adapt to new conditions
Self-motivation[a]	…driven and unlikely to give up in the face of adversity

Adapted from Petrides et al. (2016, p. 336)
Note. [a]These two independent facets feed directly into the global trait EI score without going through any factor

Petrides and Furnham (2000) proposed the label of trait EI partly to reflect the perspective that Goleman's (1995) emotional intelligence represented a popularization of ideas that had long been researched in the scientific literature on personality psychology. Since most research in the field of EI is based on self-reports, so it is, de facto, conducted within the broader domain of trait EI (Pérez et al., 2005; Petrides et al., 2016).

A short sampling domain description of each of the facets of trait EI is depicted in Table 5.2. For a more detailed description, see Roy (2015). The 15 facets of trait EI may be grouped under 4 statistical factors under the acronym WESC: well-being, emotionality, sociability, and self-control. The gold standard for the comprehensive assessment of trait EI is the Trait Emotional Intelligence Questionnaire or TEIQue, although a good alternative for adults is the Profile of Emotional Competence or PEC (Brasseaur et al., 2013).

5.5 Trait EI and Its Overlap with the Six Rings of Positive Self-Capital

Economists have long known that people are an important part of the wealth of nations, and they use the human capital construct to refer acquired and useful skills and knowledge which are a form of capital (wealth) that accounts for a great variance of economic growth in individuals and nations (Schultz, 1961). Nevertheless, "conceptually, there has not been a clear-cut definition on how human capital should be represented" (Cohen & Soto, 2007, p. 52).

But, unfortunately, psychological qualities such as human intelligence are usually neglected as a meaningful human capital indicator by economists. Nonetheless, it seems that we have been immersed for a long time in a cognitive capitalism (called the capitalism of general intellect even by Marx; see Vencellone, 2007), where the hegemony of knowledges and the increasingly immaterial and cognitive character of labor are considered as the main factors driving the new economy, but not just cognitive but also noncognitive (i.e., social and emotional), as recognized above by the World Economic Forum (2016, 2020).

Thus, a first approach to human capital that contemplates the main large areas of individual differences (i.e., intelligence and personality) could be represented in Fig. 5.1. However, in this figure we see a traditional attentional bias towards the cognitive factors of human capital. This cognitive bias, in any case, could be said to

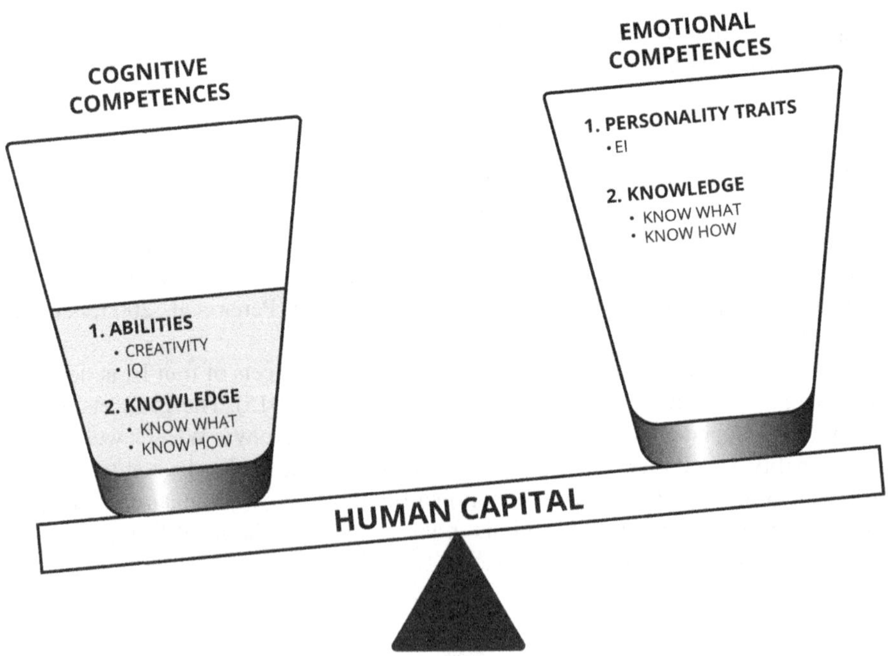

Fig. 5.1 Representation of traditional cognitive bias in the interpretation of human capital

have been unsatisfactory, since it has barely represented just half of the cognitive glass, preferring to focus more on explicit knowledge (know-what) than on procedural knowledge (know-how) and more on convergent thinking skills (IQ) than on diverting thinking skills (creativity). But this inertia has changed with the emotional *zeitgeist* of our time. It is no longer enough to reduce human capital to cognitive competence, but it is also necessary to pay attention to its emotional competence component, considering both knowledge and noncognitive skills linked to EI (Furnham, 2009; Goleman, 1995).

Given that accumulated evidence we have reviewed above has verified trait EI as a psychological variable with a high and transversal impact on personal, social, and working life, it seems more than reasonable that trait EI may be considered a fundamental asset of human capital that organizations must promote because of its potential individual, organizational, and social benefits. Thus, from this empirically based perspective, trait EI represents an advantage for individual functioning, with an impact on their well-being and performance. However, the human capital concept is not enough anymore, since it has been recently eclipsed by the positive self-capital concept as a resonance effect of positive psychology in the workplace.

The positive psychology movement initiated by Martin Seligman (Seligman & Csikszentmihalyi, 2000; Seligman, 2007) expanded in the organizational domain giving birth to the movement of the so-called positive organizational psychology. This new movement has produced two parallel approaches usually referred to as positive organizational scholarship, focused on the macro or organizational level, and positive organizational behavior, focused on the micro or individual level. Under the latter approach, positive self-capital "refers to the set of the positive features of individual functioning, which is able to support individual's health and well-being as well as to uncover their potentialities and abilities, enhancing successful performances and personal development" (Borgogni & Miraglia, 2020, p. 345).

Our time calls for our interest in this renewal of the concept of human capital as positive self-capital. Just as we did with human capital, we can recognize the remarkable weight of the trait EI within positive self-capital.

Within the positive organizational behavior approach, Luthans et al. (2007) proposed the concept of Psychological Capital or PsyCap, "which refers to a high-order construct, synergistically integrating some assets of positive self-capital" (Borgogni & Miraglia, 2020, p. 345), but not all. Although PsyCap is clearly a construct that aims to represent the concept of positive self-capital, it does not cover all key personal resources that the latter could or should cover in its comprehensive approach.

The sampling domain of positive self-capital, that is, the set of dimensions or sub-dimensions that must be part of positive self-capital, is not defined or closed yet, despite various candidate constructs have been suggested by Borgogni y Miraglia (2020): PsyCap (including self-efficacy, optimism, resiliency, and hope), core self-evaluation (including self-esteem, generalized self-efficacy, emotional stability, locus of control), and positivity (including life satisfaction, self-esteem, and optimism). Note that these three constructs and most of their constituents presumably have some overlap with trait EI. Note, moreover, although the three constructs reviewed by Borgogni and Miraglia (2020) fulfill their definition of "positive

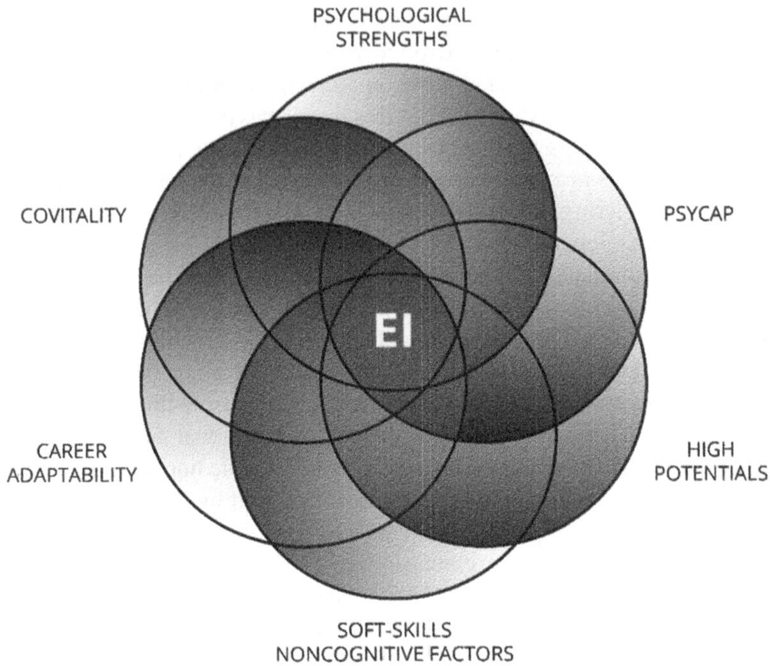

Fig. 5.2 Heuristic model of the six rings of positive self-capital, with trait emotional intelligence as the key variable with higher overlap along the rings

features of individual functioning," we believe that only the first of them (PsyCap) has a coherent theory to support it, as well as an instrument which allows its operationalization. For this reason, we believe it is justified to retain PsyCap as an essential element of positive self-capital. However, we have also identified in the literature other five factors of high order that deserve to be considered as constituent elements of positive self-capital, thus forming, together, a provisional nomological network of six factors or rings if we represent them in a circular way as interlinked sets (see Fig. 5.2).

Another related construct related to positive self-capital but omitted by Borgogni and Miraglia (2020) is the "intrapreneurial self-capital" construct (Di Fabio, 2014). It is defined as a core of individual intrapreneurial resources used to cope with career and life construction challenges and includes dimensions of core self-evaluation, hardiness, creative self-efficacy, resilience, goal mastery, decisiveness, and vigilance. This is extraordinarily interesting because it is probably here that the point of marginal utility is located in the new social contract that organizations and individuals need to agree upon. Perhaps people with high scores in intrapreneurial self-capital are those with the necessary personal resources to adapt to the new work demands and to provide the value that organizations need to keep developing in a sustainable way (Di Fabio & Duradoni, 2019). Recently, Di Fabio and Saklofske (2019) observed a high correlations ($r = 0.52$–0.70) between intrapreneurial self-capital and trait EI factors and demonstrated through hierarchical regression that trait EI explained additional variance in intrapreneurial self-capital beyond that accounted for each of the three personality trait models they measured.

According to Borgogni and Miraglia (2020), "research may aim at identifying additional positive psychological characteristics to describe positive self-capital" (p. 348). This is precisely our purpose.

On the one hand, it is to be expected that all these candidate dimensions of the still diffuse positive self-capital construct share considerable variance (overlap) with the trait EI construct. On the other hand, going back to the definition of positive self-capital offered by Borgogni and Miraglia (2020), it seems to us that at least six other higher-order constructs could be suggested as large sets defining their sampling domain. That is, under the framework of the positive self-capital concept, we understand that at least the following six constructs could be identified as key personal resources to achieve and maintain successful performances and personal development, namely, soft skills or noncognitive factors, career adaptability, covitality, psychological strengths, psychological capital, and high potentials. There is research literature supporting the link between each of these rings and occupational or career success.

All of these key personal resources (rings in Fig. 5.2) present an important theoretical and presumably empirical overlap with trait EI, to such an extent that we dare to propose trait EI as the outstanding component of positive self-capital. We suspect that there is no other high-order construct that shares the maximum possible variance with all alleged dimensions of positive self-capital and, at the same time, has a level of criterion validity comparable to that of trait EI.

Trait EI is a comprehensive, parsimonious, and pragmatic construct, whose theory is rooted in the tradition of individual differences research (Petrides, 2011; Petrides et al., 2016). From a theoretical point of view, integrating trait EI into the nomological network of positive self-capital could additionally contribute to strengthening the scientific foundation of positive self-capital. From a pragmatic point of view, the assessment of trait EI (e.g., TEIQue) as a proxy of positive self-capital might help practitioners to get a quick picture through a unique research-based construct.

In order to empirically test our proposal, future research may explore the correlations between trait EI and operationalizations of each of the six rings, as well as the moderator or mediator role of trait EI in the relationships between the six rings and job variables such as leadership, job performance, job satisfaction, or job engagement, among others.

The Soft Skills or Noncognitive Factors Ring

Wechsler (1943) introduced the concept of *non-intellective* factors as all affective and conative abilities which in any way enter into global behavior and should be part of total intelligence if it were well measured. Later, the Nobel laureate James

Heckman and colleagues (e.g., Heckman & Kautz, 2012) popularized the synonyms *noncognitive factors* or *soft skills* subsuming a diverse set of traits mainly represented by socio-emotional competences, personality, and motivational traits. Heckman and colleagues have provided robust evidence about the relevant predictive power of noncognitive factors in life outcomes such as academic performance and work success.

Although a distinction between cognitive and noncognitive factors is useful from a pragmatic point of view, to be honest it may not be convenient in all cases. In this regard, we concur with Borghans et al.'s (2008) caveat that noncognitive factors create the potential for much confusion because few aspects of human behavior are devoid of cognition, as well as many aspects of personality are influenced by cognitive processes. Nevertheless, the differentiation between cognitive versus noncognitive factors is, in a way, a parallel to the distinction in individual differences research between the fields of intelligence and personality. It should be also noted that intelligence and personality are not completely independent (orthogonal) (e.g., Román-González et al., 2018) or enough to explain the complete repertoire of differences between individuals, but the scientific study of individual differences involves essentially variables that, for the most part, can be easily distinguished as cognitive or noncognitive (Borghans et al., 2008).

The report on demanded soft skills by the Burning Glass Technologies (2015) highlighted that soft skills, often considered part of EI, are called crucial and occupy an unusual position in the debate over global workforce. Nevertheless, this report also warns that the definition of soft skills is, as the name implies, somewhat fuzzy.

According to the new rules of management of the Harvard Business Review series in 2018, despite advanced technical skills, the quick transformation of organizations is requiring human resources with soft skills such as advanced social and emotional learning skills as well as career management skills. This is in line with the World Economic Forum's (2016, 2020) reports, where EI has been identified as one of the top skills in demand by the labor market.

The Career Adaptability Ring

Adaptability to career is another of the six constructs related to professional evolution and performance in which EI is present (Coetzee & Harry, 2014; Di Fabio & Saklofske, 2014). It contains unambiguous elements that respond to the need for the development of new personal skills adapted to the demands of the organizational environment, which we find so important insofar as work provides a framework for human development and a key experience in each person's life (Savickas, 2011).

Career adaptability is the disposition to deal with predictable tasks of job preparation and participation and unpredictable adjustments caused by changes in work and working conditions (Savickas, 1997, p. 254). The link between the career adaptability and EI has increased over the last few years, being considered either as both meta-skills that can be part of models working to increase employability or as self-regulatory social resources involved in the interaction of the individual and the social environment (Coetzee, & Harry, 2014), and also in the EI's role as a mediator between self-efficacy in decision-making and the process of choosing and building careers in both work and academic settings (Di Fabio et al., 2013; Di Fabio & Palazzeschi, 2014) or even in the influence on the return to work of people over 50 (Luke et al., 2016).

The findings show correlations that leave no doubt about their high level of linkage, which in many cases has served empirically not only to verify their relationship but also to work on their development, since there is scientific literature that shows that both EI and career adaptability can be improved through training, professional orientation, and coaching programs (Di Fabio & Kenny, 2011).

The Covitality Ring

Covitality is a concept opposed to comorbidity, since concerns to the co-occurrence of various health conditions lead to some individuals tending to be generally healthy. For the first time in 2002, Weissm et al. (2002) used *covitality* as a conceptual framework to capture the tendency to positive traits of well-being, self-confidence, and general health, as well as the relations among them. Later studies (Jones et al., 2013) extended the construct to include the co-occurrence of synergy of positive psychology constructs and their relations with psychological well-being. The main advancement with the covitality construct perhaps relies on that, by examining the hypothesis that a combination of first-order positive psychology constructs map onto the second-order covitality construct, it encompasses healthy and positive functioning across life domains.

Despite the fact that most of the empirical studies on covitality have been focused at this time on adolescent samples, extending the framework to the working adult population could offer us a theoretical umbrella for linking EI and positive functioning in all life spheres. In fact, one of the psychological variables most linked to covitality is the so-called general factor of personality or Big One, considered an indicator of social efficiency, whose high overlap with trait EI has been confirmed on several occasions, finding a correlation between both variables in a range of $r = 0.65–0.85$ (Pérez-González, 2020).

In their collaboration with the largest Mutual Benefit Society in Belgium, Mikolajczak et al. (2015) split their sample of clients into two groups: (1) people with below-average trait EI and (2) people with above-average trait EI. Mikolajczak et al. (2015) found that those with below-average trait EI used medical services equivalent to an expenditure of approximately €361 per year more than those people who were above average on trait EI. When these researchers considered the whole population of Belgium, they calculated that people low on trait EI would cost Belgium insurance firms €1.8 billion per year more in partial reimbursement of medical and healthcare expenses and financial support in the event of disability, being unable to work compared to those with above-average trait EI. Mikolajczak et al.'s (2015) work suggested that improvement in EI in the general population would result in countries seeing benefits to individuals, communities, and economies. In a later econometric study of these data, Mikolajczak and Van Belleguem (2017) found that the 1% increase in the intrapersonal component of trait EI of people corresponded to 1% decrease in healthcare expenditures required annually by individuals. These findings suggest that trait EI is a favorable personal resource both for the well-being and economy of individuals and communities.

The Psychological Strengths Ring

Of the three pillars on which positive psychology bases its condition, the study of positive subjective experiences, the study of positive individual traits, and the study of institutions that foster positive experiences and traits (Seligman & Csikszentmihalyi, 2000), the character strengths and virtues classification are ascribed to the second (Peterson, & Seligman, 2004). The character strengths and virtues are distinguished on three different conceptual levels: virtues, which are the fundamental characteristics, wisdom, courage, humanity, justice, temperance, and transcendence; character strengths, which are the psychological ingredients that define the way to reach the virtues; and situational themes, the specific habits that lead people to manifest given character strengths in given situations (Peterson, & Seligman, 2004).

The framework of character strengths and virtues describe 24 character strengths: cognitive strengths (under the virtue of wisdom and knowledge: including creativity, curiosity, open-mindedness, love of learning, perspective), emotional strengths (courage: including bravery, persistence, integrity, vitality), social and community strengths (humanity, including love, kindness, and social intelligence; justice, including citizenship, fairness, and leadership), protective strengths (temperance: including forgiveness and mercy, humility and modesty, prudence, self-regulation), and spiritual strengths (transcendence: including appreciation of beauty and excellence, gratitude, hope, humor, and spirituality). Particularly, within "humanity," Peterson and Seligman (2004) included EI under the dimension of "social intelligence."

The correlation between character strengths and virtues and EI has already been partially explored (e.g., Ros-Morente et al. 2017; Zacher et al. 2013), leaving an expeditious field of research; but it is the identification or deliberate creation of institutions that facilitates good character, which Seligman and Csikszentmihalyi (2014) point to, as a third way of developing character strengths and virtues, which with the passage of time has generated greater consequences and studies and in which the relationship between character strengths and virtues and EI might be more fruitful. In order to reach an effective and sustainable adaptation to the new economic environment, paying attention to the quality of the psychological strengths and the level of well-being of a number of people who live in organizations is mandatory (e.g., Guest, 2017).

The required adaptation that organizations and individuals have made to deal with the challenges posed by the new postmodern and global service economy has been possible because of the implementation of psychological strengths or personal resources such as character strengths and EI (e.g., Di Fabio, 2017a; Luthans et al. 2004).

The twenty-first century organizations must play a new role in this new society in which there exist the need to combine well-being, profit, and sustainability (Di Fabio, 2017b), which may imply the construction of a new psychological contract which links organizations and people, not so much for the infinite guarantee of jobs but for the integral, personal, and professional development of their human teams, which guarantees commitment and results for the organization and greater future employability and well-being for the individual (Di Fabio, & Duradoni, 2019).

The Positive Psychological Capital Ring

Among the six rings that make up this metaconstruct sewn by Ariadne's thread that EI weaves, Positive Psychological Capital (PsyCap) is probably the one that, with the most determination, has positioned itself in the task of giving meaning to the new differential role that people can play in organizations (Luthans et al., 2004). The new economic and industrial model 4.0 that is emerging today generates new structures, roles, and challenges to which, once again, people will have to adapt in order to re-signify their value (Savickas & Savickas, 2019).

In this single and uncertain market of global connections, the challenge for most organizations is to respond to more complex problems in a shorter period of time, which implies the need for greater cooperation between people. The speed and power of communications networks are multiplying; however, it is the quantity of human interactions and the quality and fluidity of these that largely determine the value of the product or service that organizations place on the market (Hülsheger et al. 2009). In this environment, adaptation through psychological resources and socio-emotional skills is no longer a simple and voluntary axis of development for individuals, teams, and organizations but has transformed itself into a differential value and a strategic need to maintain and increase the competitiveness of organizations and the employability of people (Guest, 2017; Di Fabio, 2017b).

The four psychological variables which form the acronym HERO in PsyCap (Hope, Efficacy, Resilience, Optimism) make up those sorts of valuable personal resources. The inspiration that positive psychology (Seligman & Csikszentmihalyi, 2000) and positive organizational behavior bring to PsyCap determines its positive focus on the individual's psychological strengths and competences that influence work performance (Donaldson et al., 2019), although its central contribution lies, in our view, in the selection of the psychological variables that determine the new role of people in organizations. The selection criteria of these four variables constitute one by far the most original, important, and, at the same time, controversial elements of the construct (Table 5.3).

Table 5.3 Selection criteria to become part of PsyCap.

CATEGORY	POSITIVE CONSTRUCT	THEORY-BASED?	TRAIT-LIKEL RELATIVELY FIXED?	STATE-LIKEL MALLEABLE?	MEASURABLE?	RELATED TO WORK PERFORMANCE?	RELATED TO OTHER POSITIVE OUTCOMES?
SOCIAL	Gratitude	●	●	●	○	○	●
	Forgiveness	●	●	●	●	○	●
	Emotional intelligence	●	○	○	●	●	●
	Spirituality	●	●	●	●	○	●
HIGHER ORDER	Authenticity	●	●	●	●	●	●
	Courage	●	●	●	●	○	●

Adapted from Luthans et al. (2007, p. 177)

The controversial argument has to do with the "reason why" which supports the very possibility of development, malleability, growth, and management of the variables which make up the acronym HERO. An argument which places them in a territory between state and trait and which would give rise to a hybrid between the two would have the maximum percentage of the former, to be considered susceptible to modification, and the minimum of the latter, although sufficient in any case, to provide it with the necessary permanence and sustainability over time. "The positioning of PsyCap as a state-like resource is aligned with evidence from positive psychology that states and state-like characteristics still include a trait baseline (or set point)" (Luthans & Youssef-Morgan, 2017, p, 10).

It was not until 2007 (Luthans et al., 2007) that the authors of PsyCap seriously considered including EI as a positive social construct applicable to PsyCap. The result of that analysis was a conclusion that it met all the requirements, leaving open for review the controversial issue of whether EI was a trait, and therefore fixed characteristic, with little capacity for improvement, which would prevent its classification as PsyCap, or rather a state, almost a trait, which is malleable and capable of development but stable enough to be maintained over time – which is why it was considered to put this issue on hold while suggesting that EI needed more valid measurements and empirical research (Luthans et al., 2007).

Thirteen years later the case is still sub judice, and it is remarkable that, despite the long road of consistency, validity, and influence taken by EI during all this time and the increasing notoriety and robustness of PsyCap, there are a lack of studies exploring the relationships between both constructs (e.g., Pradhan et al. 2016). The theoretical and expected empirical link between EI and PsyCap may result in stronger plans to promote the development and growth of people with the necessary set of skills that the new organizational environments urgently need.

The High Potentials Ring

According to Silzer and Church (2009), the three key dimensions of potential are career dimensions (specific attributes that lead to success in a particular occupation or job), growth dimensions (growth mindset that affects development and improvement), and foundational dimensions (cognitive and noncognitive characteristics that predict success across the board). Under this framework, trait EI is located by Silzer and Church (2009) in the foundational dimensions of personality, but it could be also located in the first dimension of career dimension too. Following the Silzer and Church's (2009) framework, MacRae and Furnham (2014, 2020) developed the High Flying Personality Inventory (HFPI), which is a six-factor model of high flying personality traits including conscientiousness, adjustment, curiosity, risk approach, ambiguity acceptance, and competitiveness. These six high flying dimensions of potential are correlated with work success.

According to MacRae and Furnham (2014), specifically those with high adjustment are calm under pressure and feel less anxiety and stress, while those with lower scores in this trait tend to experience more anxiety and worry. It is pretty clear that adjustment is theoretically related to trait EI, but probably some of the rest of these high potentials may share some variance with trait EI too, since from trait EI theory (Petrides, 2010, 2011) positive correlation is expected between trait EI and conscientiousness and negative correlation is expected between trait EI and risk approach.

In a review that compared scientific research on predictors of job performance to the qualities in highest demand for the twenty-first century workforce, Hogan et al. (2013) identified three general markers of high potential or determinants of employability: social skills (rewarding to deal with), ability (able to do the job), and drive (willing to work hard). They considered social skills as a key determinant of employability, which is related to EI (Furnham, 2009). Accordingly, they later concluded that an early indicator of high potential is emotional intelligence, "which can be assessed by psychometric tests and further refined through training and development" (Chamorro-Premuzic et al., 2017).

5.6 Training EI

Around just a decade ago, Furnham (2009) concluded that few or no studies had demonstrated the efficacy of EI training. Nowadays science evolution has made its job, and the data-driven conclusion is not the same. Despite the trait nature of trait EI, against the caution expressed by Luthans et al. (2007; Luthans & Youssef-Morgan, 2017), this is not a fixed trait, but a malleable trait, and therefore a legitimate member of PsyCap and positive self-capital meeting all their selection criteria (Table 5.3).

Several comprehensive reviews on emotional education for promoting EI can be found among which we highlight two meta-analyses (Hodzic et al., 2018; Mattingly & Kraiger, 2019) and a systematic review of the literature (Kotsou et al., 2019). According these three reviews, there is enough evidence to conclude that both ability EI and trait EI can be improved through emotional education in adults.

The meta-analysis conducted by Hodzic et al. (2018) illustrated the fact that EI interventions have increased in number in recent years, providing the opportunity for a systematic analysis of outcome of EE for promoting EI. The interventions reviewed are set in different contexts (in hospitals, education, and employment) and use different theoretical models as the basis for their development; they also focus on improving different outcomes. Hodzic et al. (2018) listed those outcomes as life satisfaction, perceived health, stress reduction, emotional self-efficacy, mental health, quality of interpersonal relationships, and employability.

The meta-analyses by Hodzic et al. (2018) included 24 studies focused on interventions designed for those older than 16 years of age. The work showed a moderate effect size of EI training and showed that effect size did not change when only randomized control trials were considered. Thus, Hodzic et al. (2018) confirmed that training programs could increase EI. Moreover, they highlighted some additional findings that Pérez-González and Qualter (2018) summarized like this: (1) interventions based on the ability EI model showed significantly higher effects on outcome variables (including trait EI) than those based on trait EI models, (2) those effects on ability EI were most pronounced for the ability EI branch known as "understanding emotions," and (3) the positive changes in EI due to interventions remained over time. However, it should be noted that most of the interventions analyzed in this meta-analysis are aimed at improving EI as a personality trait (i.e., trait EI), not as ability EI. This fact implies that the meta-analysis by Hodzic et al. (2018) is primarily a collection of evidence supporting the effectiveness of trait EI training programs rather than of ability EI training programs in adults.

The Mattingly and Kraiger (2019) meta-analysis used a total of 56 samples from 50 studies with a pre-post design and 28 samples from 26 studies with a treatment control design. Coinciding with the meta-analysis by Hodzic et al. (2018), the effect size observed in the meta-analysis by Mattingly and Kraiger (2019) was approximately of Cohen's $d = 0.50$.

In the systematic review by Kotsou et al. (2019), they analyzed 46 EI intervention studies on adult populations, concluding that groups receiving EI interventions improved compared to control groups. Likewise, of the 11 studies targeting work-related outcomes, 8 reported significant results with improvement in variables such as teamwork, conflict management, coaching skills, employability and reemployment, workplace incivility, institutional climate, and job satisfaction. Kotsou et al. (2019) caveat that although there is not a minimum level of baseline EI required to obtain improvements, lower levels of baseline EI have been presented greater improvements. However, Kotsou et al. (2019) concluded that findings from their review suggest that it is possible to improve various aspects of EI in a way that often benefits psychological health, so EI interventions might be a tool to prevent and treat work burnout and chronic stress.

In an attempt to identify the active ingredients of the most effective emotional education programs, Pérez-González and Qualter (2018) observed that most of them agreed that they used three group and one individual teaching strategies, which can be summarize under the acronym DAPHnE. This acronym refers to the main kind of practices/activities (presumed or potential active ingredients/mechanisms) that contribute to changes/learning/outcomes in effective EE programs, namely, **D**iscussions in the classroom; real situational or case **A**nalyses; social **P**laying or enjoyable activities (e.g., role-playing; cooperative learning; relaxation); and **H**omework exercises (e.g., emotional diary), and successful emotional education programs used to including these four pedagogical practices that allegedly e**n**hance **E**ffectiveness (DAPHnE).

5.7 Take-Home Message

After the economic periods based on agricultural model (1850–1909), industrial model (1910–1949), and corporate model (1950–1999), we arrived to the current global economy model of the fourth industrial revolution (2000–2050), when "Life has become less predictable, less regulated, less stable, and less orderly. Post-modern social structures make it difficult, and even unwise, for individuals to commit to institutions and organisations. Instead individuals must commit to self-chosen trajectories" (Savickas & Savickas, 2019, p. 38). The leading role of the self is linked to the success of the global economy model. From that moment on, the idea of community gave way to a more independent, more anonymous, and, theoretically, more freer being, but also with greater individual burdens (Savickas, 2011). The need to develop psychological resources to face this new reality increases with globalization, the development of technology, and the fullness of economic neoliberalism at the end of the twentieth century and the beginning of the twenty-first century, as well as with the in crescendo demand for teamwork and emotional labor at any hierarchical level of the organization.

The higher complexity and variability of the so-called VUCA environment (characterized by volatility, uncertainty, complexity, and ambiguity) and the new organizational needs made the idea of hedonic organizational welfare (positive vs. negative emotions balance) as a key ingredient for work engagement. But this has recently evolved towards a comprehensive welfare of both hedonic and eudaimonic well-being, where focus on self-actualization of resources and personal strengths is a sine qua non asset (Di Fabio, 2014). This new desired well-being lies in people developing the main psychological resources necessary to face and manage their employability and career development throughout their life span under the principles of harmonization, which emphasizes the importance to act early in a primary prevention perspective, considering the importance to build personal and collective strengths, from the intrapersonal dimension (e.g., trait EI) to enhance interpersonal and community strengths to protect and sustain the environment and the relationships in/with the environment (Di Fabio & Tsuda, 2018).

In line with the idea of sustainability, a new and possible psychological contract between organizations and people is based on mutual collaboration in this construction and development work. It is a question not only of achieving a better work performance or being led in an emotionally intelligent way but also of developing the necessary skills that allow for the construction of a meaningful life and career of guarantees in accordance with the values, desires, and needs of each individual (Di

Fabio, 2014, 2017b). The longed for commitment that organizations seek to generate in their employees may pass through this process of co-construction.

As it has been seen, the development in the scientific literature of novel constructs related to positive self-capital has been especially relevant in the last three decades (e.g., Borgogni & Miraglia, 2020). Following this *zeitgeist*, we augur an equal growth in the emergence of tools and programs that help not only to enhance the individual resources necessary to face the new labor demands but also to increase those organizational resources that enable a lasting and full personal and professional self-construction, which will probably be an action in which the role of trait EI is expected to play decisive, as can be surmised from the evidence accumulated (e.g., Pérez-González et al., 2020; Petrides et al., 2016; World Economic Forum, 2016, 2020) on its pervasive incidence in multiple areas of personal and professional development and on their proven malleability through appropriate emotional education programs.

References

Alba-Juez, L., & Pérez-González, J.-C. (2019). Emotion and language 'at work': The relationship between Trait Emotional Intelligence and communicative competence as manifested at the workplace. In J. Lachlan Mackenzie & L. Alba-Juez (Eds.), *Emotion in discourse* (pp. 247–278). John Benjamins. https://doi.org/10.1075/pbns.302.10alb

Andrei, F., Siegling, A. B., Aloe, A. M., Baldaro, B., & Petrides, K. V. (2016). The incremental validity of the Trait Emotional Intelligence Questionnaire (TEIQue): A systematic review and meta-analysis. *Journal of Personality Assessment, 98*, 261–276. https://doi.org/10.12681/psy_hps.23016

Ashkanasy, N. M., & Daus, C. S. (2020). Emotional intelligence in the workplace. In B. J. Carducci & C. Nave (Eds.), *The Wiley-Blackwell encyclopedia of personality and individual differences: Vol. IV. Personality processes and individual differences* (pp. 485–490). John Wiley & Sons. https://doi.org/10.1002/9781118970843.ch345

Borghans, L., Duckworth, A. L., Heckman, J. J., & Ter Weel, B. (2008). The economics and psychology of personality traits. *Journal of Human Resources, 43*(4), 972–1059. https://doi.org/10.3368/jhr.43.4.972

Borgogni, L., & Miraglia, M. (2020). Chapter 211: Positive self-capital. In B. J. Carducci & C. Nave (Eds.), *The Wiley-Blackwell encyclopedia of personality and individual differences: Vol. III. Personality processes and individual differences* (pp. 345–349). John Wiley & Sons. Bar. https://doi.org/10.1002/9781118970843.ch236

Brasseur, S., Grégoire, J., Bourdu, R., & Mikolajczak, M. (2013). The profile of emotional competence (PEC): Development and validation of a self-reported measure that fits dimensions of emotional competence theory. *PLoS One, 8*(5), e62635. https://doi.org/10.1371/journal.pone.0062635

Burning Glass Technologies. (2015). *Blurring lines: How business and technology skills are merging to create high opportunity hybrid jobs*. http://hdl.voced.edu.au/10707/519485

Chamorro-Premuzic, T., Adler, S., & Kaiser, R. B. (2017). What science says about identifying high-potential employees. *Harvard Business Review*. Retrieved from https://hbr.org/2017/10/what-science-says-about-identifying-high-potential-employees.

Coetzee, M., & Harry, N. (2014). Emotional intelligence as a predictor of employees' career adaptability. *Journal of Vocational Behavior, 84*(1), 90–97. https://doi.org/10.1016/j.jvb.2013.09.001

Cohen, D., & Soto, M. (2007). Growth and human capital: Good data, good results. *Journal of Economic Growth, 12*, 51–76. https://doi.org/10.1007/s10887-007-9011-5

Di Fabio, A. (2014). Intrapreneurial self-capital: A new construct for the 21st century. *Journal of Employment Counseling, 51*(3), 98–111. https://doi.org/10.1002/j.2161-1920.2014.00045.x

Di Fabio, A. (2017a). Positive healthy organizations: Promoting well-being, meaningfulness, and sustainability in organizations. *Frontiers in Psychology. Organizational Psychology, 8*, 1938. https://doi.org/10.3389/fpsyg.2017.01938

Di Fabio, A. (2017b). The psychology of sustainability and sustainable development for well-being in organizations. *Frontiers in Psychology. Organizational Psychology, 8*, 1534. https://doi.org/10.3389/fpsyg.2017.01534

Di Fabio, A., & Duradoni, M. (2019). Intrapreneurial self-capital: A primary preventive resource for twenty-first century entrepreneurial contexts. *Frontiers in Psychology, 10*(MAY). https://doi.org/10.3389/fpsyg.2019.01060

Di Fabio, A., & Kenny, M. E. (2011). Promoting emotional intelligence and career decision making among Italian high school students. *Journal of Career Assessment, 19*(1), 21–34. https://doi.org/10.1177/1069072710382530

Di Fabio, A., & Kenny, M. E. (2016). Promoting well-being: The contribution of emotional intelligence. *Frontiers in Psychology, 7*. https://doi.org/10.3389/fpsyg.2016.01182

Di Fabio, A., & Rosen, M. A. (2018). Opening the black box of psychological processes in the science of sustainable development: A new frontier. *European Journal of Sustainable Development Research, 2*(4), 2–6. https://doi.org/10.20897/ejosdr/3933

Di Fabio, A., & Saklofske, D. H. (2014). Comparing ability and self-report trait emotional intelligence, fluid intelligence, and personality traits in career decision. *Personality and Individual Differences, 64*, 174–178. https://doi.org/10.1016/j.paid.2014.02.024

Di Fabio, A., & Saklofske, D. H. (2019). The contributions of personality traits and emotional intelligence to intrapreneurial self-capital: Key resources for sustainability and sustainable development. *Sustainability, 11*(5), 1240. https://doi.org/10.3390/su11051240

Di Fabio, A., & Tsuda, A. (2018). The psychology of harmony and harmonization: Advancing the perspectives for the psychology of sustainability and sustainable development. *Sustainability, 10*(12), 4726. https://doi.org/10.3390/su10124726

Di Fabio, A., & Palazzeschi, L. (2014). Incremental variance of trait emotional intelligence compared to personality traits in career decision-making difficulties. *Personality and Individual Differences, 60*(Suppl), S31. https://doi.org/10.1016/j.paid.2013.07.060

Di Fabio, A., Palazzeschi, L., Asulin-Peretz, L., & Gati, I. (2013). Career indecision versus indecisiveness: Associations with personality traits and emotional intelligence. *Journal of Career Assessment, 21*(1), 42–56. https://doi.org/10.1177/1069072712454698

Donaldson, S. I., Lee, J. Y., & Donaldson, S. I. (2019). Evaluating positive psychology interventions at work: A systematic review and meta-analysis. *International Journal of Applied Positive Psychology*, September. https://doi.org/10.1007/s41042-019-00021-8

Fiori, M., Antonietti, J. P., Mikolajczak, M., Luminet, O., Hansenne, M., & Rossier, J. (2014). What is the ability emotional intelligence test (MSCEIT) good for? An evaluation using item response theory. *PLoS One, 9*(6), e98827. https://doi.org/10.1371/journal.pone.0098827

Furnham, A. (1992). *Personality at work: The role of individual differences in the workplace*. Routledge.

Furnham, A. (2001). *The psychology of behaviour at work. The individual in the organization.* Oxford University Press.

Furnham, A. (2009). The importance and training of emotional intelligence at work. In *Assessing emotional intelligence* (pp. 137–155). Springer. https://doi.org/10.1007/978-0-387-88370-0_8

Goleman. (1995). *Emotional intelligence: Why it can matter more than IQ*. Bantam Books.

Guest, D. E. (2017). Human resource management and employee well-being: Towards a new analytic framework. *Human Resource Management Journal, 27*(1), 22–38. https://doi.org/10.1111/1748-8583.12139

Harms, P. D., & Credé, M. (2010). Emotional intelligence and transformational and transactional leadership: A meta-analysis. *Journal of Leadership & Organizational Studies, 17*(1), 5–17. https://doi.org/10.1177/1548051809350894

Heckman, J. J., & Kautz, T. (2012). Hard evidence on soft skills. *Labour Economics, 19*(4), 451–464. https://doi.org/10.1016/j.labeco.2012.05.014

Hodzic, S., Scharfen, J., Ripoll, P., Holling, H., & Zenasni, F. (2018). How efficient are emotional intelligence trainings: A meta-analysis. *Emotion Review, 10*(2), 138–148. https://doi.org/10.1177/1754073917708613

Hogan, R., Chamorro-Premuzic, T., & Kaiser, R. B. (2013). Employability and career success: Bridging the gap between theory and reality. *Industrial and Organizational Psychology, 6*(1), 3–16. https://doi.org/10.1111/iops.12001

Hülsheger, U. R., Anderson, N., & Salgado, J. F. (2009). Team-level predictors of innovation at work: A comprehensive meta-analysis spanning three decades of research. *Journal of Applied Psychology, 94*(5), 1128–1145. https://doi.org/10.1037/a0015978

Jones, C. N., You, S., & Furlong, M. J. A. (2013). Preliminary examination of covitality as integrated well-being in college students. *Social Indicators Research, 111*, 511–526. https://doi.org/10.1007/s11205-012-0017-9

Joseph, D. L., & Newman, D. A. (2010). Emotional intelligence: An integrative meta-analysis and cascading model. *Journal of Applied Psychology, 95*(1), 54–78. https://doi.org/10.1037/a0017286

Joseph, D. L., Jin, J., Newman, D. A., & O'Boyle, E. H. (2015). Why does self-reported emotional intelligence predict job performance? A meta-analytic investigation of mixed EI. *Journal of Applied Psychology, 100*(2), 298–342. https://doi.org/10.1037/a0037681

Kotsou, I., Mikolajczak, M., Heeren, A., Grégoire, J., & Leys, C. (2019). Improving emotional intelligence: A systematic review of existing work and future challenges. *Emotion Review, 11*(2), 151–165. https://doi.org/10.1177/1754073917735902

Luke, J., McIlveen, P., & Perera, H. N. (2016). A thematic analysis of career adaptability in retirees who return to work. *Frontiers in Psychology, 7*(February), 1–12. https://doi.org/10.3389/fpsyg.2016.00193

Luthans, F., & Youssef-Morgan, C. M. (2017). Psychological capital: An evidence-based positive approach. *Annual Review of Organizational Psychology and Organizational Behavior, 4*, 339–366. https://doi.org/10.1146/annurev-orgpsych-032516-113324

Luthans, F., Luthans, K. W., & Luthans, B. C. (2004). Positive psychological capital: Beyond human and social capital. *Business Horizons, 47*(1), 45–50. https://doi.org/10.1016/j.bushor.2003.11.007

Luthans, F., Youssef, C. M., & Avolio, B. J. (2007). Psychological capital: Developing the human competitive edge. In *Psychological capital: Developing the human competitive edge*. https://doi.org/10.1093/acprof:oso/9780195187526.001.0001

MacCann, C., Jiang, Y., Brown, L. E. R., Double, K. S., Bucich, M., & Minbashian, A. (2020). Emotional intelligence predicts academic performance: A meta-analysis. *Psychological Bulletin, 146*(2), 150–186. https://doi.org/10.1037/bul0000219

MacRae, I., & Furnham, A. (2014). *High potential: How to spot, manage and develop talented people at work*. Bloomsbury Publishing.

MacRae, I., & Furnham, A. (2020). A psychometric analysis of the high potential trait inventory (HPTI). *Psychology, 11*(08), 1125–1140. https://doi.org/10.4236/psych.2020.118074

Martins, A., Ramalho, N., & Morin, E. (2010). A comprehensive meta-analysis of the relationship between emotional intelligence and health. *Personality and Individual Differences, 49*, 554–564. https://doi.org/10.1016/j.paid.2010.05.029

Matthews, G., Deary, I. J., & Whiteman, M. C. (2003). *Personality traits*. Cambridge University Press.

Mattingly, V., & Kraiger, K. (2019). Can emotional intelligence be trained? A meta-analytical investigation. *Human Resource Management Review, 29*(2), 140–155. https://doi.org/10.1016/j.hrmr.2018.03.002

Mavroveli, S., Petrides, K. V., Rieffe, C., & Bakker, F. (2007). Trait emotional intelligence, psychological Well-being, and peer-rated social competence in adolescence. *British Journal of Developmental Psychology, 25*, 263–275. https://doi.org/10.1348/026151006X118577

Miao, C., Humphrey, R. H., & Qian, S. (2017). A meta-analysis of emotional intelligence effects on job satisfaction mediated by job resources, and a test of moderators. *Personality and Individual Differences, 116*, 281–288. https://doi.org/10.1016/j.paid.2017.04.031

Miao, C., Humphrey, R., & Qian, S. (2018a). Emotional intelligence and authentic leadership: A meta-analysis. *Leadership and Organization Development Journal, 39*(5), 679–690. https://doi.org/10.1108/LODJ-02-2018-0066

Miao, C., Humphrey, R. H., & Qian, S. (2018b). A cross-cultural meta-analysis of how leader emotional intelligence influences subordinate task performance and organizational citizenship behaviour. *Journal of World Business, 53*(4), 463–474. https://doi.org/10.1016/j.jwb.2018.01.003

Mikolajczak, M., & Van Bellegem, S. (2017). Increasing emotional intelligence to decrease healthcare expenditures: How profitable would it be? *Personality and Individual Differences, 116*, 343–347. https://doi.org/10.1016/j.paid.2017.05.014

Mikolajczak, M., Avalosse, H., Vancorenland, S., Verniest, R., Callens, M., van Broeck, N., Fantini-Hauwel, C., & Mierop, A. (2015). A nationally representative study of emotional competence and health. *Emotion, 15*(5), 653–667. https://doi.org/10.1037/emo0000034

Moroń, M., & Biolik-Moroń, M. (2020). Trait emotional intelligence and emotional experiences during the COVID-19 pandemic outbreak in Poland: A daily diary study. *Personality and Individual Differences, 168*, 110348. https://doi.org/10.1016/j.paid.2020.110348

O'Boyle, E. H., Humphrey, R. H., Pollack, J. M., Hawver, T. H., & Story, P. A. (2011). The relation between emotional intelligence and job performance: A meta-analysis. *Journal of Organizational Behavior, 32*, 788–818. https://doi.org/10.1002/job.714

O'Connor, P. J., Hill, A., Kaya, M., & Martin, B. (2019). The measurement of emotional intelligence: A critical review of the literature and recommendations for researchers and practitioners. *Frontiers in Psychology, 10*, 1116. https://doi.org/10.3389/fpsyg.2019.01116

Perera, H. N., & DiGiacomo, M. (2013). The relationship of trait emotional intelligence with academic performance: A meta-analytic review. *Learning and Individual Differences, 28*, 20–33. https://doi.org/10.1016/j.lindif.2013.08.002

Pérez, J. C., Petrides, K. V., & Furnham, A. (2005). Measuring trait emotional intelligence. In R. Schulze & R. D. Roberts (Eds.), *Emotional intelligence: An international handbook* (pp. 181–201). Hogrefe & Huber.

Pérez-González, J. C. (2020). Chapter 211: General personality factor. In B. J. Carducci & C. Nave (Eds.), *The Wiley-Blackwell encyclopedia of personality and individual differences: Vol. III. Personality processes and individual differences* (pp. 195–200). John Wiley & Sons. https://doi.org/10.1002/9781118970843.ch210

Pérez-González, J. C., & Sánchez-Ruiz, M. J. (2014). Trait emotional intelligence anchored within the big five, big two and big one frameworks. *Personality and Individual Differences, 65*, 53–58. https://doi.org/10.1016/j.paid.2014.01.021

Pérez-González, J.-C., Saklofske, D. H., & Mavroveli, S. (2020). Editorial: Trait emotional intelligence: Foundations, assessment, and education. *Frontiers in Psychology*. https://doi.org/10.3389/fpsyg.2020.00608

Pérez-González, J.-C., & Qualter, P. (2018). Emotional intelligence and emotional education in school years. In L. Dacree Pool & P. Qualter (Eds.), *An introduction to emotional intelligence* (pp. 81–104). Wiley.

Peterson, C., & Seligman, M. E. P. (2004). *Character strengths and virtues: A handbook and classification*. American Psychological Association. Oxford University Press.

Petrides, K. V. (2010). Trait emotional intelligence theory. *Industrial and Organizational Psychology: Perspectives on Science and Practice, 3*, 136–139. https://doi.org/10.1111/j.1754-9434.2010.01213.x

Petrides, K. V. (2011). Ability and trait emotional intelligence. In T. Chamorro-Premuzic, A. Furnham, & S. von Stumm (Eds.), *The Blackwell-Wiley handbook of individual differences*. Wiley.

Petrides, K. V., & Furnham, A. (2000). On the dimensional structure of emotional intelligence. *Personality and Individual Differences, 29*(2), 313–320.

Petrides, K. V., Mikolajczak, M., Mavroveli, S., Sánchez-Ruiz, M.-J., Furnham, A., & Pérez-González, J.-C. (2016). Developments in trait emotional intelligence research. *Emotion Review, 8*, 335–341. https://doi.org/10.1177/1754073916650493

Petrides, K. V., Pérez-González, J. C., & Furnham, A. (2007a). On the criterion and incremental validity of trait emotional intelligence. *Cognition and Emotion, 21*(1), 26–55. https://doi.org/10.1080/02699930601038912

Petrides, K. V., Pita, R., & Kokkinaki, F. (2007b). The location of trait emotional intelligence in personality factor space. *British Journal of Psychology, 98*(2), 273–289. https://doi.org/10.1348/000712606X120618

Petrides, K. V., Sanchez-Ruiz, M.-J., Siegling, A. B., Saklofske, D. H., & Mavroveli, S. (2018). Emotional intelligence as personality: Measurement and role of trait emotional intelligence in educational contexts. In K. Keefer, J. Parker, & D. Saklofske (Eds.), *Emotional intelligence in education. The springer series on human exceptionality*. Springer. https://doi.org/10.1007/978-3-319-90633-1_3

Pradhan, R. K., Jena, L. K., & Bhattacharya, P. (2016). Impact of psychological capital on organizational citizenship behavior: Moderating role of emotional intelligence. *Cogent Business and Management, 3*(1), 1–17. https://doi.org/10.1080/23311975.2016.1194174

Rauthmann, J. F., Gallardo-Pujol, D., Guillaume, E. M., Todd, E., Nave, C. S., Sherman, R. A., … Funder, D. C. (2014). The situational eight DIAMONDS: A taxonomy of major dimensions of situation characteristics. *Journal of Personality and Social Psychology, 107*(4), 677. https://doi.org/10.1037/a0037250

Román-González, M., Pérez-González, J.-C., Moreno-León, J., & Robles, G. (2018). Extending the nomological network of computational thinking with non-cognitive factors. *Computers in Human Behavior, 80*, 441–459. https://doi.org/10.1016/j.chb.2017.09.030

Ros-Morente, A., Alsinet Mora, C., Torrelles Nadal, C., Blasco Belled, A., & Jordana Berenguer, N. (2017). An examination of the relationship between emotional intelligence, positive affect and character strengths and virtues. *Anales de Psicología, 34*(1), 63. https://doi.org/10.6018/analesps.34.1.262891

Roy, S. R. (2015). *Promoting trait emotional intelligence in leadership and education*. IGI Global.

Sánchez-Álvarez, N., Extremera, N., & Fernández-Berrocal, P. (2016). The relation between emotional intelligence and subjective well-being: A meta-analytic investigation. *The Journal of Positive Psychology, 11*, 276–285. https://doi.org/10.1080/17439760.2015.1058968

Savickas, M. L. (1997). Career adaptability: An integrative construct for life-span, life-space theory. *Career Development Quarterly, 45*(3), 247–259. https://doi.org/10.1002/j.2161-0045.1997.tb00469.x

Savickas, M. L. (2011). New questions for vocational psychology: Premises, paradigms, and practices. *Journal of Career Assessment, 19*(3), 251–258. https://doi.org/10.1177/1069072710395532

Savickas, M. L., & Savickas, S. (2019). A history of career counselling. In *International handbook of career guidance* (pp. 25–43). Springer. https://doi.org/10.1007/978-3-030-25153-6_2

Schultz, T. W. (1961). Investment in Human Capital. *The American Economic Review, 51*(1), 1–17. https://www.jstor.org/stable/1818907

Seligman, M. E. P., & Csikszentmihalyi, M. (2014). Positive psychology: An introduction. In *Flow and the foundations of positive psychology*. Springer. https://doi.org/10.1007/978-94-017-9088-8_18

Seligman, M. E. (2007). Coaching and positive psychology. *Australian Psychologist, 42*(4), 266–267. https://doi.org/10.1080/00050060701648233

Seligman, M. E., & Csikszentmihalyi, M. (2000). Positive psychology: An introduction. *American Psychologist, 55*(1), 5–14. https://doi.org/10.1007/978-94-017-9088-8_18

Siegling, A. B., Vesely, A. K., Saklofske, D. H., Frederickson, N., & Petrides, K. V. (2015). Incremental validity of the Trait Emotional Intelligence Questionnaire–Adolescent Short Form (TEIQue–ASF). *European Journal of Psychological Assessment, 97*, 525–535. https://doi.org/10.1027/1015-5759/a000267

Silzer, R., & Church, A. H. (2009). The pearls and perils of identifying potential. *Industrial and Organizational Psychology, 2*(4), 377–412. https://doi.org/10.1111/j.1754-9434.2009.01163.x

Stough, C., Saklofske, D. H., & Parker, J. D. (2009). A brief analysis of 20 years of emotional intelligence: An introduction to assessing emotional intelligence: Theory, research, and applications. In *Assessing emotional intelligence* (pp. 3–8). Springer. https://doi.org/10.1007/978-0-387-88370-0_1

Van der Linden, D., Schermer, J. A., de Zeeuw, E., Dunkel, C. S., Pekaar, K. A., Bakker, A. B., Vernon, P. A., & Petrides, K. V. (2018). Overlap between the general factor of personality and trait emotional intelligence: A genetic correlation study. *Behavior Genetics, 48*, 147–154. https://doi.org/10.1007/s10519-017-9885-8

Vercellone, C. (2007). From formal subsumption to general intellect: Elements for a Marxist reading of the thesis of cognitive capitalism. *Historical Materialism, 15*, 13–36. https://doi.org/10.1163/156920607X171681

Wechsler, D. (1943). Non-intellective factors in general intelligence. *The Journal of Abnormal and Social Psychology, 38*(1), 101–103.

Weiss, A., King, J. E., & Enns, R. M. (2002). Subjective well-being is heritable and genetically correlated with dominance in chimpanzees (Pan troglodytes). *Journal of Personality and Social Psychology, 83*(5), 1141–1149. https://doi.org/10.1037/0022-3514.83.5.1141

World Economic Forum. (2016). *The future of jobs. Employment, skills and workforce strategy for the fourth industrial revolution*. Retrieved (January 2021) http://www3.weforum.org/docs/WEF_Future_of_Jobs.pdf

World Economic Forum. (2020). *The future of jobs report 2020*. Retrieved (January 2021) from: https://www.weforum.org/reports/the-future-of-jobs-report-2020

Zacher, H., McKenna, B., & Rooney, D. (2013). Effects of self-reported wisdom on happiness: Not much more than emotional intelligence? *Journal of Happiness Studies, 14*(6), 1697–1716. https://doi.org/10.1007/s10902-012-9404

Zeidner, M., Matthews, G., & Roberts, R. D. (2012). The emotional intelligence, health, and Well-being nexus: What have we learned and what have we missed? *Applied Psychology: Health and Well-Being, 4*(1), 1–30. https://doi.org/10.1111/j.1758-0854.2011.01062.x

Zysberg, L., & Zisberg, A. (2020). Days of worry: Emotional intelligence and social support mediate worry in the COVID-19 pandemic. *Journal of Health Psychology, 18*. https://doi.org/10.1177/1359105320949935

Chapter 6
Sustainability in Indian Organizations

Alpana Vaidya

6.1 What Is Sustainability?

There is no universally agreed upon definition of sustainability. Etymologically, the word sustainability refers to something that can be sustained for a period of time. It refers also to something that can be supported, tolerated or confirmed over time and that can be stated with certainty. It concerns building on the present in such a way as not to put the future at risk (Di Fabio, 2017b).

Sustainability can be seen as the process by which something is kept at a certain level. Usually the word sustainability is used in connection with protection of natural resources thereby maintaining ecological balance. In short, sustainability is defined as the processes and actions through which humankind avoids the depletion of natural resources to keep an ecological balance so that society's quality of life doesn't decrease. This approach also takes into consideration meeting the needs of the present generation without compromising the ability of future generations to meet their needs (Di Fabio, 2017b; Di Fabio & Rosen, 2018).

From the traditional point of view, a product is sustainable if it uses increasingly smaller amounts of material; if it is based on renewable and non-polluting processes and materials; if it is not toxic; and if it is easy to maintain, process, dismantle, demolish, dispose of and recycle (Di Fabio, 2017b). In politics, technology, economy and ecology, sustainability is about balancing current aims with future aims without jeopardizing the latter (Di Fabio, 2017b; Di Fabio & Maree, 2016). According to Brundtland Report (1987) and Harris (2003), sustainable development was based on the three "Es", namely, economy, equity and ecology, highlighting the right of present as well as future generations to enjoy the environment and natural resources.

A. Vaidya (✉)
Symbiosis College of Arts and Commerce, Pune, India
e-mail: alpana.vaidya@symbiosiscollege.edu.in

A. Di Fabio (ed.), *Cross-cultural Perspectives on Well-Being and Sustainability in Organizations*, https://doi.org/10.1007/978-3-030-86709-6_6

There are three pillars of sustainability, namely, economic, environmental and social—also known informally as profits, planet and people (cited in Mitchell Grant, 2020). The definition of each one is given below.

The word economic sustainability refers to practices that support long-term economic growth without negatively impacting social, environmental and cultural aspects of the community. – University of Mary Washington, Economic Sustainability.

Environmental sustainability refers to the protection of the environment. For example, it is concerned with whether environmental resources will be protected and maintained for future generations (Tejvan Pettinger, 2018).

Social sustainability refers to "a process for creating sustainable successful places that promote wellbeing, by understanding what people need from the places they live and work. Social sustainability combines design of the physical realm with design of the social world – infrastructure to support social and cultural life, social amenities, systems for citizen engagement, and space for people and places to evolve (cited from https://www.esg.adec-innovations.com/about-us/faqs/what-is-social-sustainability/)". From a business perspective, social sustainability is about understanding the impacts of corporations on people and society. Decisions taken by organizations have to take into consideration their impact on environment, society and people. For example, instead of concentrating on short-term gains, they encourage business enterprise to consider diverse and long-term gains. For example, zero waste packaging, reduction in carbon emission, improving work conditions for workers, deciding on paying minimum wages to be paid, use of less toxic chemicals and the like.

Regarding psychological sustainability and sustainable development, it represents new field of research in psychology. It addresses the issues of sustainability and how psychologists can make contribution in this field. This innovative psychological perspective enlarges the concept of sustainability. It is not only limited to ecological and socioeconomic environment, but it also seeks to improve the quality of life of each human being with and in the environments (Di Fabio, 2017a, b; Di Fabio & Rosen, 2018). This definition of sustainability tries to go beyond traditional framework which is based on three "Es" (economy, equity, ecology) (Burtland Report, 1987; Harris, 2003), "avoiding" something in terms of exploitation, depletion and irreversible alteration.

Modern view of sustainability centres around promotion and enrichment of well-being of individual and society. It is based on a new framework which has focussed on regenerating resources and on a positive approach (Di Fabio, 2017a, b). This approach is based on new keywords such as promotion, enrichment, growth and flexible change (Di Fabio, 2017a, b). In short, according to modern view, any project is sustainable if it is accessible, is de-constructable and is aimed at promoting individual and organizational well-being (Di Fabio, 2016).

In order to promote psychological sustainability.;, organizations need to make efforts towards providing psychologically safe work place. For example, studies have shown that it is necessary for organizations to provide a psychologically safe workplace to improve employee engagement (Kompaso & Sridevi, 2010). From a positive psychology point of view, the four factors in a healthy organization that need to be considered are the individual, the group, the organization and the inter-organizational processes (Henry 2005). Thus, psychological sustainability sees sustainability in terms of the development of well-being and satisfaction of the employees at work place.

6.2 Why Sustainability in Indian Organizations?

This section deals with why it is necessary to understand the sustainability in Indian organization. What is the situation with respect to environmental sustainability, public health and psychological sustainability in Indian organizations? What are the challenges that Indian companies and Indian government are facing with regard to environmental pollution, public health and psychological well-being of the employees? First let's try to understand why it is important to understand sustainability in Indian organizations.

India is one of the fastest growing economies in the world. Most of the Indian companies seem to be well aware of sustainable development as a multidimensional and multidisciplinary concept having a special relevance in the context of environment, allocation and utilization of natural or environmental resources in particular. But still there is lot of scope for improvement. The Indian Constitution 1950 is based on the principles of social and economic justice, and these principles of constitution are implemented by the states through many socially sensitive policies, including affirmative actions. In the 1970s, India also began to focus more sharply on safeguarding the environment. Several laws were put in place, and the Ministry of Environment and Forests came into being in 1985. In the seven decades since Indian independence, the country's large public sector enterprises, nationalized banks and state-owned insurance companies have consistently implemented initiatives directed towards economic growth and distributive justice. The country has also signed on to most major international conservation treaties related to habitat, species and environment (Deepika Krishan, 2019).

But sustainability efforts of India suffered as the Indian economy was liberalized in the 1990s. With liberalization, private corporations were allowed to set up businesses in India, and this leads to rapid GDP growth. This helped India to cut multi-dimensional poverty in half. But this development was at the cost of environment. For example, seven of the ten most polluted cities in the world are in India, and protected forest lands have been opened up for mining, road and rail construction, oil drilling and pipeline and canal construction. Despite economic progress, poverty remains a major challenge. According to the United Nations Development Programme's 2018 Multidimensional Poverty Index, a staggering 364 million Indians continue to experience acute deprivations in health, nutrition, schooling and sanitation. Inequality has worsened dramatically over time (Deepika Krishan, 2019).

Environmental Sustainability in Indian Organization

Indian companies are facing various challenges regarding environmental sustainability. There are a lot of polluting companies in India. For example, there has been 136% increase in the number of grossly polluting units of industry in India between 2011 and 2018. These include pulp and paper mills; distilleries; sugar mills; textile units; tanneries; thermal power plants; food, dairy, and beverage units; chemical units; and slaughterhouses (cited in Kiran Panday, 2019).

Another challenge towards sustainability is India's coal-run thermal energy plants. They are among the most inefficient in the world, and many don't meet the already relatively insufficient Indian standards for air pollution, water consumption or waste disposal.

Another problem area that the Indian government urgently needs to address is the Indian automobile industry, which currently contributes 7.1% of India's GDP. The government is moving towards stricter emission norms and is giving incentives towards the faster adoption and manufacturing of hybrid and electric vehicles. However, progress has been slow. Large-scale green public transport projects haven't taken off in India. As of May 2019, India had only 280,000 electric vehicles, which is far from the government's target of 15 million to 16 million by 2020. In the budget of 2019, the government has made allocations to support e-mobility (Deepika Krishan, 2019).

According to a survey of Indian executives by the Global Compact Network India—one of the country's largest corporate sustainability initiatives—Indian business leaders are focussed primarily on regulatory compliance, not on core sustainability (Deepika Krishnan, 2019). This is because 90% of CEOs surveyed reported that they face no pressure at all from investors to move the needle on sustainability. There is need for more pressure from all the sides for India to make rapid progress towards sustainable development.

The overall amount of green lending in India is still small compared to other types of lending. There is no pressure on banks to use environmental, social and governance criteria while financing polluting projects or projects such as large dams, which cause disruption (Deepika Krishnan (2019).

This shows that in a developing economy, it's impractical to stop new units from launching, and as a result, balancing growth with sustainability is a big challenge.

Public Health in India

When one looks at public health in India, it has been found that providing public healthcare facilities for all is a big challenge. The healthcare system in India suffers from inadequate funding. There are several structural problems too, like the lack of integration between disease control, family welfare programmes, non-communicable diseases control and other programmes in the social sectors. There has been poor accountability and commitment towards universal healthcare (Suresh Kumar, 2019). A lot needs to be done in the area of public healthcare.

Psychological Sustainability in Indian Organizations

Regarding psychological sustainability in Indian organization, it has been found that it is a relatively new area for research and a lot needs to be done in this area. Even at the community level, community mental health and creating mental health awareness programmes are necessary. At organizational level too, attention needs to be paid to the mental health and well-being of the employees.

India has seen a significant rise in the number of organizations over the past years. Organizations play an important role in the development and enhancement of a society. The performance at an individual level can impact an organization's effectiveness. Studies have shown that most of the research on well-being has been from a Western framework (Agarwal et al., 2011; Joshanloo & Weijers, 2014).

Review of literature on well-being and sustainability in Indian organizations has shown that in India, which is the most overworked country in the world, work-related stress can have serious consequences. Portrayed as the "silent killer", stress has been reportedly termed as the root of depression and workplace anxiety in India (Business Insider India 25th September 2019 post). According to a recent LinkedIn report, the majority of the workforce in India is stressed due to the workload, fear of job loss and office politics (Business Insider India 25th September 2019 post).

A 2016 survey of 200, 000 professionals employed across 30 Indian firms found that 46% are reported suffering from extreme stress as a consequence of their work. Pressure related to their jobs had caused at-risk individuals to contemplate suicide (Richa & Kala Vijayraghavan, 2016).

A recent study by People Matters 2019 pointed out that on a scale of one to five, Indian organizations rated their satisfaction with current wellness initiatives at three, indicating the average results of wellness programmes. This shows that a high number of organizations do not offer personalized wellness activity resulting in less employee adoption and engagement. This could be attributed to lack of awareness in defining well-being programmes selecting the best tools for assessment and measuring the cost-effectiveness of these interventions (Sethi, 2019).

It has also been found that in India, only 10% of the 1.7 million registered companies run a formal mental health support programme (Sethi, 2019).

A report based on the views of 12,500 corporate employees from 150 companies across 18 broad sectors like media, telecom, KPO, etc. found that about 42.5% employees from private sectors suffer from depression or general anxiety disorder compared to government employees (ASSOCHAM study, 2015). It was also seen that 38.5% of corporate employees sleep less than 6 hours a day due to high levels of stress that arise due to tough targets set by employers. Around 48% of respondents said they felt fatigue on a regular basis due to general anxiety, and about 27%

of participants in the survey admitted to having regular headaches (ASSOCHAM study, 2015) (cited in Sethi, 2019).

A study on the workplace motivators and employee satisfaction found that front-line sales workers in India were less satisfied with the organization as the mean satisfaction score was 2.04 out of 4. A major reason for this was an incongruence between an employee's expectations and what they were getting. The study indicated that employee's satisfaction is significantly influenced by compensations along with working conditions and support from management (Mishra & Gupta, 2009). A study conducted by Gargi and Ghazi (2018) found that people look more from the work front than just subjective well-being and hedonic happiness point of view.

Limited budgets have been a major challenge for companies to drive workplace wellness initiatives (Hemant Sethi, 2019). Though there has been a spike in the investments in health and wellness programmes, there still is a lot more scope for corporate India. A high number of companies do not offer personalized wellness programmes/activities resulting in less employee adoption and engagement.

The liberalized Indian economy demands a more focussed and efficient workplace culture (Singh, 2011). The human values are the most important aspect that influences organization goals and policies (Pareek, 2002). Indian organizations need to create an attractive and motivating working environment (Agarwal et al., 2014).

In view of the challenges regarding sustainability (environmental, public health and psychological sustainability), efforts are being made in India at the government level, NGO level and corporate level to meet these challenges and to create societal awareness and consumer awareness. As a result of these efforts, the situation is slowly changing.

6.3 Efforts by NGOs, Corporate and Government Towards Sustainable Development

The following section highlights how efforts are being made by NGOs including the Indian companies and Indian government in attaining sustainability (environmental, public health and psychological sustainability).

Few years ago, the terms like climate change, environmental sustainability and circular economy were not very popular in India. Also, awareness about eco-friendly products and the sustainability of them was limited to only certain section of society. But now people and the organizations are increasingly becoming aware about the sustainability, environmental protection, social awareness and social inclusion. Consumers are also becoming aware about these things and are, accordingly, changing their lifestyle and trying to have a minimalistic living. In a recent study conducted in North America, Europe and Asia of 6000 consumers, it has been found that more than 80% of survey participants said they felt it was "important or extremely important" for companies to design environmentally conscious products (Andrew Martinez, 2019).

There are many organizations including NGOs that are making essential contributions to the environment and society towards the sustainability of India. They are making efforts to facilitate the conditions under which humans and nature can co-exist while still satisfying the social, health and environmental needs of present and future generations.

Highlights of some Indian NGOs and social enterprise which are balancing their development with minimal or no impact on Mother Earth is given below.

There are a lot of organizations that are striving to reduce pollution or the other major negative impacts on our non-renewable resources and are working for improved standards of living with sustainable solutions, thereby making an effort towards green environment, for example, NavAlt Solar & Electric Boats Pvt. Ltd. based in Kochi, Avani, an NGO based in Kumaon region of the Uttarakhand. Avani is a community built on the principles of sustainability and local empowerment. They have various initiatives which help to protect the environment to providing employment opportunities to local people, thereby providing opportunity for rural development. Greenobazaar in Gujarat is an eco-friendly bazaar which provides organic food and eco-friendly products at door steps of the customers. Daily Dump is a design company in Karnataka established in 2006. It is based on PM of India Mr. Narendra Modi's Swachh Bharat, Clean India Mission. At this company they use design to help imagine alternative scenarios that can help change behaviour. The objective of this company is to reduce waste, improve material recovery and enable better livelihoods, and this is done through the voluntary collective action of urban citizens. Aadhan is a social enterprise located in NCR, Delhi, and is in the recycled infrastructure space, which recycles old shipping containers into an eco-friendly, mobile infrastructure. They design buildings which can be sent to any location in the country. These structures are made from shipping containers retired from their tenure at sea and other environmentally friendly materials, emphasizing a culture of sustainability. Aspartika Biotech company situated in Bengaluru uses locally generated waste and by-products of agro-industries and palm oil industries to develop value-added products like omega-3 fatty acids for human and animal application and feed additives for aqua, poultry cattle and swine. Oorja Energy Engineering is a clean-tech heating and cooling company in Hyderabad which has focussed on providing sustainable solutions for industrial and commercial heating and cooling. They provide solutions based on solar and energy efficiency that are economically viable for customers. Their products either eliminate or greatly reduce the need for fossil fuel consumption for these processes in industries and commercial establishments. Oizom is an environmental solution company based in Gujarat which started with a life-centric approach; Oizom builds data-driven solutions for natural resources like air, water, soil, energy, etc. Oizom Instruments aims to make the environment understandable and predictable. With a dense network of Oizom Instruments and machine learning data science, Oizom strives to create a sustainable living environment. S4S Technologies located in Mumbai is a food preservation company that invents new food processing machines. They sell these machines to farmers or use these machines at their own facility to produce the best quality processed food. S4S Technologies work with a range of partner organizations to create a sustainable supply of processed food products. Their technology includes a Solar Conduction Dryer, Haldi Tech, SmartDry and Frost Dry.

Along with NGOs, various Indian companies are making significant contribution towards sustainability. For example, Hero MotoCorp which is the world's largest two-wheeler company believes in innovation at the core of its philosophy. It has been at the forefront of designing and developing technologically advanced motorcycles and scooters for customers around the world. With over 90 million satisfied customers across the globe, it continues to champion socioeconomic progress and empowerment through its range of products and services. Hero MotoCorp's manufacturing facilities are based on the core principle of sustainable development, as the company remains committed to maintaining the highest ecological standards. As a responsible corporate citizen, Hero MotoCorp has always gone beyond its business objectives to ensure a green, safe and happy environment for the society at large. Hero MotoCorp's corporate social responsibility (CSR) platform has an initiative—the "Happy Earth" programme for the environment. This programme aims to protect, conserve, restore and optimally use environmental resources through various initiatives including tree plantation drives, water conservation programmes and conservation of wildlife. Additionally, under its community care programme, Hero MotoCorp has been engaged in the socioeconomic development of more than 100 villages in several parts of India through various initiatives including skill development of the youth, promotion of sports, health camps and water harvesting projects. Thus, the company has carried out various activities right from environmental sustainability to water conservation and forest conservation.

Indian Oil Corporation Limited (IOCL), d/b/a IndianOil, is an Indian public sector oil and gas company headquartered in New Delhi. It is the largest commercial oil company in the country, with a net profit of ₹19,106 crore for the financial the year 2016–2017. Indian Oil Corporation under the flagship scheme of the government of India, i.e. Pradhan Mantri Ujjwala Yojana, has provided LPG connections to Below Poverty Line (BPL) families since its launch in May 2016. The objective of the scheme is to provide clean cooking fuel to the BPL families to eliminate use of biomass-like agricultural residue, wood, cow dung, etc. The clean fuel has helped in eliminating domestic pollutions, thus improving the health of women from BPL families. As per study by the World Health Organization, about one million people die every year in India, which can be attributable to diseases related to indoor air pollution. Pradhan Mantri Ujjwala Yojana (PMUY, 2019) aims to safeguard the health of women and children by providing them with cleaner cooking fuel, i.e. LPG. Under this scheme, Indian Oil contributes 20% of 2% of previous year's profit towards release of one-time grant to BPL families in rural areas for release of new LPG connections. During 2018–2019, under CSR, LPG connections were released to about 5.33 lakh families.

The foundation of Axis Bank has been built on its core values—customer centricity, ethics, transparency, teamwork and ownership—and by doing the right thing for its stakeholders.

Axis Bank is guided by a strategy that focusses on achieving sustainable and profitable growth for the organization while catalysing positive economic, social and environmental value creation for its stakeholders in the short and long term. Recognizing the increasing concern of global warming and its ill effects on the environment, the bank commenced undertaking green energy initiatives in its operations in order to reduce its carbon footprint. Towards this end, the bank has installed around 5 MW of rooftop solar plants in various branches and offices since inception of the project in the FY 2014–2015. This includes the largest installation of 1.27 MW at Axis House, Noida, at one single location. Further, the bank has set up a 2 MW solar plant by acquiring 8.59 acres of land at Sonalwadi Village, Sangola Taluka, Solapur, through open-access system which can generate 28 lakhs electricity units per annum, and aims to reduce 2296 tons of carbon emissions annually.

At Reliance, they believe in human capital and knowledge economy. Therefore, for the human capital and knowledge variable, awareness training in Six Sigma has been imparted to nearly 25% supervisory staff, and the company wants to achieve the target of nearabout 40% in the coming 3 years. Training is also imparted to the food handlers in the canteens to ensure food and water safety. The company is following a comprehensive community development programme which includes education, healthcare, disaster relief and skills training, and like in the community health centres, the company has created a society named the Dahej Health & Welfare Society (DHWS) to run a 50-bedded hospital for secondary-level healthcare facilities at Dahej in partnership with the government of Gujarat. All raw materials are transported in a secure manner in order to make sure the safety of customers, carriers, suppliers, distributors and contractors. The company ensures adherence to EUREACH regulations which aimed at ensuring that every chemical substance that is manufactured in or imported into Europe is safe to use. Dividend payment is made as per the shareholders aspirations, and company's policy to pay sustainable dividend is linked to long-term growth objectives of the company.

In order to have a better and in-depth understanding of environmental conservation and its importance, Apeejay Tea has started Nature Clubs, a sustainability programme for school children of our tea gardens in Assam. In the Q1 of 2015, Apeejay Tea announced its initiative towards reducing the impact of human elephant conflict in Assam with many first-of-a-kind ideas built in and implemented by the World Wide Fund for Nature (WWF). Along with the efforts at corporate level and government level, efforts must be made in creating awareness among consumers regarding sustainable development.

Wipro, an IT company through its project Readiness designed for the fresh graduates joining the company, prepares them to live in the real-life lab of operating project environment. For the safety of female employees, there is a campus safety training programme designed by the company. For the sales team, a training programme is available with the benefits covering IP, FCPA, antitrust, anti-competitive acts, discrimination and work place behaviour. Wipro Cares – a non-for-profit trust was formally established in 2002. This is an employee led initiative, through which unique opportunity is provided for Wipro employees to contribute to social causes and to community needs. For example, employees engage meaningfully with disadvantaged communities who are proximate to Wipro locations. In addition to this, access to education, education for children with disabilities, community ecology program, community healthcare program, and the like programs are supported by Wipro Cares. Suppliers are also expected to follow sustainability standards. From the socioeconomic point of view of local material procurement, 50% of the construction material is procured from the local suppliers. Nearly the 80% suppliers are based in India (Wipro Sustainability Report, 2013). Dividends are paid to the shareholders every year, and the rate of payment is increasing year by year.

Tata Consultancy Services (TCS) invests in the training of science graduates to transform them into software professionals, and this training programme includes computing, core technology skills, TCS processes and tools, communication and team skills, business literacy and cultural awareness in addition to hands-on project experience. The company has a focus on the areas of education and skill building, health, environment and affirmative action, and it is involved in providing exemplary services with the assistance of Andhra Pradesh government to digitize their operations and to offer an end-to-end IT solution to fruitfully automate the NREGA scheme in the state of Andhra Pradesh. In the area of supply chain sustainability, TCS has initiated a network flow based "Optimized Empty Tank Container Repositioning System". This has resulted into rise in the revenue by 12%, margin rose by 10% for the next three years, and a considerable amount of carbon emissions was avoided (TCS Consultancy Report 2013). Dividends are paid every year with minimum payout ratio of 30%.

The government has taken some steps to force improvements by defining standards, imposing a coal tax on the industry and putting the Ministry of New and Renewable Energy to work on funding, promoting and researching cleaner energy. However, in 2018, India met only 6.3% of the country's target for the production of wind energy and 5.86% of the target for solar. Coal continues to pollute, and gas and hydropower are running far below capacity.

The government has also created eco-sensitive zones, and these zones tightened the benchmarks and efforts that are being made to monitor corporate environmental performance.

There are encouraging signs that financial markets are making moves to encourage sustainability. State Bank of India, the country's largest bank, has obtained lines of credit of more than $1 billion from the World Bank and Germany's KfW for projects in solar energy and affordable housing. And as of March 2019, the State Bank of India had disbursed (pdf) $4.3 billion towards 656 completed renewable energy projects and had an affordable housing portfolio of $37 billion. Similar examples exist among Indian private banks.

The government has also introduced highly publicized government awards for companies that implement sustainability programmes, thereby recognizing efforts of the corporate towards sustainability and encouraging them to sharpen their focus.

International trends in sustainable and green lending are helping drive the corporate sustainability agenda in India, but more can be done.

Also, green bonds in India have grown into more than $6 billion market. Green "masala bonds", issued outside India but in rupees, have been listed on international exchanges since 2015. And in 2017, the Securities and Exchange Board of India finalized norms for the issuing and listing of these and other environmentally focussed bonds. Both public and private sector corporations in India are now using such bonds to raise finance, especially for renewable energy projects. For example, the Securities and Exchange Board of India took a step in the right direction by issuing a paper in 2012 asking the top 100 listed companies in India (based on market capitalization) to release annual business responsibility reports. In 2017, this was extended to the top 500 companies. The reporting format is influenced by the standards set by the Global Reporting Initiative and based on locally developed principles of sustainability. However, the progress in this direction is very slow, and there are no regulator audits of the reports and no accountability for projects listed in them.

Indian stock exchanges have been responding to the global trend towards sustainable business. The two major stock exchanges in India—the Bombay Stock Exchange and the National Stock Exchange—have launched sustainability-based indexes (the S&P BSE Carbonex, the S&P BSE Greenex and the NIFTY100 ESG). However, media and investors have not yet shown much interest. More corporations must embrace holistic sustainability.

As a result of government initiatives and efforts, many leading Indian businesses like Tata, Reliance, Mahindra and Birla, to name a few, have made great progress on sustainability, spearheaded by visionary CEOs and supported by capable senior teams. They've implemented major internal campaigns to reduce and recycle and to influence employee behaviour. Their efforts have also enabled the creation of green supply chains, expanding the idea of sustainability to many smaller companies. The idea of a "circular economy", in which waste becomes a resource, is also catching on. However, it's still limited to only the large, professionally managed companies. Regarding inequality in all aspects of the business, distribution of profits, pay to employees, manage supply chains, etc., these issues are rarely addressed.

India has a strong legacy of civil society movements. This shows how efforts are being made by Indian NGOs, social enterprise, companies and the government in making India a better place to live by their contribution towards environmental sustainability measures.

Regarding public health, efforts are being made by the Indian government by recently adopting the National Health Policy in March 2017 (NHP, 2017). This is in line with United Nations Sustainable Development Goals (SDGs). Health is a core dimension of the Sustainable Development Goals (SDGs); the NHP 2017 aims to "attain the highest possible level of health and well-being for all at all ages through a preventive and promotive health care orientation in all developmental policies, and universal access to good quality health care services without anyone having to face financial hardship as a consequence". NHP has also highlighted seven priority areas, namely, air pollution, better solid waste management, water quality, occupational safety, road safety, housing, vector control and reduction of violence and urban stress.

Since the Sustainable Development Goals have to be implemented by 2030, the Indian government is taking efforts by formulating national policies and regulations. The government is also roping in corporations and business to complement these actions.

Study has indicated that that 60% of the top 100 companies incorporate SDGs into their responsible business actions (Futurescape Newsletter rankings). All the top ten companies map their business goals with SDGs (Majumdar & Rana, 2020).

On an average, companies map 12 SDGs with a low of 1 to a maximum of 17. Of the companies that mapped their SDG goals, 88% were in the private sector and 68% were in the manufacturing companies. In terms of focus, the SDGs 8 (decent work), 6 (clean water and sanitation), 12 (responsible consumption and production), 13 (climate action), 3 (good health and well-being) and 4 (quality education) occupy the top position with over 82% companies reporting so. On the other hand, SDG 14 (life below water) was mapped by the least number of companies—32% (Futurescape Newsletter).

Mapping of top 100 companies identified that the prominent SDGs were SDG 8 (decent work and economic growth), SDG 9 (industry, innovation and infrastructure) and SDG 13 (climate action), with over 70% of the companies indicating it as an action item. Other important ones were SDGs 1 (no poverty), 6 (clean water and sanitation), 16 (peace, justice and strong institutions) and 4 (quality education). Both manufacturing and service companies focussed on SDGs 8, 13, 9 and 1—similar to the overall results. The pattern repeats itself across all industries. Efforts are being made in India to provide quality healthcare facilities to all by establishing primary healthcare centres, establishing medical colleges and also introducing various changes in National Health Policy.

Regarding psychological well-being, Indian companies assume that the concept of sustainable development is significantly correlated with employees' well-being. With innovation and changing technology, mergers and acquisitions and organizational restructuring, the resultant stresses and strains have become part of human life. They not only affect individual's life but also affect professional, social or personal activities. In order to save organizations from the negative performance, it is necessary to develop programmes that will promote complete well-being of employee. Various Indian companies are making efforts to develop wellness programmes for the employees. Some of the examples are given below.

Infosys Technologies which is a leading IT sector provider in India has adopted a wellness initiative named Health Assessment and Lifestyle Enrichment (HALE). The initiative primarily consists of a portal for online dissemination of information on the various stress-related issues and a hotline whereby employees can directly receive counselling. This initiative can help the HR track patterns and take suitable actions to deal with the prevalent issues in the organization. The company also has its in-house doctors for regular check-ups as well as professional counsellors offering psychological counselling to its employees.

Wipro is another leading IT and BPO sector that has created a supportive environment and culture focused on wellness. Its most successful initiative has been once called "MITR" (meaning "friend" in Hindi). In this program, 28 employees, all volunteers, were trained to counsel fellow employees to manage stress. "MITR" signifies to employees that they do have a friend in Wipro and they should be able to confide and share their problems. "We at Mitr can help employees cope with their problems", says Wipro's head of compensation and benefits, Anil Jalali.

Indian start-up WeWork initiated a programme called "right to disconnect". The start-up floated policies which allowed employees to feel free and switch off on weekends and non-working hours from 9 pm to 7 am.

Other global giants like Hindustan Unilever and Pidilite Industries have been strongly opposing the late working policy to support employees.

A study showed that MNCs in Jamshedpur are making strategic moves in retaining their employees. Obtained results showed that the implementation of three R's-respect, recognition and rewards will increase employee retention. Similarly, conducting frequent reviews and not blaming the employees for policy errors also leads to employee retention (Nafisa Shahin, 2014).

With the outbreak of global pandemic COVID-19, life of people has changed. With the new norm of social distancing, use of online platforms has become common in everyday life. At Microsoft team's workplace, skype for online meetings has become commonplace.

According to a report by JLL, 66% of employees in India immediately shifted to working from home after the outbreak of the pandemic. While 30% of Indian working population has said that the lockdown has given them an enhanced opportunity to balance their work and life, as much as 41% of them are missing the professional environment (Sanjeev Sinha, 2020). Regarding physical health and well-being amidst the global pandemic COVID-19, efforts are being made by the Indian companies to promote health and well-being of the employees. Majority of the Indian companies view that setting up and designing activities to help reduce health risks and will enhance their ability to function. For example, in line with the current situation, Indian companies are rethinking their employee-wellness strategies and customizing programmes to accommodate the technological change in people's habits. In addition to these measures, the companies are advising employees to adopt to simple health measures, making employees understand how to deal with work from home and adapt to uncertain situations and the like. Companies like Vedanta are putting employee well-being ahead of productivity issues and offer psychological counselling helplines to its employees facing anxiety and stress-related problems along with meditation sessions and a series of engagement activities to keep them motivated and engaged.

In view of the diverse and changing environment of organizations, Indian companies are making efforts to attract and motivate employees at workplace. They are trying to create an effective organizational culture and develop a mechanism of collaboration, trust and reciprocal accountability by setting interactive associations with employees. All these efforts will promote psychological well-being and will help improve quality of life, thereby promoting psychological sustainability.

In order to meet the challenges of sustainability, the Indian government has taken various measures by adopting regulations, standards, subsidies, initiatives and public investment in green technologies. Some of the government initiatives are mentioned below.

NGOs have played a critical role in the enactment of many important legislative changes. Several NGOs provide consulting services to corporations and help companies implement their CSR activities, and they have the necessary skills to create an independent green rating system for India. It won't be easy, but with support from the government, it can be done (Deepika Krishnan, 2019).

Governments in India have always been alerting to popular trends and have responded in the past with many pro-people policies. Awareness regarding sustainability must reach to common Indian. The Indian media also has to play a major role in creating a public opinion. Media should give coverage to companies that have taken major initiatives towards sustainability. Digital media company should make digital platforms available and tell stories of sustainability and sell sustainable products online. Currently it is reaching out to audience of 30 million per month. Still such efforts are not enough.

At school level right from primary school, social and environmental responsibility is part of the core curriculum. Engineering and design schools must increase their focus on sustainable design. And more than 3500 business schools in India should offer courses that teach students the practical aspects of building and running sustainable businesses.

India's sustainable future also requires efforts from research firms, innovation hubs, incubators, social scientists, auditors, lawyers, analysts and talent search companies. In this direction, consulting companies and industry associations have started efforts, for example, they have started offering training programmes, advisory services and green product certification. Such industry-led efforts need to be accelerated.

6.4 Conclusions

Despite government efforts, results have shown that corporate sustainability isn't yet a fully realized idea in India. Several leading companies consider social and environmental issues to be as important to their business as financial growth, but many other companies do not consider it as important and most aren't thinking of sustainability as a core strategic priority.

In Indian organizations, sustainability is understood mainly in terms of environmental sustainability such as reductions of waste, water, emissions, circular economy and commitment to renewable energy.

Corporate social responsibility (CSR) is contributing to charitable causes or taking up projects that have larger social implications. Indian consumers are ready to get connected to these causes or initiatives initially, but they want to ensure whether the real on-ground action is promoted by the brand over time.

For business to grow in India, it is necessary for the corporations to embrace sustainability as an essential part of the corporate agenda. Collaborative efforts must be made by the government, financial markets, CEOs, media and civil society in creating awareness and commitment towards the sustainable development and well-being.

The Indian government is making efforts to maintain balance between the two by making efforts to achieve the goal of sustainability and well-being in organization. In order to achieve this goal, various provisions and policy measures and incentives are given to the companies.

It is believed that progress of company depends upon the quality of its employees. If the employees are healthy and have mental well-being, then they will be able to make contribution towards the sustainable development of the company. In view of this, the companies have realized that well-being is not only a central dimension of sustainable development but also a key contributor to sustainability.

References

Agarwal, B., Mandhanya, Y., & Mishra, D. P. (2014). Analyzing the impact of HRD culture on managerial effectiveness. *Vishwakarma Business Review, 4*(2), 68–73.

Agrawal, J., Murthy, P., Philip, M., Mehrotra, S., Thennarasu, K., John, P. J., & Isaac, M. (2011). Socio-demographic correlates of subjective well-being in urban India. *Social Indicators Research, 101*(3), 419–434.

Andrew, M. (2019). Most Consumers want sustainable product packaging. *Business Daily*, June 4, 2019. https://www.businessnewsdaily.com/15087-consumers-want-sustainable-products.html

Brundtland Report. (1987). *Our common future*. Butterworth.

Deepika, K. (2019). Making Indian business sustainable. *Energy and Sustainability.* https://www.strategy-business.com/article/Making-Indian-businesses-sustainable?gko=f72e8

Di Fabio, A. (2016). Positive relational management for healthy organizations: Psychometric properties of a new scale for prevention for workers. *Frontiers in Psychology, 7*, 1523. https://doi.org/10.3389/fpsyg.2016.01523

Di Fabio, A., & Maree, J. G. (2016). A psychological perspective on the future of work: Promoting sustainable projects and meaning-making through grounded reflexivity. *Counseling: Giornale Italiano di Ricerca e Applivazioni, 9*. https://doi.org/10.14605/CS931619

Di Fabio, A. (2017a). Positive Healthy Organizations: Promoting well-being, meaningfulness, and sustainability in organizations. *Frontiers in Psychology. Organizational Psychology, 8*, 1938. https://doi.org/10.3389/fpsyg.2017.01938

Di Fabio, A. (2017b). Frontiers in psychology. *Organizational Psychology, 8*, 1534. https://doi.org/10.3389/fpsyg.2017.01534

Di Fabio, A., & Rosen, M. A. (2018). Opening the black box of psychological processes in the science of sustainable development: A new frontier. *European Journal of Sustainable Development Research, 2*(4), 47. https://doi.org/10.20897/ejosdr/3933

Gargi, S., & Ghazi, S. (2018). Index of Psychological well-being at work—Validation of tool in the Indian organizational context. *Vision: The Journal of Business Perspective, 22*(2), 1–11. https://doi.org/10.1177/0972262918766134

Harris, J. M. (2003). Sustainability and sustainable development. *International Society for Ecological Economics., 1*, 1–12. https://sustainabledevelopment.un.org/content/documents/5987our-common-future.pdf

Henry, J. (2005). The healthy organization. In A. S. G. Antoniou & C. L. Cooper (Eds.), *Research companion to organizational health psychology* (pp. 382–391). Edward Elgar.

Joshanloo, M., & Weijers, D. (2014). Aversion to happiness across cultures: A review of where and why people are averse to happiness. *Journal of Happiness Studies: An Interdisciplinary Forum on Subjective Well-Being, 15*(3), 717–735. https://doi.org/10.1007/s10902-013-9489-9

Kompaso, S., & Sridevi, S. (2010). Employee engagement: The key to improving performance. *International Journal of Business Management, 5*(2). https://doi.org/10.5539/ijbm.v5n12p89

Majumdar, U., & Rana N. (2020). Responsible Business Rankings India's Top companies for sustainability and CSR 2020. *Futurscape*. Retrieved from https://www.futurescape.in/

Mishra, S., & Gupta, B. (2009). Work place motivators and employee's satisfaction: A study of retail sector in India. *Indian Journal of Industrial Relations, 44*(3), 509–519.

Mitchell. (2020). *Sustainability*. Sustainability Definition (investopedia.com)

Nafisa, S. (2014). Role of employee retention practices in Indian industry – A study of select MNCs in Jamshedpur. *International Journal of Engineering and Management Research, 4*(4), 206–213.

National Health Policy. (2017). *Ministry of Health and Family Welfare Government of India*. Retrieved from https://main.mohfw.gov.in/documents/policy

Pandey, K. (2019). Grossly polluting industries more than doubled in 8years: SOE in figures. *Down To Earth*. Retrieved from https://www.downtoearth.org.in/news/pollution/grossly-polluting-industries-more-than-doubled-in-8-years-soe-in-figures-64962

Pareek, U. R. (2002). *Effective organisations: Beyond management to institution building*. Oxford and IBH.

Pradhan Mantri Ujjwala Yojana (PMUY). (2019). *CSR Project By: Indian Oil Corporation Ltd (IOCL)*. Retrieved from https://csrbox.org/India_CSR_Project_Indian-Oil-Corporation-Ltd-(IOCL)-PNG-and-LPG-Scheme-for-BPL-Families-pan-india_10572

Sethi, H. (2019). Indian Business should make workplace wellbeing a priority. *Business World People Inventiva*. Retrieved September 30, 2021, from http://bwpeople.businessworld.in/article/Indian-Businesses-Should-Make-Workplace-Wellbeing-A-Priority/04-09-2019-175674/

Singh, A. K. (2011). HRD practices & managerial effectiveness: Role of organisation culture. *Indian Journal of Industrial Relations, 47*(1), 138–148.

Sinha, S. (2020). How the workplaces in India will change in Post Covid-19 world. *Financial Express*, July, 3rd, 2020.

Suresh, J. (2019). Can current Indian health system achieve health related SDG'S? *Open Journal of Pediatrics and Child Health, 4*(1), 013–030. https://doi.org/10.17352/2640-7612.000017

Tejvan, P. (2018). Environmental sustainability – Definition and issues. *Economics Help*. https://www.economicshelp.org/blog/143879/economics/environmental-sustainability-definition-and-issues/

Wipro Sustainability Report. (2013). *Wipro Sustainability Report 2012–2013*. Retrieved from https://www.wipro.com/content/dam/nexus/en/sustainability/archieves/wipro-sustainability-report-2012-13.pdf

Chapter 7
Human Capital Sustainability Leadership Scale: Psychometric Proprieties of the Malaysian Version

Chua Bee Seok, Pan Lee Ching, and Rosnah Ismail

7.1 Introduction

As the industry becomes more competitive, human capital sustainability is privilege to management. Managing human capital effectively is an essential strategy to retain talent because organizations may face difficulty to withstand stiff competition (Mackipere & Yip, 2008). Leaders are responsible for organizations' success through collective work with employees. Leaderships are important because, through leader's guidance, they are responsible to identify issues and overcome it for stakeholders' interests including organizations, employees, clients, and investors (Ulrich & Smallwood, 2012). There are various leadership instruments available in the literature. Examples include authentic leadership, spiritual leadership, ethical leadership, and servant leadership (Yukl et al., 2011). Nonetheless, looking into the lens of business complexity, it is highly recommended for leaders to possess more than one leadership style for managing a comprehensive leadership context. The recent research area of the Psychology of Sustainability and Sustainable Development (Di Fabio, 2017a, b; Di Fabio & Rosen, 2018) underlined the importance of sustainability of human resources and flourishing of workers and organizations. For enhancing sustainability of human resources, it is also critical to valorize processes of harmonization (Di Fabio & Tsuda, 2018) to promote well-being in different environments. In this perspective, Di Fabio and Peiró (2018) answered to the call for new instruments regarding leadership, introducing the Human Capital Sustainability Leadership Scale (HCSLS). This newly established construct integrates with other leadership behavior that is essential from a psychological

C. B. Seok · P. L. Ching
Universiti Malaysia Sabah, Kota Kinabalu, Malaysia

R. Ismail (✉)
University of Cyberjaya, Cyberjaya, Malaysia

A. Di Fabio (ed.), *Cross-cultural Perspectives on Well-Being and Sustainability in Organizations*, https://doi.org/10.1007/978-3-030-86709-6_7

perspective. The human capital sustainability leadership focuses on employees' well-being, embraces change, values morality, and designs a robust strategy for the sustainability of the organizations and employees. This scale has four subdimensions: ethical, sustainable, mindful, and servant leadership.

Ethical leadership emphasizes morale as being caring, fair, and balanced decision-making, as well as practicing integrity (Brown & Trevino, 2006; Di Fabio & Peiró, 2018; Yukl et al., 2011). Malaysian culture underlines the value of morale and integrity not just in life but also at work. The study by Madanchian et al. (2018) highlighted that ethical leadership have direct and positive effect on leadership effectiveness among SMEs in Malaysia.

Leadership for sustainability should be able to foresee the work pattern and problem solver, acquire emotional intelligence, and adapt to changes (Metcalf & Benn, 2012) and that the person shall continue to learn, actively engage with dynamic society, develop human capital with fairness, and lead other leaders (Di Fabio & Peiró, 2018).

Since organizations are constantly changing, leaders should be prone to manage this complexity. Nonetheless, some leadership may be static when it comes to problem-solving; leaders tend to provide past solutions to present issues (King & Haar, 2017). That means, to depend on experience to solve a problem is an unlikely ideal. This is where mindful leadership plays a role. The leadership style is keen to observe changes and create a strategy from the observation (Di Fabio & Peiró, 2018). A rigorous approach is done in the Western context, but the mindful context is closely associated with Eastern practice (Ehrlich, 2017).

The servant leaders' priority is always not about themselves but other people, they are focusing on objectives and on subordinate's development and recognition (Di Fabio & Peiró, 2018; Luthans & Avolio, 2003). At this point, it is fair to claim that servant leader belief is when employees' justice is protected; they would perform optimum in exchange of their leaders' support (Van Dierendonck, 2010). In Malaysia's setting, employers that practice servant leadership are able to gain trust from the employees and, as a result, reduce the intention to resign (Ng et al., 2016).

Based on the leadership literature, Eastern studies are underrepresented with respect to studies conducted in Western countries (Madanchian et al., 2018). Malaysia practices high power distance between leaders and subordinates, although it shows that leadership style is essential, but very small studies could affirm the Malaysian leadership context (Madanchian et al., 2018). This may occur due to a lack of leadership instrument fitness to Malaysia's work setting. Thus, it is essential to investigate the Human Capital Sustainability Leadership Scale (HCSLS) adjusted according to the Eastern context.

The scale (Di Fabio & Peiró, 2018) is newly constructed in Western context; therefore this study explored east context. Thus, this paper aims to analyze the psychometric properties of the Human Capital Sustainability Leadership Scale in Malaysia workplace setting. According to the original scale, this study hypothesized that each of these dimensions, ethical, sustainable, mindful, and servant leadership can be placed under a higher construct called human capital sustainability leadership. As the HCSLS was originally tested with the Flourishing Scale in terms of concurrent validity, this study would do similarly.

7.2 Method

Respondents

A sample of 305 (169 females, 136 males) employees was selected randomly using a convenience sampling method from public and private organization sectors in Kota Kinabalu, Sabah, Malaysia. The respondent ages ranged from 18 to 56 years old with a mean of 33.0 years old, and their average tenure with their current employer was 7.46 years ($SD = 7.46$). In this study, the sample size was determined based on the expense of data collection and the sufficient statistical power, to conduct a multivariate statistical technique. There were 16 constructs for the Human Capital Sustainability Leadership Scale and 305 cases involved. The ratio of cases to free parameters was 31:1, which revealed the sample size of the current study is sufficient and should not be an issue.

Measures

The Human Capital Sustainability Leadership Scale (HCSLS)

The Human Capital Sustainability Leadership Scale (HCSLS) is a newly developed scale by Di Fabio and Peiró (2018) to measure the construct of human capital sustainability leadership style that has its roots in the existing definition of the leadership term but aims to achieve higher integration of different functions. The human capital sustainability leadership is a higher-order construct with the four first-order constructs: ethical leadership, sustainable leadership, mindful leadership, and servant leadership. The HCSLS consisted of 16 items with a 5-point Likert scale response format (1 = strongly disagree, 2 = disagree, 3 = neither agree nor disagree, 4 = agree, 5 = strongly agree). The Malaysian version of the HCSLS (by Chua Bee Seok et al., 2020) was obtained through the back translation method. The examples of the items are as follows: "My leader keeps his/her promise to his/her collaborators" (ethical leadership); "My leader develops, rather than exhaust, the human resources that work with him/her" (sustainable leadership); "My leader is aware of the strengths and the limitations of his/her collaborators" (mindful leadership); and "My leader encourages his/her collaborators when he/she realizes that they encounter difficulties" (servant leadership). The Cronbach's alpha coefficients in the Di

Fabio and María Peiró (2018) study were .80 (ethical leadership); .86 (sustainable leadership); .83 (mindful leadership); .86 (servant leadership); and .94 for the total scale.

The Flourishing Scale

The Flourishing Scale is designed to measure important aspects of human functioning such as meaning and purpose in life, positive relationships, optimism, and self-esteem. The scale provides a single psychological well-being score. It consists of eight items, and the responses for the items were given on seven-point scales ranging from 1 = strongly disagree to 7 = strongly agree. Originally, the scale was called Psychological Well-Being; however, the scale was changed to flourishing as the content went beyond psychological well-being. All items are phrased in a positive direction, and high scores indicate that respondents view themselves in positive terms in important areas of functioning. The examples of the items are as follows: "I am a good person and live a good life" and "I am engaged and interested in my daily activities." The internal consistency for the Italian version of the Flourishing Scale was Cronbach's alpha coefficient = 0.88, and the internal consistency for the Malaysian version of this scale was Cronbach's alpha coefficient = 0.93 (Chua Bee Seok et al., 2020).

Data Analysis

The data of the current study was analyzed using the IBM SPSS AMOS 23 Program. To determine the sample size, Tabachnick and Fidell (2001) suggested the minimum sample size should be 200. To conduct the confirmatory factor analysis, as suggested by Kline (2005), the ratio of cases to free parameters should be 10:1. The confirmatory factor analysis (CFA) was conducted to confirm the higher-order structure with the four first-order constructs: ethical leadership, sustainable leadership, mindful leadership, and servant leadership. To estimate the fit of empirical data to the model, we compared the results of the current study to the recommended four model fit indices suggested by Hu and Bentler (1998): Comparative Fit Index (CFI) >0.90; Tucker-Lewis Index (TLI) >0.90; Root Mean Square Error of Approximation (RMSEA) 0.05–0.08; and Chi-Square Statistic value (CMIN/DF) <5.0. The chi-square value (χ^2) and its associated degree of freedom (df) were also reported. The reliability of the Human Capital Sustainability Leadership Scale (HCSLS) Malaysian version was verified using Cronbach's alpha coefficient. Concurrent validity was examined through correlations of the Malaysian version of HCSLS with the Flourishing Scale.

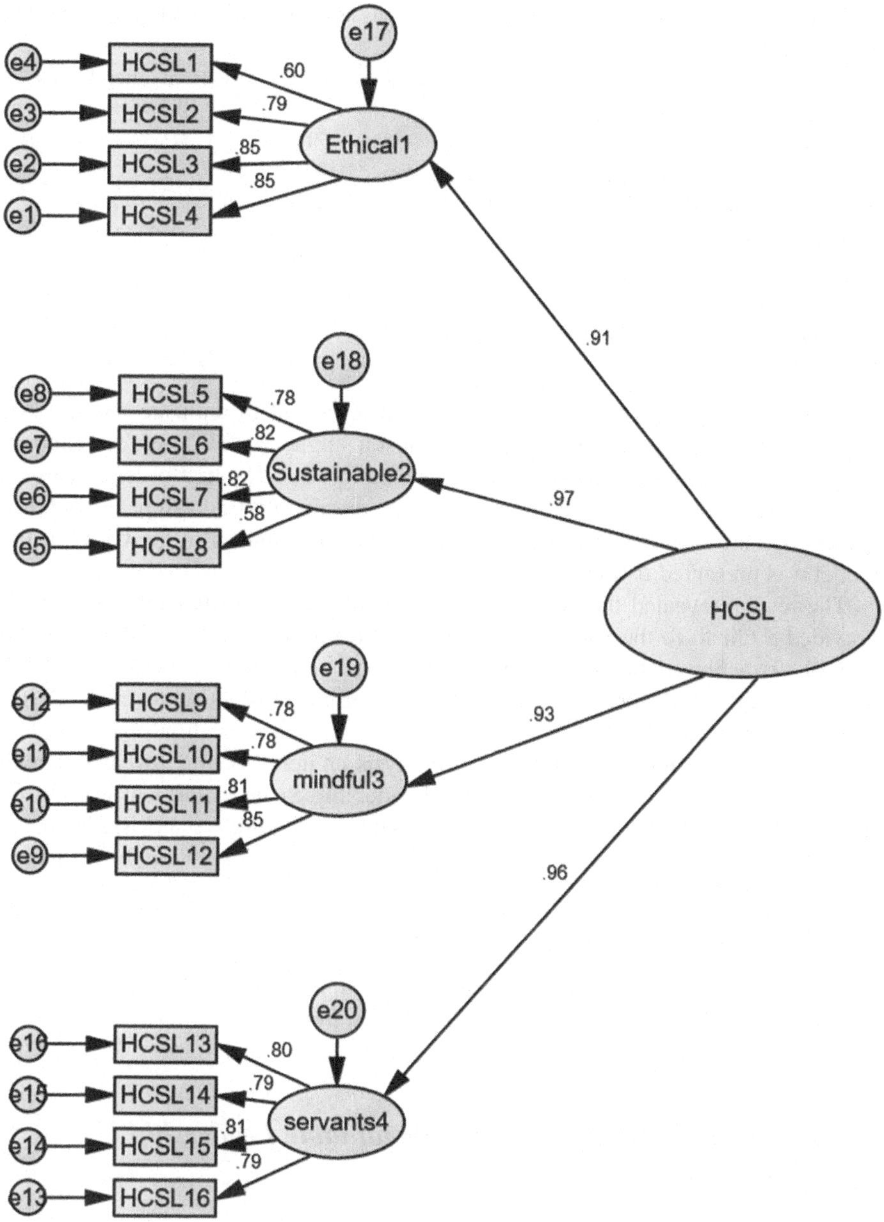

Fig. 7.1 Second-order confirmatory factor analysis of the Human Capital Sustainability Leadership Scale (HCSLS)

Table 7.1 Model fit statistics of confirmatory factor analysis for the Human Capital Sustainability Leadership Scale (HCSLS)

The fit indices	Recommended fit	Model fit	Model fit (after deleted item HCSL8)
Comparative Fit Index (CFI)	>0.90	.91	.92
Tucker-Lewis Index (TLI)	>0.90	.89	.90
Root Mean Square Error of Approximation (RMSEA)	0.05–0.08	.10	.10
Chi-Square Statistic value (CMIN/DF)	<5.0	4.36	4.24

Results

The confirmatory factor analysis (CFA) was carried out to replicate the analysis conducted by Di Fabio and Peiró (2018) to test a single second-order factor model of HCSLS with the four first-order factors (see Fig. 7.1). The first-order constructs model is as follows: (1) ethical leadership, (2) sustainable leadership, (3) mindful leadership, and (4) servant leadership. The model fit statistics of the CFA for the model was presented in Table 7.1.

The results revealed that the model of Human Capital Sustainable Leadership provided a fair fit to the data. The CMIN/DF ratio (4.36), the CFI (0.91), and the TLI (0.89) achieved the recommended model fit indices, whereas the RMSEA (0.10) was close to the criteria value of 0.08. According to MacCallum et al. (1996), RMSEA in the range of 0.05 to 0.10 was considered an indication of fair fit, and values above 0.10 indicated poor fit. The analysis on items loading revealed that all items of the Human Capital Sustainable Leadership Scale loaded sufficiently on their hypothesized latent variable, except item HCSL8 (loading = 0.58). Standardized regression weights for the factor of ethical leader ranged from 0.60 to 0.85; sustainable leader (0.58–0.78); mindful leader (0.78–0.85); and servant leader (0.79–0.81) (refer to Fig. 7.1).

We tested again the second-order factor model of HCSLS after deleting item HSCL8. The result showed a better model fit with the CMIN/DF ratio (4.24), the CFI (0.92), the TLI (0.90) and the RMSEA (0.10) (refer to Fig. 7.1).

Reliability of the Human Capital Sustainable Leadership Scale (HCSLS)

The reliability of the Malay version of the Human Capital Sustainable Leadership Scale (HCSLS) was tested by using a method of internal consistency Cronbach's alpha with a criterion of 0.70, indicating good reliability (Kock and Sexton 2017). The result showed Cronbach's alpha for the total scale of HCSLS was .96, and the internal consistency for all the dimensions of HCSLS were range from alpha coefficient 0.83 to 0.88 (refer Table 7.2).

Table 7.2 Internal consistency Cronbach's alpha for the Human Capital Sustainable Leadership Scale (HCSLS)

Dimension	Number of items	Items	Cronbach's alpha coefficients
Total scale	16 items		.96
Ethical leader	4 items	1, 2, 3, 4	.86
Sustainable leader	4 items	5, 6, 7, 8	.83
Mindful leader	4 items	9, 10, 11, 12	.88
Servant leader	4 items	13, 14, 15, 16	.88

Table 7.3 Correlations between Human Capital Sustainable Leadership Scale (HSCLS) and Flourishing Scale

Scale	Total HSCLS	Ethical	Sustainable	Mindful	Servant	Flourishing
Ethical	.88**					
Sustainable	.93**	.81**				
Mindful	.92**	.71**	.77**			
Servant	.92**	.71**	.80**	.84**		
Flourishing	.24**	.20**	.24**	.22**	.20**	-

**Correlation is significant at the 0.01 level (two-tailed)

The Validity of the Human Capital Sustainable Leadership Scale (HCSLS)

The results of correlations between HSCLS and the Flourishing Scale (FS) showed the HSCLS showed good concurrent validity (refer to Table 7.3).

7.3 Discussions

This study examined the psychometric proprieties of the Human Capital Sustainability Leadership Scale (HCSLS) in east context and assessed the suitability of the factorial model for employees. The employees are currently working in both public and private work settings in Sabah, Malaysia.

Prior item deletion, the model is deem fit if based on Hu and Bentler (1998) recommendation. Nevertheless, the fitness improved after one item is deleted. The result comprehends the idea of Kenny and McCoach (2003) whereby additional construct or item that degrades the model fitness creates biasness. Rigdon (1996) claimed that Comparative Fit Index (CFI) and Root Mean Square Error of Approximation (RMSEA) contain advantages and disadvantages when it comes to model fitness determination. Thus, it is ideal to yield more than one criteria of which appears in Hu and Bentler (1998) suggestion. On the first hand, McDonald and Marsh (1990) and Marsh and Balla (1994) are concern about the sample size that may lead to biasness of the model. But Cudeck and Henly (1991) denied the claim. Perhaps future study may take caution in regard to sample size.

This study hypothesized that these four leadership styles – ethical, sustainable, mindful, and servant leadership – can be merged as a construct, the human capital sustainability leadership. The results show that the dimensions have a positive and significant score for reliability and convergent tests. This suggested that the translated version of HCSLS is suitable for Eastern setting especially Malaysia. And also, the instrument is deemed fit for Malaysian employers as this study proves that the HCSLS has a satisfying model fitness when operating with working people. The good score may be related to the positive impact that organizations and employees could acquire from the leadership style. For example, Malaysian employers can act more effectively when being ethical to their employees (Madanchian et al., 2018). Being a servant leader could eliminate the intention to quit among the employees (Ng et al., 2016). The leadership style enriches the mutual trust between employers and employees that form smooth operation.

Di Fabio and Peiró (2018) claimed Human Capital Sustainability Leadership Scale (HCSLS) has a positive relationship with the Flourishing Scale. This study showed results in a similar direction. Leaders with higher human capital sustainability leadership seem to perceive themselves as more flourishing in terms of psychological well-being in important areas such as presence of purpose, optimism, and relationships. From the results of the present study on the psychometric properties of the Malaysian version of the HCSLS, it is possible to conclude that this version of the instrument is practical and usable in east countries too, specifically Malaysia. Past studies revealed that ethical and servant leadership styles produce a positive impact like increase commitments (Madanchian et al., 2018; Ng et al., 2016).

This study contributes a deem fit HCSLS instrument in Malay version for Malaysian employers to apply multiple leadership styles for organization sustainability. Future research could consider using HCSLS instrument among Malaysian employers as the instrument permits to detect a higher-order construct including more than one leadership style. Individuals who are responsible for human capital development strategy could consider HCSLS for needs analysis and relative trainings.

Apart from that, practicing more than one leadership style also helps employers to be versatile and act accordingly. For instance, a leader has to supervise more than an individual. Different individual has own advantages and disadvantages. Being mindful of employee personality would ease collaboration.

The study had the limit to be carried out only among employees in Sabah, Malaysia. It could be important to extend the research on the psychometric properties of the HCSLS to other states that have larger-scale industry especially in Selangor, Pulau Pinang, and Johor for providing new insight about the leadership. Due to high complexity in a larger firm, the employers may have a different perspective towards the desired leadership style while managing the firm.

7.4 Conclusion

The original version of HCSLS that is formed in the Western context (Di Fabio & Peiró, 2018) could not affirm whether it is well fit to its counterpart, Eastern. After running a psychometric test, the HCSLS shows satisfying goodness fit for employers that serve in both the public and private sectors. Also, the translated instrument is reliable and valid to be used in Malaysia working setting. The opportunity to have also the Malaysian version of the HCSLS opens interesting perspectives of research and intervention for healthy people and healthy organizations (Di Fabio et al., 2020) in a cross-cultural perspective, following the framework of both the Psychology of Sustainability and Sustainable Development (Di Fabio, 2017a, b; Di Fabio & Rosen, 2018) and the Psychology of Harmonization (Di Fabio & Tsuda, 2018).

References

Brown, M. E., & Trevino, L. K. (2006). Ethical leadership: A review and future directions. *The Leadership Quarterly, 17*(6), 595–616. https://doi.org/10.1016/j.leaqua.2006.10.004

Chua Bee Seok, Pan Lee Ching, & Rosnah Ismail. (2020). Exploring the role of Malaysian Student's Intrapreneurial Self-Capital in the relationship between satisfaction with life, academic performance, and flourishing. *Sustainability, 12*, 508. https://doi.org/10.3390/su12020580

Cudeck, R., & Henly, S. J. (1991). Model selection in covariance structures analysis and the "problem" of sample size: A clarification. *Psychological Bulletin, 109*(3), 512–519. https://doi.org/10.1037/0033-2909.109.3.512

Di Fabio, A. (2017a). Positive Healthy Organizations: Promoting well-being, meaningfulness, and sustainability in organizations. *Frontiers in Psychology. Organizational Psychology, 8*, 1938. https://doi.org/10.3389/fpsyg.2017.01938

Di Fabio, A. (2017b). *Frontiers in Psychology. Organizational Psychology, 8*, 1534. https://doi.org/10.3389/fpsyg.2017.01534

Di Fabio, A., & Peiró, J. (2018). Human capital sustainability leadership to promote sustainable development and healthy organizations: A new scale. *Sustainability, 10*(7), 2413. https://doi.org/10.3390/su10072413

Di Fabio, A., Cheung, F., & Peiró, J.-M. (2020). Editorial Special Issue Personality and individual differences and healthy organizations. *Personality and Individual Differences, 166*. https://doi.org/10.1016/j.paid.2020.110196

Di Fabio, A., & Rosen, M. A. (2018). Opening the black box of psychological processes in the science of sustainable development: A new frontier. *European Journal of Sustainable Development Research, 2*(4), 47. https://doi.org/10.20897/ejosdr/3933

Di Fabio, A., & Tsuda, A. (2018). The psychology of harmony and harmonization: Advancing the perspectives for the psychology of sustainability and sustainable development. *Sustainability MDPI, 10*(12), 4726. https://doi.org/10.3390/su10124726

Ehrlich, J. (2017). Mindful leadership. *Organizational Dynamics, 46*(4), 233–243. https://doi.org/10.1016/j.orgdyn.2017.05.002

Hu, L. T., & Bentler, P. M. (1998). Fit indices in covariance structure modeling: Sensitivity to under parameterized model misspecification. *Psychological Methods, 3*(424), 453.

Kenny, D. A., & McCoach, D. B. (2003). Effect of the number of variables on measures of fit in structural equation modeling. *Structural Equation Modeling: A Multidisciplinary Journal, 10*(3), 333–351. https://doi.org/10.1207/s15328007sem1003_1

King, E., & Haar, J. M. (2017). Mindfulness and job performance: A study of Australian leaders. *Asia Pacific Journal of Human Resources, 55*(3), 298–*319*. https://doi.org/10.1111/1744-7941.12143
Kline, R. B. (2005). Principles and Practice of Structural Equation Modeling (2nd Edition ed.). New York: The Guilford Press.
Kock, N., & Sexton, S. (2017). Variation sharing: A novel numeric solution to the path bias underestimation problem of PLS-based SEM. *International Journal of Strategic Decision Sciences, 8*(4), 46–68.
Luthans, F., & Avolio, B. J. (2003). Authentic Leadership: A Positive Developmental Approach. In: Cameron, K.S., Dutton, J.E. and Quinn, R.E., Eds., Positive Organizational Scholarship, Barrett-Koehler, San Francisco, 241–261.
MacCallum, R. C., Browne, M. W., Sugawara, H., & M. (1996). Power analysis and determination of sample size for covariance structure modeling. *Psychological Methods, 1*(2), 130–149.
Madanchian, M., Hussein, N., Noordin, F., & Taherdoost, H. (2018). The impact of ethical leadership on leadership effectiveness among SMEs in Malaysia. *Procedia Manufacturing, 22*, 968–974. https://doi.org/10.1016/j.promfg.2018.03.138
Makipere, K., & Yip, G. (2008). Sustainable leadership. *Business Strategy Review, 19*(1), 64–67. https://doi.org/10.1111/j.1467-8616.2008.00521.x
Marsh, H. W., & Balla, J. (1994). Goodness of fit in confirmatory factor analysis: The effects of sample size and model parsimony. *Quality & Quantity, 28*(2), 185–217. https://doi.org/10.1007/bf01102761
McDonald, R. P., & Marsh, H. W. (1990). Choosing a multivariate model: Noncentrality and goodness of fit. *Psychological Bulletin, 107*(2), 247–255. https://doi.org/10.1037/0033-2909.107.2.247
Metcalf, L., & Benn, S. (2012). Leadership for sustainability: An evolution of leadership ability. *Journal of Business Ethics, 112*(3), 369–384. https://doi.org/10.1007/s10551-012-1278-6
Ng, X. L., Choi, S. L., & Soehod, K. (2016). The effects of servant leadership on employee's job withdrawal intention. *Asian Social Science, 12*(2), 99. https://doi.org/10.5539/ass.v12n2p99
Rigdon, E. E. (1996). CFI versus RMSEA: A comparison of two fit indexes for structural equation modeling. *Structural Equation Modeling: A Multidisciplinary Journal, 3*(4), 369–379. https://doi.org/10.1080/10705519609540052
Tabachnick, B. G., & Fidell, L. S. (2001). Using Multivariate Statistics. 4th Edition, Allyn and Bacon, Boston.
Ulrich, D., & Smallwood, N. (2012). What is leadership? *Advances in Global Leadership, 9–36*. https://doi.org/10.1108/s1535-1203(2012)0000007005
Van Dierendonck, D. (2010). Servant leadership: A review and synthesis. *Journal of Management, 37*(4), 1228–1261. https://doi.org/10.1177/0149206310380462
Yukl, G., Mahsud, R., Hassan, S., & Prussia, G. E. (2011). An improved measure of ethical leadership. *Journal of Leadership & Organizational Studies, 20*(1), 38–48. https://doi.org/10.1177/154805181142935

Chapter 8
Leadership for Sustainable Organizational Health and Well-Being: Indigenous Community Perspectives

Christopher Pitt, Kerrie E. Doyle, Elias Mpofu, Paul Saunders, Robert William Doyle, and Lauren Parkinson Zarb

8.1 Introduction

Historically, leadership has been underpinned by the ideals of competitiveness and individualism, with underlying racial and gender bias which favour white masculinity (Liu, 2017). Increasingly, inclusivity has been recognized as being essential to organizational competitiveness in the globalized world economies. For that reason, business organizations need to acknowledge and accept the need for a leadership culture which values inclusivity (Di Fabio, 2017a, b, c). This need for inclusive workplaces is particularly compelling in societies with Indigenous communities, who contribute to the vitality of business entities by virtue of their unique cultural heritages. Yet, few works have considered models of leadership which are inclusive of Indigenous community values, despite organizational well-being being closely associated with an alignment of employee values and organizational climate (Di Fabio, 2017b; Freegard et al., 2004). Conversely, dissonances between employee values and organizational climate are associated with high employee dissatisfaction and staff turnover (Di Fabio, 2017b; Price & Whiteley, 2014; Santos et al., 2012) and poor satisfaction by service users (Douglas et al., 2015; Padma & Ahn, 2020; Sezgen et al., 2019). Indigenous community employees are vulnerable to poor

C. Pitt (✉) · K. E. Doyle · P. Saunders · R. W. Doyle
Western Sydney University, Sydney, NSW, Australia
e-mail: c.pitt@westernsydney.edu.au

E. Mpofu
Western Sydney University, Sydney, NSW, Australia

University of North Texas, Denton, TX, USA

University of Johannesburg, Johannesburg, South Africa

L. P. Zarb
La Trobe University, Melbourne, VIC, Australia

A. Di Fabio (ed.), *Cross-cultural Perspectives on Well-Being and Sustainability in Organizations*, https://doi.org/10.1007/978-3-030-86709-6_8

quality of work life in organization with less inclusivity, such as those that assume a predominately white culture world view (Asmar et al., 2009). We seek to address this need, by proposing a model Indigenous leadership which harness inclusive workplace cultures aimed at promoting sustainable organizational health and well-being.

In business organizations, 'sustainability' refers to organizational longevity (profitability) while considering environmental and social capital issues (Sustainability Accounting Standards Board: SASB, 2019). Putting *cultural* sustainability into the conversation of business organizations would add to community inclusiveness, business leadership practices and increase employee well-being, which would in turn enhance the business organization's health and well-being (Di Fabio, 2017a, b). In other words, the well-being of organizations and their employees are indissolubly intertwined (Di Fabio, 2017b).

The United Nations Department of Economic and Social Affairs (UNDESA) defines Indigenous peoples as those members of communities that have retained social, cultural, economic and political characteristics distinct from those of the dominant societies in which they live, as a function of being colonized by another people/peoples (Hohmann & Weller, 2018). There are 476 million Indigenous peoples in 90 countries, or approximately 6% of the global population, which includes a sizable workforce. UNDESA's 17 Sustainable Development Goals recognize the ongoing disparities between population groups in countries, including Indigenous populations (UN Department of Economics and Social Affairs, 2015). Goal 4 recognizes quality education as a human right, and Goal 8 affirms that governments are charged with promoting sustained, inclusive, sustainable economic growth that leads to productive and decent work for all citizens (UNDESA, 2015). Yet Indigenous community members experience low work participation rates compared to mainstream community members (see Altman & Daly, 2018: Daly, 2018; Holcombe & Kemp, 2020; Liu et al., 2020).

Australia Indigenous people make up approximately 3.3% of the total population (ABS, 2019), while their participation in the formal employment sector is less than 1% of employable persons (ABS, 2016; Staines, 2018). Indigenous families are three times more likely to live in poverty, with unemployment rates twice that of urban non-Indigenous people and even higher in remote areas (McLachlan et al., 2013). These disparities continue, despite interventions to promote Indigenous community well-being by organizations such as the Australian Commonwealth Aboriginal and Torres Strait Islander Workforce Strategy (APSC, 2019). As an example, Indigenous people continue to face significantly worse outcomes in employment rates (Dunnett, 2021).

Improvements in Indigenous organizational work well-being should move beyond the deficit approach of oversimplifying Indigenous employment issues, ascribing the need for education and training (Hunter & Gray, 2017). Also, Indigenous organizational work well-being goes beyond geolocation issues, presumably to be solved by relocation of Indigenous workers to employment hubs that would increase their employment potential (Biddle, 2010; Biddle & Markham, 2018; Brereton & Parmenter, 2008; Jordan, 2018). For instance, the 3% of people who comprise the Australian Indigenous populations are scattered across the continent, located in metropolitan, urban, regional, rural and remote communities (Kingsley et al., 2013). Increasing workplace inclusivity practices which adopt indigenist approaches would improve work participation of Indigenous community employees and make for more inclusive workplace cultures.

Inclusive Leadership for Sustainable Organizational Health and Well-Being Inclusive leadership refers to organizational frameworks based on social justice and social inclusion (Randel et al., 2018) that recognize and act on exclusive practices (Shields, 2003), foster accountability for inclusive behaviours (McNeil, 2000) and reinforce the leader's responsibility for inclusive practices (Ryan, 2006a, b). As a business practice, inclusive leadership aims to create cultural value and richness for the organization by enhancing employee organizational citizenship, in order to achieve more complete involvement of diverse individuals in the workplace (Shore et al., 2011). This model of leadership provides the opportunity for organizational members to reach their full potential, by increasing employees' perception of being esteemed members of a work group or organization (Shore et al., 2011). Organizations have long been perceived to bear some responsibility for negative environmental and societal impacts (see, e.g. Dunphy et al., 2003). Workplaces with poor organizational health and well-being experience decreased productivity as a result of increased absenteeism (where employee takes paid or unpaid leave) and presenteeism (where the employee attends work but due to illness or injury is not fully productive) (APSC, 2018). Conversely, employees with high organizational citizenship identified with the values within their work organization are willing to go the extra mile for the betterment of their organization (Gao, & He, 2017; Jiang et al., 2017; Younas et al., 2020). Healthy organizational workplaces which value inclusiveness use targeted system-level interventions in their quest to create sustainable and effective cultural, social and physical aspects which promote employee well-being (Brand et al., 2017). With Indigenous communities, this entails adopting indigenist approaches.

Indigenist Approaches Indigenist model of leadership are characterized by egalitarian and holistic work environments, which recognize the role of community engagement, long-term orientations and cultural authenticity (Julien et al., 2010; Haar et al., 2019). A review of ten global leadership studies highlighted successful leaders within Indigenous-led organization to display behaviours focussed upon effective management of time and resources, with detailed planning (Park et al., 2018). For example, Palmer and Masters (2010) observed Maori values of kinship, compassion, caring and support to underpin their work organization followership, resulting in lower staff absences.

Around the world, Indigenous leaders and followers within business organizations are commonly a minority (Asmar et al., 2009). This might inadvertently result in a lack of cultural awareness by business organization leaders, subjecting Indigenous employees to unwanted workplace estrangement (Leitch, 2017), resulting from a lack of cultural fluency by the management. Cultural fluency is defined as the ability of a person to recognize and adapt to various cultures (Mourey et al., 2015). Being culturally fluent requires workplace managers and employees to understand the often entrenched cultural biases and the challenges of different behavioural requirements (Leitch, 2017).

Next, we review leading theoretical approaches and empirical research to embed a sustainable indigenist organizational culture. These are important for understanding the indigenist culture influences on leadership styles.

8.2 Theories and Models on Indigenist-Oriented Organization Leadership

Indigenist-oriented models of leadership prioritize egalitarian and holistic community – focussed approaches to leadership (Ruwhiu & Elkin, 2016). Leading types of such leadership include transformational, authentic, aspirational and servant. Transformational leadership is one of the most studied and contested styles of leadership, characterized by a leadership style where leaders influence followers by aligning self-interest to that of the organization (Muchiri et al., 2012; Schaubroeck et al., 2007). A transformational leadership style that privileges the voices of Indigenous leaders and followers has been shown to foster an inclusive environment for Indigenous followers (Doyle & Hungerford, 2015; Santamaría et al., 2015). However transformational leadership has been criticized for its primary focus on leadership effectiveness rather than followers' need satisfaction (Van Dierendonck et al., 2014). The strong moral dimensions found in newer leadership styles have been said to address deficiencies present in transformational leadership (Banks et al., 2016).

Authentic leadership is a type of transformational leadership which prioritizes the importance of self-awareness, demonstrated by an ability to honour their own core values, beliefs, strengths and weaknesses (Gardner et al., 2011). In its development, authentic leadership theory drew heavily from Kernis's (2003) theory of optimal self-esteem. Optimal self-esteem is characterized by four key elements: self-awareness, unbiased processing, relational authenticity and authentic behaviour/action. A central to the theory of authentic leadership is recognition of the self-referential nature of authenticity (Avolio & Gardner, 2005). Authenticity is associated with self-aware, transparent leader who is invested in the well-being of the workplace collective (Walumbwa et al., 2008). Authentic leadership has a cascading effect, in promoting positive behaviours at both team and individual levels (Hirst et al., 2016).

Aspirational leadership emphasizes the importance of leaders looking forward and having a future plan for the organization and the broader community (Koning & Waistall, 2012; see Fig. 8.1). A leader adopting an aspirational model will display behaviours focussed on creating a culture of wellness and sustainability, through displaying humility and actively listening to team members (Donnelly, 2017).

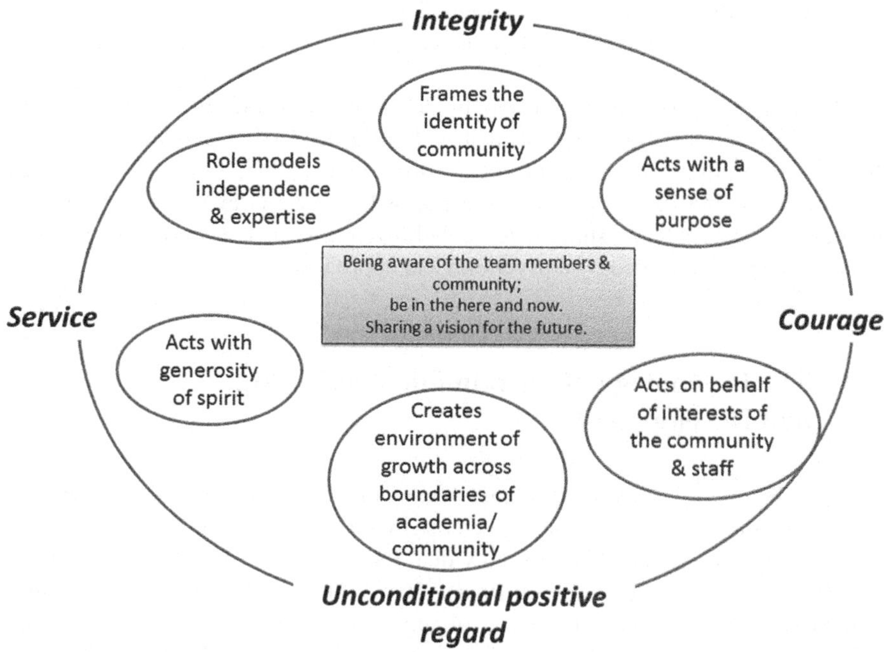

Fig. 8.1 Model of indigenist aspirational servant leadership. (Adapted from Doyle et al. (2020, p151))

Aspirational leadership behaviours can be embedded within an organization through organization-wide communication, stakeholder engagement and policies promoting aspirational leadership (Penttilä, 2014). The model of aspirational servant leadership aims to support individuals and celebrate everyone's success to create community cohesion and reduce social exclusion (Doyle et al., 2020).

A servant model of leadership is able develop organizational inclusion through empowering and developing people, displaying humility and authenticity and interpersonal acceptance and providing direction and stewardship (Gotsis & Grimani, 2016), dependent on the skills and attitudes of the leader. A central tenant of servant leadership is focussed on promoting inclusiveness, integrity, humility, servanthood, goal setting and shared decision-making (Gotsis & Grimani, 2016; Winston & Fields, 2015). The servant leader displays the following qualities:

- Character traits – Concerned with undertaking a servant attitude
- People orientation – Concerned with how a leader interacts with their team
- Task orientation – Focussed on how the leader completes work tasks
- Process orientation – Pertains to how a leader impacts the broader organization (Page & Wong, 2000)

Servant leadership is about demonstrating the behavioural skills of standing back, forgiveness, courage, empowerment, accountability, authenticity, humility and stewardship (Dierendonck & Nujiten, 2011). Aspects of servant leadership are merged with an indigenist leadership that considers cultural as an important skill in a leader's tool kit.

Doyle et al. (2020) piloted a hybrid model of servant leadership and aspirational leadership, and findings suggest that a hybrid leadership model aligns with Indigenous cultural community protocols, professional ethics and a need to create a safe space for Indigenous peoples (see Fig. 8.1). Doyle et al.'s (2020) indigenist aspirational servant leadership model includes values such as integrity and includes behaviours best described by a psychological approach.

8.3 The Psychology of Sustainable Leadership and Development

The psychology of sustainability and sustainable development is a new research area within the field of sustainability science that focusses upon improving the quality of life and well-being of every individual for the benefit of the broader community (Di Fabio, 2017a, Di Fabio & Rosen, 2018). The psychology of sustainability and sustainable development requires organizations to incorporate behaviours and systems that promote well-being and positive relationships (Di Fabio, 2017a).

Applied to organizational leadership, sustainable development requires the engagement and collaboration of employees across an organization at all levels (Di Fabio, 2017b). With Indigenous communities, the sustainability and sustainable development of organizations would depend on finding meaning within their work (Doyle et al., 2020). Meaning within work is defined as employees obtaining a sense of direction, belonging and significance through their work (Schnell et al., 2013). Organizations promote meaning within their work by finding a work fit between employees and their role, outlining the significance of employees' tasks and promoting a socio-moral climate (Schnell et al., 2013). Indigenist organizations would implement sustainable organization development initiatives that require inclusive managerial styles for staff well-being. For instance, organization leadership would seek to facilitate the development of the interpersonal and intrapersonal talents of all employees for the benefit of the broader organization and the general community (see also Di Fabio & Rosen, 2018), improving the quality of life of each person they interact with.

An inclusive managerial practice grounded in indigenist leadership approach is harmonization (Doyle et al., 2020). Within a work context, harmonization is focussed upon the development of positive relationships with one's self, place of work and community (Cortini et al., 2019; Di Fabio & Tsuda, 2018). Harmonization is an important facet of sustainable organization development in directing leaders to better balance their own behaviour and their external relationship (Di Fabio & Tsuda, 2018), sustaining work community relationships. Harmonization indigenist business leadership practices are likely by leaders with emotional intelligence (see also Baron et al., 2018; Miao et al., 2018; Tuleja, 2014).

8.4 Emotional Intelligence and Indigenist Leadership

Emotional intelligence refers to the capacity of a person to comprehend and respond to others within a social environment, with mindfulness (Prati et al., 2003). Emotional intelligence has only been shown to be effective in improving organizational commitment and job satisfaction when paired with a transformational leadership style (Lam & O'Higgins, 2012). Mindfulness is defined as the state of being focussed and absorbing what actions are occurring in the present situation (Brown & Ryan, 2003). Within a work context, leaders who score highly on the trait of mindfulness are characterized with positive affect and self-efficacy beliefs (Carleton et al., 2018; Klockner, 2017). Sinclair (2015) and Di Fabio (2017b) have proposed a series of guiding principles that constitute a framework for the tools a mindful leader utilizes, such as the following:

- Placing emphasis upon people and their happiness
- Focussing upon behaviour rather than talk
- Demonstrating leadership
- Implementing mindful behaviour in everyday situations

These practices align with indigenist organization leadership practices. Arendt et al. (2019) evaluated the practical influence of mindful leadership through evaluating the perceptions of 34 leaders and 98 followers and found that leaders with higher self-rated mindfulness scores correlated higher with the needs of followers. The work of Arendt et al. (2019) highlighted the association between leaders' mindfulness behaviour and followers' perceptions. Ashton and Scott (2017) reported leadership inclusivity to empower employees, providing social environments of trust and respect that fostered collaboration and commitment to the team.

8.5 Practices in Sustainable Indigenist-Oriented Leadership

Sustainable indigenist-oriented organizational leadership has a community focus, emphasizing co-ownership of work place practices (see Ruwhiu & Elkin, 2016; Julien et al., 2010). For instance, Indigenous leaders are frequently viewed as Indigenous people first and leaders second (Evans & Sinclair, 2016). Thus, a core challenge that Indigenous leaders face is identifying who they represent and whom they are speaking for in mainstream organizations with less inclusivity. Table 8.1 summarizes the differences between Indigenous and mainstream leadership approaches.

Community-Oriented Practices

Collectiveness identity in a work culture reduces Indigenous staff turnover (Warren & Quine, 2013). Moreover, indigenist-oriented leadership enhanced Indigenous employees' work resilience by proving a broader workplace family for them (Henry

Table 8.1 Organization leadership values: Indigenous vs mainstream

Indigenous leaders	Mainstream leaders
Community-focussed approach to leadership (Julien et al., 2010; Ruwhiu & Elkin, 2016; Stewart, & Warn, 2017)	Individualistic (Julien et al., 2010)
Holistic approach to leadership (Warren & Quine, 2013; Henry & Wolfgramm, 2018)	Hierarchical leadership structure (Henry & Wolfgramm, 2018)
Shared leadership (Henry & Wolfgramm, 2018)	Focus on person as employee, separation between employee and family (Julien et al., 2010)
Cultural authentic leadership (Haar et al., 2018)	
Collectiveness identity (Stewart, & Warn, 2017)	

& Wolfgramm, 2018). Workplace family is important in collectivist cultures and an essential quality of ethical leadership based on power sharing with the followers (Lubogoyi et al., 2018). For instance, Ibrahim and Mayende (2018) considered ethical leaders' behaviours among Ugandan cultures, also collectivists, to include morality, honestly, fairness and power sharing.

Holistic Practices

Holistic leadership is concerned with the organization as a whole, rather than a siloed approach. Holistic leaders are self-directed and *other*-focussed and seek their own personal leadership development (Dhiman, 2017). This style of leadership can suit collectivist cultures. For example, Kiruhi (2017) reports 25 leaders in Kenya successfully adopted a holistic leadership approach to increase employment recruitment and retention in a way that best suits communities. Chai et al. (2016) describe an Indigenous Korean leadership model based on Confucian values where loyalty and trust are exchanged with job security (see also Chang, 2012), where leaders took a holistic approach to organization management/leadership, promoting self-reflection and sensitive caring for the health and well-being of employees through a sense of shared leadership.

Shared Leadership Practices

Shared leadership is an indigenist organization leadership practice. It implies a distributive model of functional leadership (Goldsmith, 2010). There are three common practices in this leadership model – shared, lateral modes of influence; consideration of team performance and success; and leadership roles and influence dispersed across team members (Pearce & Conger, 2003). Hans and Gupta (2018) considered shared leadership practices to create psychological safe environments that allow peers to demonstrate self-efficacy. Sharing leadership requires the leader to score highly on emotional intelligence and to be able to match tasks to team members' creativity (Thrassou et al., 2018). While shared leadership practices can work well on collective and working networks (see Cullen-Lester & Yammarino, 2016) such as projects (Scott-Young et al., 2019) or design teams (Novoselich & Knight, 2018), it is less appropriate when task complexity increases (Müller et al., 2018). Shared leadership would enhance Indigenous employee retention when used as a development tool for team members (see also Pitelis & Wagner, 2019; Van De Mieroop et al., 2020). Ultimately, indigenist organization leadership is about cultural authenticity.

Cultural Authenticity

Culturally authenticity organization leadership practices are those that have a communal identity focus (Carr & MacLachlan, 1997), through work meaningfulness (Di Fabio, 2017b). In the context of indigenist organization business practices, cultural authenticity refers to the values of the following:

- Spirit and integrity
- Cultural continuity
- Equity
- Respect
- Responsibility (NHMRC, 2018)

Indigenist leadership practices with cultural authenticity create a sense of belongingness, which is an essential aspect of organizational health and well-being (Carr & MacLachlan, 1997; Doyle et al., 2020). Culturally authentic leaders are knowledgeable, proactive and supportive when implementing organizational change for sustainability (Gifford et al., 2017; Torres et al., 2017). The Australian Closing the Gap Report (COA, 2020) proposes authentic culture solutions aimed to supporting Indigenous peoples' self-determination and culturally safe employment. For this strategy to be successful as an organizational leadership strategy requires an understanding of applied cultural authenticity for social outcome-oriented workplace practices.

Social Outcome-Oriented Practices

Social outcome-oriented leadership practices by government and non-government organizations are aimed at promoting harmony (Di Fabio & Tsuda, 2018). Their significance is grounded in the concept of social responsibility, as proposed by the International Organization for Standardization (ISO) in ISO 26000 (ISO, 2018). Applied to indigenist-oriented organization leadership, social outcome-oriented leadership aims to balance human rights and labour practices as a social responsibility. For instance, in Australia, the Australian Leadership Index (ALI, 2019) reported effective leadership to be premised on high ethical standards, transparent organizational structures and responsiveness to the interests of society and the people they serve by the organizations (ALI, 2019). With implementation, social outcome-oriented leadership would be indigenist in its governance structures.

8.6 Issues for Research and Practice

We believe that organizations and leaders do not need to be Indigenous to utilize indigenist-oriented models – in the same way that non-Indigenous people utilize indigenist research methods when researching Indigenous topics, organizations with a view to increased sustainability would do well to consider a model that fosters values that empower a collectivist people even in the absence of Indigenous people. For instance, supporting Australian Indigenous leaders requires business organizations to support Indigenous leaders negotiate and navigate their own way through the many layers of management and/or into positions of leadership. This would lay the ground for the dissemination and utilization of indigenist leadership across business organizations.

Nonetheless, there is a dire need for research on best practices that support the engagement of Indigenous staff by applying indigenist, strength-based approaches (Doyle et al., 2020; Doyle & Hungerford, 2015), of which we considered a good number in this chapter. A strength-based approach to leadership focusses on community and not merely the individual (Doyle et al., 2020). Indigenist-oriented models of leadership need to foster the strengths of Indigenous cultures and identities, by undertaking a community approach to management (Doyle & Hungerford, 2015; Ruwhiu & Elkin, 2016; Walker et al., 2014).

Indigenous populations are not homogenous across communities, and such diversity needs to be recognized and addressed in workplace practices for organizational health and well-being (Doyle, 2012; Doyle et al., 2020; Walker et al., 2014). The diversity in the world's Indigenous populations makes it impossible to have one framework that fits all Indigenous people. For this reason, we recommend organizational leaders to:

- Understand the organization's mission, vision and values and measure the organizational capability and capacity for leadership change
- Educate organization on indigenist leadership models and embed cultural competency principles in all facets of the organization
- Become familiar with the history and culture of contemporary Indigenous communities and privilege the knowledges of local people by engaging authentically with community leaders and elders
- Include Indigenous leaders on organizational boards as mentors in indigenist leadership

Indigenist models of leadership and governance are human rights oriented, employing a stewardship approach to organization development and management. Social outcome-oriented leadership practices allow for inclusive work practices, such as integrity, reciprocity and respect, which are important pillars of social responsibility-oriented leadership.

8.7 Conclusion

This chapter has considered indigenist leadership to be a vehicle of organizational sustainability by demonstrating leadership practices through an indigenist lens. Leadership practices that are holistic with grounded authentic leaders able to role model shared leadership practice are matched to aspects of indigenist leadership. Organizations implementing inclusive leadership for sustainability need to consider how best to manage competing world views, applying strength-based approaches. Organizations need to consider cultural sustainability as a core component, and part of this sustainability needs to privilege indigenist leadership as a vehicle for producing workspaces that are welcoming and affirming to people from collectivist backgrounds. Utilizing indigenist-oriented leadership approaches will increase organizational health and well-being through higher organization citizenship behaviours of Indigenous employees, and following the axiom of getting it right for Indigenous people means getting it right for all people.

References

Altman, J., & Daly, A. E. (2018). *Do fluctuations in the Australian macroeconomy influence Aboriginal employment status?* Centre for Aboriginal Economic Policy Research (CAEPR), The Australian National University.

Ann Roche, M., Haar, J. M., & Brougham, D. (2018). Māori leaders' well-being: A self-determination perspective. Leadership (London, England), 14(1), 25–39.

Arendt, J. F., Verdorfer, A. P., & Kugler, K. G. (2019). Mindfulness and leadership: Communication as a behavioral correlate of leader mindfulness and its effect on follower satisfaction. *Frontiers in Psychology, 10*, 667–683.

Ashton, A. S., & Scott, N. (2017). International retirement migration and Thai stakeholders' views: A Japanese case study. *Journal of Place Management and Development, 10*(1), 7–22.

Asmar, C., Mercier, O., & Page, S. (2009). 'You do it from your core': Priorities, perceptions and practices of research among Indigenous academics in Australian and New Zealand universities. In A. Brew & L. Lucas (Eds.), *Academic research and researchers* (pp. 146–160). Open University Press and McGraw-Hill.

Australian Bureau of Statistics. (2019). *Estimates and projections, Aboriginal and Torres Strait Islander Australians, 2006 to 2031. ABS cat. no. 3238.0*. ABS.

Australian Leadership Index. (2019). *Australian Leadership Index: 2019 National Survey Report*. Retrieved from https://www.australianleadershipindex.org/wp-content/uploads/2020/06/2002-SALI-Report-ACCESSIBLE-Final.pdf.

Australian Public Service Commission (APSC). (2019). *An Evaluation of the Commonwealth Aboriginal and Torres Strait Islander Employment Strategy*. Commonwealth of Australia.

Australian Public Service Commission (APSC). (2018). *How employee health and wellbeing affects organisational productivity*. Australian Government, Retrieved from https://www.apsc.gov.au/how-employee-health-and-wellbeing-affects-organisational-productivity

Avolio, B. J., & Gardner, W. L. (2005). Authentic leadership development: Getting to the root of positive forms of leadership. *The Leadership Quarterly 16*, 315–338.

Banks, G. C., McCauley, K. D., Gardner, W. L., & Guler, C. E. (2016). A meta-analytic review of authentic and transformational leadership: A test for redundancy. *The Leadership Quarterly, 27*(4), 634–652.

Baron, L., Rouleau, V., Grégoire, S., & Baron, C. (2018). Mindfulness and leadership flexibility. *Journal of Management Development, 37*(2), 165–177.

Biddle, N., & Markham, F. (2018). *Indigenous identification change between 2011 and 2016: Evidence from the Australian Census Longitudinal Dataset*. Centre for Aboriginal Economic Policy Research, ANU.

Biddle, N. (2010). Proximity to labour markets: Revisiting Indigenous employment through an analysis of Census place of work data. *Australian Journal of Labour Economics, 13*(2), 175–189.

Brand, S. L., Thompson Coon, J., Fleming, L. E., Carroll, L., Bethel, A., & Wyatt, K. (2017). Whole-system approaches to improving the health and wellbeing of healthcare workers: A systematic review. *PloS one, 12*(12), e0188418.

Brereton, D. J., & Parmenter, J. (2008). Indigenous employment in the Australian mining industry. *Journal of Energy and Natural Resources Law, 26*(1), 66–90.

Brown, K. W., & Ryan, R. M. (2003). The benefits of being present: Mindfulness and its role in psychological well-being. *Journal of Personality and Social Psychology, 84*, 822–848.

Carleton, E. L., Barling, J., & Trivisonno, M. (2018). Leaders' trait mindfulness and transformational leadership: The mediating roles of leaders' positive affect and leadership self-efficacy. *Canadian Journal of Behavioural Science, 50*(3), 185–194.

Carr, S. C., & MacLachlan, M. (1997). Motivational Gravity. In D. Munro, J. F. Schumaker, & S. C. Carr (Eds.), *Motivation and culture*. Routledge.

Chai, D. S., Jeong, S., Kim, J., Kim, S., & Hamlin, R. G. (2016). Perceived managerial and leadership effectiveness in a Korean context: An Indigenous qualitative study. *Asia Pacific Journal of Management, 33*(3), 789–820.

Chang, S. (2012). Study on human resource management in Korea's chaebol enterprise: A case study of Samsung Electronics. *International Journal of Human Resource Management, 237*, 1536–1561.

Commonwealth of Australia, Department of the Prime Minister and Cabinet. (2020). Closing the Gap Report 2020. https://ctgreport.niaa.gov.au/sites/default/files/pdf/closing-the-gap-report-2020.pdf

Cortini, M., Converso, D., Galanti, T., Di Fiore, T., Di Domenico, A., & Fantinelli, S. (2019). Gratitude at work works! A mix-method study on different dimensions of gratitude, job satisfaction, and job performance. *Sustainability, 11*(14), 3902.

Crawford, J. A., & Kelder, J. A. (2019). Do we measure leadership effectively? Articulating and evaluating scale development psychometrics for best practice. *The Leadership Quarterly, 30*(1), 133–144.

Cullen-Lester, K. L., & Yammarino, F. J. (2016). Collective and network approaches to leadership: Special issue introduction. *The Leadership Quarterly, 27*(2), 173–180. https://doi.org/10.1016/j.leaqua.2016.02.001

Daly, A. E. (2018). *The employment of Aboriginal Australians in the labour market*. Centre for Aboriginal Economic Policy Research (CAEPR), The Australian National University.

Dhiman, S. (2017). *Holistic leadership: A new paradigm for today's leaders*. Springer.

Di Fabio, A. (2017a). Positive healthy organizations: Promoting well-being, meaningfulness, and sustainability in organizations. *Frontiers in Psychology, 8*, 1938.

Di Fabio, A. (2017b). The psychology of sustainability and sustainable development for well-being in organizations. *Frontiers in Psychology, 8*, 1534.

Di Fabio, A. (2017c). Positive Healthy Organizations: Promoting well-being, meaningfulness, and sustainability in organizations. *Frontiers in Psychology, 8*, 1938.

Di Fabio, A., & Rosen, M. A. (2018). Opening the black box of psychological processes in the science of sustainable development: A new frontier. *European Journal of Sustainable Development Research, 2*(4), 47.

Di Fabio, A., & Tsuda, A. (2018). The psychology of harmony and harmonization: Advancing the perspectives for the psychology of sustainability and sustainable development. *Sustainability, 10*(12), 4726.

Dierendonck, D. V., & Nuijten, I. (2011). The servant leadership survey: Development and validation of a multidimensional measure. *Journal of Business Psychology, 26*, 249–267.

Donnelly, L. F. (2017). Aspirational characteristics for effective leadership of improvement teams. *Paediatric Radiology, 47*(1), 17–21.

Douglas, J. A., Douglas, A., McClelland, R. J., & Davies, J. (2015). Understanding student satisfaction and dissatisfaction: an interpretive study in the UK higher education context. *Studies in Higher Education, 40*(2), 329–349.

Doyle, K. (2012). Modes of colonisation and patterns of contemporary mental health: Towards an understanding of Canadian Indigenous. *Australian Indigenous and Maori peoples, Indigenous and Islander Health Worker Journal, 35*(1), 20–37.

Doyle, K., & Hungerford, C. (2015). Leadership as a personal journey: An Indigenous perspective. *Issues in Mental Health Nursing, 36*(5), 336–345.

Doyle, K. E., Hungerford, C., Pitt, C., Saunders, P., & Wilkey, K. (2020). Indigenist leadership in academia: Towards an aspirational model of mindful servant leadership. In J. Frawley, G. Russell, & J. Sherwood (Eds.), *Cultural competence and the higher education sector: Perspectives, policies and practice* (pp. 137–158). Wiley.

Dunnett, A. (2021). *'They think they know us better': Aboriginal experiences of education, health and employment in contemporary South Australia* (unpublished doctoral dissertation). John Moores University Liverpool, UK.

Evans, M., & Sinclair, A. (2016). Containing, contesting, creating spaces: Leadership and cultural identity work among Australian Indigenous arts leaders. *Leadership, 12*(3), 270–292.

Freegard, H., Hoogland, L., & Marquis, R. (2004). Cultures that support caregiver retention in residential aged care. *Geriaction, 22*(1), 3–12.

Gao, Y., & He, W. (2017). Corporate social responsibility and employee organizational citizenship behavior. *Management Decision, 55*(2), 1–12.

Gardner, W. L., Cogliser, C. C., Davis, K. M., & Dickens, M. P. (2011). Authentic leadership: A review of the literature and research agenda. *The Leadership Quarterly, 22*(6), 1120–1145.

Gifford, W., Graham, I. D., Ehrhart, M. G., Davies, B. L., & Aarons, G. A. (2017). Ottawa Model of Implementation Leadership and Implementation Leadership Scale: mapping concepts for developing and evaluating theory-based leadership interventions. *Evidenced Based Practice Journal of Healthcare Leadership, 9*, 15–23.

Goldsmith, M. (2010). Sharing leadership to maximize talent. *Harvard Business Review*. Retrieved from https://hbr.org/2010/05/sharing-leadership-to-maximize/

Gotsis, G., & Grimani, K. (2016). The role of servant leadership in fostering inclusive organizations. *Journal of Management Development, 35*(8), 985–1010.

Griffiths, A., Dunphy, D., & Benn, S. (2003). Building corporate sustainability. Training and Development in Australia, 30(2), 20–22.

Haar, J., Roche, M., & Brougham, D. (2019). Indigenous insights into ethical leadership: A study of Māori leaders. *Journal of Business Ethics, 160*, 621–640.

Hans, S., & Gupta, R. (2018). Job characteristics affect shared leadership. *Leadership & Organization Development Journal, 39*(6), 730–744. https://doi.org/10.1108/LODJ-03-2018-0101

Hirst, G., Walumbwa, F., Aryee, S., Butarbutar, I., & Chen, C. J. H. (2016). A Multi-level Investigation of Authentic Leadership as an Antecedent of Helping Behavior. *Journal of Business Ethics, 139*(3), 485–499.

Henry, E., & Wolfgramm, R. (2018). Relational leadership – An Indigenous Maori perspective. *Leadership, 14*(2), 203–219.

Hohmann, J., & Weller, M. (Eds.). (2018). *The UN declaration on the rights of indigenous peoples: A commentary*. Oxford University Press.

Holcombe, S., & Kemp, D. (2020). From pay-out to participation: Indigenous mining employment as local development? *Sustainable Development, 28*(5), 1122–1135.

Hunter, B., & Gray, M. (2017). *Occupational mobility of Indigenous and other Australians, 2006-11*. CAEPR Working Paper.

Ibrahim, A. M., & Mayende, S. T. (2018). Ethical leadership and staff retention in Uganda's health care sector: The mediating effect of job resources. *Cogent Psychology, 5*(1), 1466634.

International Organisation for Standardization (ISO). (2018). *Discovering ISO 26000*. Switzerland: iso.org. ISBN 978-92-67-10973-2

Jiang, W., Zhao, X., & Ni, J. (2017). The impact of transformational leadership on employee sustainable performance: The mediating role of organizational citizenship behavior. *Sustainability, 9*(9), 1567.

Jordan, K. (2018). Aboriginal and Torres Strait Islander employment policy and Welfare to Work: The Community Development Programme and the need for new narratives, new alliances and new institutions. *Australian Journal of Social Issues, 53*(3), 239–261.

Julien, M., Wright, B., & Zinni, D. M. (2010). Stories from the circle: Leadership lessons learned from Aboriginal leaders. *The Leadership Quarterly, 21*(1), 114–126.

Kernis, M. (2003). Toward a Conceptualization of Optimal Self-Esteem. *Psychological Inquiry, 14*(1), 1–26.

Kingsley, J., Townsend, M., Henderson-Wilson, C., & Bolam, B. (2013). Developing an exploratory framework linking Australian Aboriginal peoples' connection to country and concepts of wellbeing. *International Journal of Environmental Research and Public Health, 10*(2), 678–698.

Kiruhi, T. M. (2017). Accelerating Africa's renaissance using contextual holistic leadership development. In K. Patterson & B. Winston (Eds.), *Leading an African renaissance. Palgrave studies in African leadership*. Palgrave Macmillan. https://doi.org/10.1007/978-3-319-40539-1_7

Klockner, K. (2017). Developing organisational resilience: Organisational mindfulness and mindful organising. *The Australian Journal of Emergency Management, 32*(4), 47–51.

Koning, J., & Waistell, J. (2012). Identity talk of aspirational ethical leaders. *Journal of Business Ethics, 107*(1), 65–77.

Lam, C., & O'Higgins, E. R. E. (2012). Enhancing employee outcomes: The interrelated influences of managers' emotional intelligence and leadership style. *Leadership and Organization Development Journal, 33*(2), 149–174.

Leitch, A. (2017). Indigenous public service leadership and issues of cultural fluency. *Australian Journal of Public Administration, 76*(4), 403–405.

Liu, W. M. (2017). White Male Power and Privilege: The Relationship Between White Supremacy and Social Class. *Journal of Counseling Psychology, 64*(4), 349–358.

Liu, Y., Daff, S., & Pearson, C. (2020). Shaping sustainable employment and social consequences of Indigenous Australians in a remote region. *Sustainability, 12*(21), 9054.

Lubogoyi, B., Kasekende, F., Kagaari, J., Ngoma, M., Munene, J. C., & Bakunda, G. (2018). Stewardship behaviour and perceived goal congruence in local governments in Uganda. *Leadership & Organization Development Journal, 39(6), 712–729.*

McLachlan, R., Gilfillan, G., & Gordon, J. (2013). *Deep and persistent disadvantage in Australia*. Productivity Commission Staff Working Paper.

McNeil, L. M. (2000). *Contradictions of school reform: Educational costs of standardized testing*. Routledge.

Miao, C., Humphrey, R. H., & Qian, S. (2018). The relationship between emotional intelligence and trait mindfulness: A meta-analytic review. *Personality and Individual Differences, 135*, 101–107.

Mourey, J. A., Lam, C. P., & Oyserman, D. (2015). Consequences of Cultural Fluency. *Social Cognition, 33*(4), 308–344.

Muchiri, M., Cooksey, R. W., & Walumbwa, F. O. (2012). Transformational and social processes of leadership as predictors of organisational outcomes. *Leadership & Organization Development Journal, 33*(7), 662–683.

Müller, E., Pintor, S., & Wegge, J. (2018). Shared leadership effectiveness: Perceived task complexity as moderator. *Team Performance Management: An International Journal., 24*(5/6), 298–315.

National Health and Medical Research Council (NHMRC). (2018). *Ethical conduct in research with Aboriginal and Torres Strait Islander Peoples and communities: Guidelines for researchers and stakeholders*. Canberra.

Novoselich, B. J., & Knight, D. B. (2018). Shared leadership in capstone design teams: Social network analysis. *Journal of Professional Issues in Engineering Education and Practice, 144*(4), 04018006.

Padma, P., & Ahn, J. (2020). Guest satisfaction & dissatisfaction in luxury hotels: An application of big data. *International Journal of Hospitality Management, 84*, 102318.

Page, D., & Wong, P. (2000). A conceptual framework for measuring servant leadership. In S. Adjibolooso (Ed.), *The human factor in shaping the course of history and development* (pp. 69–110). American University Press.

Palmer, F. R., & Masters, T. M. (2010). Māori feminism and sport leadership: Exploring Māori women's experiences. *Sports Management Review, 13*(4), 331–344.

Park, S., Jeong, S., & Jang., Yoon, S.W., & Lim, D.H. (2018). Critical review of global leadership literature: Toward an integrative global leadership framework. *Human Resource Development Review, 17*(1), 95–120.

Pearce, C. L., & Conger, J. A. (2003). All those years ago. In C. L. Pearce & J. A. Conger (Eds.), *Shared leadership: Reframing the hows and whys of leadership* (pp. 1–18). Sage.

Penttilä, V. (2014). Aspirational talk in strategy texts: A longitudinal case study of strategic episodes in corporate social responsibility communication. *Business & Society, 59*(1), 67–97.

Pitelis, C. N., & Wagner, J. D. (2019). Strategic shared leadership and organizational dynamic capabilities. *The Leadership Quarterly, 30*(2), 233–242.

Prati, L. M., Douglas, C., Ferris, G. R., & Ammeter, A.P.& Buckley, M.R. (2003). Emotional intelligence, leadership effectiveness, and team outcomes. *International Journal of Organizational Analysis, 11*, 21–40.

Price, C., & Whiteley, A. (2014). Corporate culture and employee identity: Cooption or commitment through contestation? *Journal of Change Management, 14*(2), 210–235.

Randel, A. E., Galvin, B. M., Shore, L. M., Ehrhart, K. H., Chung, B. G., Dean, M. A., & Kedharnath, U. (2018). Inclusive leadership: Realizing positive outcomes through belongingness and being valued for uniqueness. *Human Resource Management Review, 28*(2), 190–203.

Ruwhiu, D., & Elkin, G. (2016). Converging pathways of contemporary leadership: In the footsteps of Māori and servant leadership. *Leadership, 12*(3), 308–323.

Ryan, J. (2006a). Inclusive leadership and social justice for schools. *Leadership and Policy in Schools, 5*(1), 3–17.

Ryan, J. (2006b). *Inclusive leadership*. Jossey-Bass.

Santamaría, A. P., Webber, M., & Santamaría, L. J. (2015). Effective school leadership for Māori achievement: Building capacity through Indigenous, national, and international cross cultural collaboration. In N. D. Erbe & A. H. Normore (Eds.), *Cross-Cultural Collaboration and Leadership in Modern Organizations, USA* (pp. 99–119). Business Science Reference.

Santos, A., Hayward, T., & Ramos, H. M. (2012). Organizational culture, work and personal goals as predictors of employee well-being. *Journal of Organizational Culture, Communications and Conflict, 16*, 25–48.

Schaubroeck, J., Cha, S. E., & Lam, S. S. K. (2007). Embracing transformational leadership: Team values and the impact of leader behaviour on team performance. *Journal of Applied Psychology, 92*(4), 1020–1030.

Schnell, T., Höge, T., & Pollet, E. (2013). Predicting meaning in work: Theory, data, implications. *The Journal of Positive Psychology, 8*(6), 543–554.

Scott-Young, C. M., Georgy, M., & Grisinger, A. (2019). Shared leadership in project teams: An integrative multi-level conceptual model and research agenda. *International Journal of Project Management, 37*(4), 565–581.

Sezgen, E., Mason, K. J., & Mayer, R. (2019). Voice of airline passenger: A text mining approach to understand customer satisfaction. *Journal of Air Transport Management, 77*, 65–74.

Shields, C. (2003). *Good intentions are not enough: Transformative leadership for communities of difference*. Scarecrow Press.

Shore, L. M., Randel, A. E., Chung, B. G., Dean, M. A., Holcombe Ehrhart, K., & Singh, G. (2011). Inclusion and diversity in work groups: A review and model for future research. *Journal of Management, 37*(4), 1262–1289.

Sinclair, A. (2015). Possibilities, purpose and pitfalls: Insights from introducing mindfulness to leaders. *Journal of Spirituality, Leadership and Management, 8*(1), 3–11.

Staines, Z. (2018). Ground-level impacts of remote employment policy: Social disadvantage under the Community Development Programme. *Journal of Australian Political Economy, 82*, 107–132.

Stewart, J., & Warn, J. (2017). Between two worlds: Indigenous leaders exercising influence and working across boundaries. *Australian Journal of Public Administration, 76*(1), 3–17.

Sustainability Accounting Standards Board. (2019). 2019 Building Resilience SASB, https://www.sasb.org/wp-content/uploads/2020/12/SASB-Annual-Report-2019-122320.pdf

Thrassou, A., Orfanos, D., & Tsoukatos, E. (2018). Linking motivational leadership with creativity. In *Innovation and capacity building* (pp. 77–108). Palgrave Macmillan.

Torres, E. M., Ehrhart, M. G., Beidas, R. S., Farahnak, L. R., Finn, N. K., & Aarons, G. (2017). Validation of the Implementation Leadership Scale (ILS) with supervisors' self-ratings. *Community Mental Health Journal, 54*(1), 49–53.

Tuleja, E. A. (2014). Developing cultural intelligence for global leadership through mindfulness. *Journal of Teaching in International Business, 25*(1), 5–24.

United Nations Department of Economics and Social Affairs. (2015). *17 sustainable development goals, 17 partnerships*. Retrieved from https://sustainabledevelopment.un.org/content/documents/211617%20Goals%2017%20Partnerships.pdf

Van De Mieroop, D., Clifton, J., & Verhelst, A. (2020). Investigating the interplay between formal and informal leaders in a shared leadership configuration: A multimodal conversation analytical study. *Human Relations, 73*(4), 490–515.

Van Dierendonck, D., Stam, D., Boersma, P., de Windt, N., & Alkema, J. (2014). Same difference? Exploring the differential mechanisms linking servant leadership and transformationalleadership to follower outcomes. *The Leadership Quarterly, 25*(3), 544–562.

Walker, M., Fredericks, B., Mills, K., & Anderson, D. (2014). "Yarning" as a method for community-based health research with Indigenous women: The Indigenous women's wellness research program. *Health Care for Women International, 35*(10), 1216–1226.

Walumbwa, F. O., Avolio, B. J., Gardner, W. L., Wernsing, T. S., & Peterson, S. J. (2008). Authentic Leadership: Development and Validation of a Theory-Based Measure. *Journal ofManagement, 34*(1), 89–126.

Warren, E., & Quine, J. (2013). A holistic approach to supporting the learning of young Indigenous students: One case study. *The Australian Journal of Indigenous Education, 42*, 12–23.

Winston, B., & Fields, D. (2015). Seeking and measuring the essential behaviours of servant leadership. *Leadership & Organization Development Journal, 36*(4), 413–434.

Younas, A., Wang, D., Javed, B., & Konte, A. G. (2020, January). *Inclusive leadership and change-oriented organizational citizenship behavior: role of psychological safety.* In Proceedings of the 2020 4th International Conference on Management Engineering, Software Engineering and Service Sciences (pp. 169–173).

Part II
Widening the Perspectives for Innovation: Theoretical Approaches and Empirical Research

Chapter 9
Positive and Negative Affects and Meaning at Work: Trait Emotional Intelligence as a Primary Prevention Resource in Organizations for Sustainable and Positive Human Capital Development

Annamaria Di Fabio and Maureen E. Kenny

9.1 Introduction

The twenty-first century is characterized by insecurity and continuous change in liquid societies (Bauman, 2000) with constant transitions growing exponentially (Blustein et al., 2019). These characteristics of the current world of work threaten the well-being of workers (Di Fabio & Kenny, 2016b). The Psychology of Working Framework (PWF; Blustein, 2006) and the Psychology of Working Theory (PWT; Duffy et al., 2016) highlight the contributions of decent work for individual, organizational, and societal well-being. The current unstable working context of the twenty-first century calls for building individual psychological resources, which can serve a protective psychological function and foster well-being.

The construct of well-being is itself multifaceted. The literature distinguishes between hedonic and eudaimonic well-being with hedonic well-being entailing the achievement of pleasure and the avoidance of pain (Watson et al., 1988). Affectively, well-being comprises both positive and negative affect (Watson et al., 1988) and a cognitive component related to the evaluation of satisfaction in one's life (Diener et al., 1985). Eudaimonic well-being refers to the attainment of meaning, flourishing, self-realization, and optimal functioning, focusing on building resources and strengths that go beyond the goals of pleasure attainment and pain avoidance (Ryan & Deci, 2001; Waterman et al., 2010). Three complementary models have been developed to define and explain eudaimonic well-being (Vázquez et al., 2006):

A. Di Fabio (✉)
Department of Education, Languages, Intercultures, Literatures and Psychology (Psychology Section), University of Florence, Florence, Italy
e-mail: annamaria.difabio@unifi.it

M. E. Kenny
Lynch School, Boston College, Boston, MA, USA

A. Di Fabio (ed.), *Cross-cultural Perspectives on Well-Being and Sustainability in Organizations*, https://doi.org/10.1007/978-3-030-86709-6_9

Ryff's multidimensional model of psychological well-being (Ryff, 1995), Seligman's model of the three ways to happiness (Seligman, 2002), and Deci and Ryan's motivational model of well-being (Deci & Ryan, 2000). Ryff's (1995) multidimensional model comprises six dimensions of well-being: self-acceptance, a positive attitude about oneself; positive relations with others in terms of the establishment of good relationships; autonomy as independence from others in thought and action; environmental mastery in terms of the perceived ability to manage environment to reach personal desires and values; purpose in life as the search for meaningful aims; and personal growth in terms of continued growth and personal development. Seligman's (2002) model contains three dimensions: pleasant life, relative to the appreciation of basic pleasures and relationships, natural environment, and bodily needs; good life, achieved through discovering one's own unique virtues and strengths and employing them creatively to enhance one's own life; and meaningful life, in which a deep sense of fulfilment is found by using one's own unique strengths for an aim greater than oneself. Finally, Deci and Ryan's (2000) motivational model considers healthy psychological functioning based on two dimensions: an adequate satisfaction of basic psychological needs (e.g., in terms of autonomy, relatedness, competence) and a system of intrinsic objectives in line with one's own interests and values.

Consistent with the Psychology of Working (Blustein, 2006; Duffy et al., 2016), work can be central to well-being, as work satisfies not only economic needs but also psychological needs for relationship, self-determination, social contribution, meaning, identity, and optimally to the human experience of fulfillment. Work may thus entail both hedonic and eudaimonic components of well-being. Satisfaction with work captures the cognitive aspect of hedonic well-being, reflecting people's cognitive evaluations of their work experience (Gagné & Houlfort, 2007). Hedonic well-being in the workplace also includes the experience of positive and negative affect as specifically related to the work context (Di Fabio & Gori, 2021). Work is also integral to eudaimonic well-being in its focus on meaning, fulfillment, identity, and social contribution.

Meaningful work is a multidimensional construct, which, according to Roberson (1990), includes the centrality of work, working values, and intrinsic orientation. May et al. (2004) underlined the importance of worthwhile, valuable, and meaningful work activities. The definition of meaningful work by Arnold et al. (2007) emphasizes "finding a purpose in work that is greater than the extrinsic outcomes of the work" (p. 195). Meaningful work has also been defined as including the experience of meaning, pride, engagement, and challenge in one's work (Britt et al., 2001), as well as good earnings and reputation (Oscós-Sánchez et al., 2008). An analysis of the literature thus suggests that the construct of meaningful work comprises various aspects: meaning attributed to tasks, pride, engagement, challenge, identity linked to the job role, centrality of work, work values, intrinsic work orientation, spirituality, good earnings, and reputation.

Aligned with an emphasis on eudaimonic well-being, Steger et al. (2012) proposed a multidimensional and comprehensive definition of the construct of meaningful work, encompassing the positive meaning associated with one's work and the perception that work gives extensive advantages for oneself and the nearby environment. Steger et al. (2012) developed the *Work and Meaning Inventory* (WAMI) to evaluate this definition. The WAMI assesses the following three dimensions: positive meaning at work, meaning making through work, and greater good motivation. Positive meaning at work entails the meaning and relevance that people positively ascribe to their work; meaning making focuses on the extent to which offers the opportunity to construct meaning in one's professional and personal life, contributing to personal growth and awareness, and the capacity to make meaningful sense of the world; and greater good motivation refers to a desire to introduce positive differences through one's work in the world and for others.

Although a relationship between hedonic and eudaimonic well-being is well established, the literature has been mixed on the nature of this relationship, with debate on whether positive affect determines the realization of life meaning or whether positive affect is a consequence of meaning. Fisher (2014) underlined that hedonic pleasure has been predicted by eudaimonic behaviors in many studies (Kashdan et al., 2008, Sheldon et al., 1996; Steyer et al., 2008; Waterman, 2008) but also notes that positive affect emerged as a determinant of meaning in other studies (King et al., 2006). With regard to the role of positive affect as a predictor, King et al. (2006) completed six studies examining the relationship of positive affect (PA) with meaning in life (MIL). In Study 1, PA emerged as the strongest predictor of MIL. Similar results emerged in Study 2 regarding the context of MIL and PA in a daily experience. Furthermore daily PA also predicted global MIL. Study 3 found no prospective relationship between measures of MIL and PA over a 2-year period. Study 4 and Study 5 demonstrated that positive mood concepts (PA) enhanced MIL. Study 6 distinguished between meaningful and meaningless activities and found that positive mood (PA) could predispose individuals to experience meaningful life, also increasing sensitivity to the meaning relevance of a situation. While existing research has examined the relationship between hedonic and eudaimonic well-being broadly and with specific attention to the relationship between affect meaning in life, the contribution of positive or negative affect to meaning in work or the psychological resources that influence the relationship between affect and work meaning have not been examined. As presented below, we propose that emotional intelligence may be one such psychological resource.

From a strength-based prevention (Di Fabio & Saklofske, 2021) and primary prevention perspective (Di Fabio & Kenny, 2016a, b; Hage et al., 2007; Kenny & Hage, 2009), building psychological resources and strengths through early intervention is important to enhance well-being and prevent distress (Di Fabio & Kenny, 2011, 2016a, b; Di Fabio & Saklosfke, 2014a, b). Emotional intelligence represents a promising strength-based factor since it has been amenable to specific trainings (Di Fabio & Kenny, 2011; Vesely et al., 2016) and has been associated with well-being and health (Austin et al., 2005; Di Fabio & Kenny, 2016b, 2019; Di Fabio & Saklofske, 2014b; Furnham & Petrides, 2003; Mikolajczak & Van Bellegem, 2017; Petrides et al., 2017). Two models of emotional intelligence are highlighted in the literature (Stough et al., 2009). The ability-based model focuses on the processing of emotional information (Mayer & Salovey, 1997), and the trait emotional intelligence models of Bar-On (1997) and Petrides and Furnham (2001) focus on the self-evaluation of one's own emotional abilities (Bar-On, 1997; Petrides & Furnham, 2001). The comprehensive trait emotional intelligence model of Petrides and Furnham (2001) defines trait emotional self-efficacy as a constellation of self-perceptions of emotions aligned with one's personality.

Both trait and ability-based models have been studied in relation to well-being and health. Furnham and Petrides (2003) examined trait emotional intelligence as a predictor of the affective and cognitive components of hedonic well-being in conjunction with measures of personality traits and cognitive ability among participants (mean age 19.79 years) in the United Kingdom. The results of a three-step hierarchical regression with trait emotional intelligence at the first step, cognitive ability at the second step, and Big Five personality traits as the third step showed that trait emotional intelligence contributed most to happiness explaining over 50% of variance. Furthermore, in the presence of trait emotional intelligence, neither cognitive skills nor personality traits were significantly associated with happiness. A further study (Mavroveli et al., 2007) investigated the relationship between trait EI and four distinct criteria (coping strategies, depressive thoughts, frequency of somatic complaints, peer nominations) among Dutch adolescents. The results of the study showed positive relationships of trait emotional intelligence with adaptive coping styles and negative associations with depressive thoughts and frequency of somatic complaints.

Di Fabio and Saklofske (2014b) assessed the relationship of life satisfaction as a component of hedonic well-being with both ability-based (Mayer et al., 2002) and trait (Bar-On, 1997; Petrides & Furnham, 2000) models of emotional intelligence, controlling for the effects of fluid intelligence and personality traits among Italian high school students. That study found that trait emotional intelligence, according to the Petrides and Furnham (2001) model, explained a greater percentage of variance in life satisfaction than trait emotional intelligence according to the Bar-On (1997) model, after controlling for the effects of fluid intelligence and personality traits. Ability-based emotional intelligence did not contribute significant variance in explaining life satisfaction.

In a subsequent study, Di Fabio and Kenny (2016b) extended the study by Di Fabio and Saklofske (2014b) adding positive and negative affect, in addition to life satisfaction, as indicators of hedonic well-being, and added meaning in life as an indicator of eudaimonic well-being. In this study (Di Fabio & Kenny, 2016b) of Italian high school students, trait emotional intelligence, according to the Petrides and Furnham (2000, 2001) model, explained a greater percentage of variance in both hedonic and eudaimonic well-being, assessed through the Italian version (Di Fabio, 2014) of the Meaningful Life Measure (MLM, Morgan & Farsides, 2009), than self-reported emotional intelligence according to Bar-On (1997), after controlling for the effects of fluid intelligence and personality traits. Also in this case, ability-based emotional intelligence did not contribute significant variance.

Di Fabio and Kenny (2019) next examined the relationships between trait emotional intelligence according to the Petrides and Furnham (2001) model and both hedonic well-being (positive affect, negative affect, and life satisfaction) and eudaimonic well-being (meaning in life and flourishing), controlling for the effects of personality traits among Italian university students. In this study flourishing was added as an indicator of eudaimonic well-being. The results showed that trait emotional intelligence explained a percentage of incremental variance beyond that accounted for by personality traits in relation to both hedonic and eudaimonic well-being. In particular the contribution of trait emotional intelligence was significant in explaining eudaimonic well-being, especially flourishing. Overall, the results of the studies reported above reveal that trait emotional intelligence according to the Petrides and Furnham (2001) model is associated with many facets of well-being and represents a psychological resource that could be fostered to promote well-being.

Emotional intelligence has also been studied as a mediator in the relationship between individual differences and health. For example, a study by Johnson et al. (2009) found that trait emotional intelligence mediated the relationship between the Big Five personality traits and overall health among university students at the University of Manchester. Additional individual difference factors have been found to mediate the relationship between well-being and health. Schutte (2014), for example, examined general self-efficacy as a mediator between increases in positive affect and positive change in work and relationship satisfaction and mental health in Australian and Canadian adults. The findings confirmed the mediating role of general self-efficacy in the relationships between positive affect and work and relationship satisfaction (as part of flourishing) and between positive affect and mental health.

The current study extends prior research examining the relationship between hedonic and eudemonic well-being and existing research examining emotional intelligence and well-being by assessing meaningful work as a dimension of eudaimonic well-being. The study also builds on the limited existing research examining emotional intelligence as a mediator of individual differences and well-being. Based on evidence that emotional intelligence can mediate the relationship between personality and health, we hypothesize that the individual difference variable of emotional intelligence could mediate the relationship between affect and eudaimonic well-being.

9.2 Aim and Hypotheses

Based on and extending the existing literature, the present study analyzed the relationships among positive affect and negative affect, meaningful work, and trait emotional intelligence, with a focus on the mediating role of trait emotional intelligence in the relationship between both positive affect and negative affect and meaningful work.

Specifically for the model with positive affect, we hypothesized (H1) that (1) positive affect will be positively correlated with both trait emotional intelligence and meaningful work; (2) trait emotional intelligence and meaningful work will be positively correlated; and (3) trait emotional intelligence will mediate the effect of positive affect on meaningful work.

For the model with negative affect, we hypothesized (H2) that (1) negative affect will be inversely correlated with both trait emotional intelligence and meaningful work; (2) trait emotional intelligence and meaningful work will be positively correlated; and (3) trait emotional intelligence will mediate the effect of negative affect on meaningful work.

9.3 Method

Participants

Participants were 222 workers from a multinational pharmaceutical company (63.5% men, 36.5% women). The mean age is 39.27 (*SD* = 7.12). Workers were recruited in their organization and were asked to participate in this research. They filled in the questionnaires voluntarily, and they did not receive any kind of reward.

Measures

Positive and Negative Affect Schedule (PANAS) To evaluate positive affect (PA) and negative affect (NA), the Positive and Negative Affect Schedule (PANAS, Watson et al., 1988) in the Italian version by Terraciano et al. (2003) was used. The PANAS includes 20 adjectives: 10 for positive affect (PA; e.g., determined, active, enthusiastic) and 10 for negative affect (NA; e.g., irritable, nervous, upset). The participant indicates how they generally feel on average on a Likert scale from 1 = very slightly or not at all to 5 = extremely. The Cronbach's alpha coefficients for the present study were .72 for positive affect and .83 for negative affect.

Trait EI Questionnaire Short Form (TEIQue-SF) To evaluate trait EI, the Trait EI Questionnaire Short Form (TEIQue-SF, Petrides, 2009) in the Italian version by Di Fabio and Palazzeschi (2011) was used. This questionnaire includes 30 items with a 7-point Likert scale response format (1= *completely disagree* to 7 = *completely agree*). Examples of items are as follows: "I feel that I have a number of good qualities" for well-being; "On the whole, I'm able to deal with stress" for self-control; "I often pause and think about my feelings" for emotionality; and "I'm usually able to influence the way other people feel" for sociability. The Cronbach's alpha coefficients in the present study were .80 for well-being, .81 for self-control, .83 for emotionality, and .82 for sociability.

Work as Meaning Inventory To evaluate work meaning, the *Work as Meaning Inventory* (WAMI, Steger et al., 2012) in the Italian version by Di Fabio (2018) was used. The WAMI assesses meaningful work in terms of the degree to which people find their work to have significance and purpose, the contribution work can provide in finding greater meaning in life, and the desire and means for one's work to make a positive contribution to the greater good. The instrument is composed of ten items with a 7-point Likert response format ranging from 1 = *strongly disagree* to 7 = *strongly agree* and represents three dimensions. Examples of items are as follows: "I have a good sense of what makes my job meaningful" for positive meaning; "I view my work as contributing to my personal growth" for meaning making through work; and "I know my work makes a positive difference in the world" for greater good motivations. The Cronbach's alpha coefficients for the current study were .82 for positive meaning, .79 for meaning making through work, .74 for greater good motivations, and .80 for the total score.

Procedure and Data Analysis

Trained psychologists administered the questionnaire to small groups of workers. With the aim of controlling possible effects of presentation of the instruments, the order of administration was counterbalanced. The research was carried out respecting the requirements of privacy and informed consent of Italian law. Descriptive statistics, Pearson's r correlations, and mediation analysis using the bootstrap method (Hayes, 2018) were effected. A simple mediation model to examine the effect of one independent variable (X) on a dependent variable (Y) through a mediator variable (M) was verified.

9.4 Results

Means, standard deviations, and correlations for the PANAS, TEIQue, and WAMI are shown in Table 9.1.

Table 9.1 Means, standard deviations, and correlations for the PANAS, TEIQue, and WAMI

	M	*SD*	1	2	3	4
1. PANAS PA	41.44	4.94	–			
2. PANAS NA	16.27	7.10	−.41**	–		
3. TEIQue Total	144.86	14.63	.33**	−.23**	–	
4. WAMI	54.23	7.38	.29**	−.27**	.51**	–

Note. $N = 222$. * < .05, ** < .01

The first mediation analysis used positive affect as independent variable, meaningful work as dependent variable, and TEIQue total as mediator.

Positive affect was directly associated with meaningful work and also indirectly associated with meaningful work through TEIQue total. As can be seen in Fig. 9.1, participants who had higher positive affect perceived themselves to have more emotional intelligence (a = .33) and, in turn, perceived themselves to have more meaningful work (b = .47). A bias-corrected bootstraps confidence interval for the indirect effect (ab = .16) based on 5000 bootstrap samples was entirely above zero (.09 to .43).

The effect of positive affect on meaningful work was reduced after controlling for emotional intelligence (TEIQue) but remained statistically significant (path c' in Fig. 9.1; $p < .001$): these results therefore indicated a partial mediation model with R^2 mediator effect size of .15.

A second mediation analysis was carried out, using negative affect as independent variable, meaningful work as dependent variable, and TEIQue total as mediator.

Negative affect was inversely associated with meaningful work and also indirectly associated with meaningful work through TEIQue total. As appears in Fig. 9.2, participants who had higher negative affect perceived themselves to have less emotional intelligence (a = −.23), but trait emotional intelligence seems able to reduce this effect (b = .47). A bias-corrected bootstraps confidence interval for the indirect effect (ab = −.11) based on 5000 bootstrap samples was entirely below zero (−.20 to −.04).

The effect of negative affect on meaningful work was reduced after controlling for emotional intelligence but remained statistically significant (path c' in Fig. 9.1; $p < .001$): these results therefore indicated a partial mediation model with R^2 mediator effect size of .12.

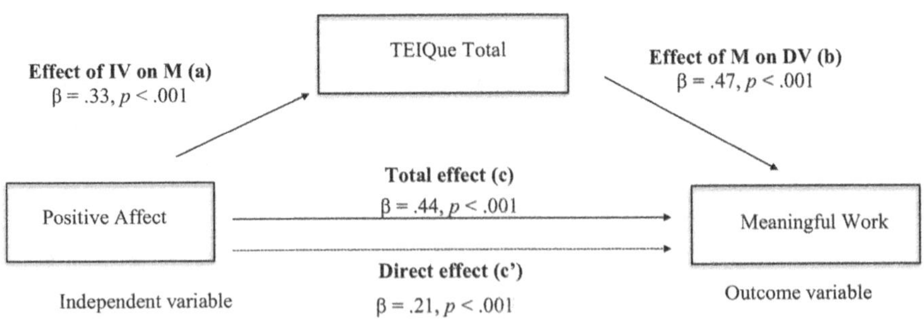

Fig. 9.1 Relationship between positive affect (PANAS) and meaningful work (WAMI) with trait emotional intelligence (TEIQue) total score as mediator. R^2 mediator effect size = .15

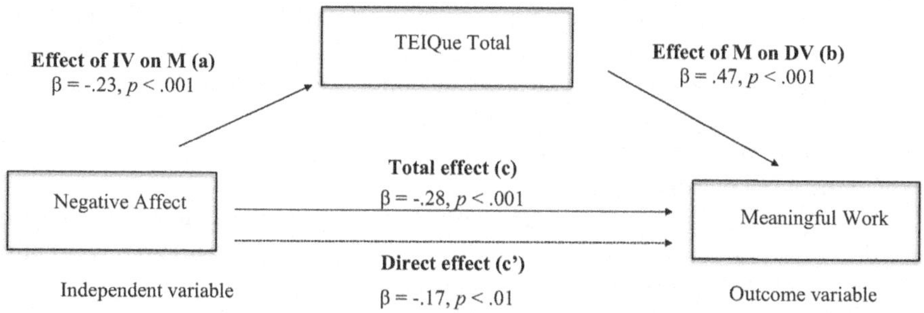

Fig. 9.2 Relationship between negative affect (PANAS) and meaningful work (WAMI) with trait emotional intelligence (TEIQue) total score as mediator. R^2 mediator effect size = .12

9.5 Discussion

This study examined, among Italian workers, the contribution of trait emotional intelligence as a mediator in the relationships between positive affect and meaningful work and also in the relationship between negative affect and meaningful work. The first hypothesis was confirmed. Individuals with higher positive affect report higher emotional intelligence and meaningful work. Moreover the contribution of positive affect on meaningful work was partially mediated by trait emotional intelligence. Participants with a higher level of positive affect were more likely to have greater trait emotional intelligence, which was associated with meaningful work.

The second hypothesis was also confirmed. Individuals with higher negative affect report lower meaningful work. Moreover negative affect appears to be negatively related to meaningful work, but trait emotional intelligence seems able to reduce this effect. Trait emotional intelligence seems to be a protective factor for meaningful work, because it reduces the effect of negative affect in undermining meaningful work.

The findings are consistent with prior research showing that components of hedonic and eudaimonic well-being are related but distinct (Di Fabio & Kenny, 2016b; Fisher, 2014). This study extended those findings with attention to meaningful work as an indicator of eudaimonic well-being. Although the current study does not directly address the debate in the literature as to whether hedonic well-being influence hedonic well-being or vice versa (Fisher, 2014; King et al., 2006), we sought to assess whether the positive and negative affect components of hedonic well-being contribute to work meaning through the mediator of trait emotional intelligence. Our findings confirm previously established relationships between trait EI and both hedonic and eudaimonic well-being (Austin et al., 2005; Di Fabio & Kenny, 2016b, 2019; Di Fabio & Saklofske, 2014b; Furnham & Petrides, 2003; Petrides et al., 2017). Also consistent with prior research (Di Fabio & Kenny, 2019), trait EI demonstrated a robust relationship with eudaimonic well-being in this study. The role of trait EI as a mediator (Johnson et al., 2009) between individual differences and health observed in prior research was extended in the current study by observing trait EI as a mediator between positive and negative affect and work meaning. The contribution of trait EI as a partial mediator suggests that EI can bolster the effects of positive affect on work meaning. Trait EI also mediates the relationship between negative affect and work meaning, indicating that higher levels of trait emotional intelligence can also reduce the adverse effects of negative affect in explaining work meaning.

These results are promising with regard to advancing understanding of the relationship between general well-being and well-being at work (Di Fabio & Kenny, 2016b; Fisher, 2014). The identification of trait EI as a factor that might mitigate the adverse effects of negative affect on work meaning and add to the positive contributions of positive affect suggests that the promotion of emotional intelligence may be an important focus for primary prevention and intervention, reinforced by evidence that EI is modifiable through intervention (Di Fabio & Kenny, 2011). Although the effects of positive and negative affect on work meaning remain after accounting for the mediating role of emotional intelligence, the contribution of affect is reduced and the role of EI is stronger than affect. The findings point to the contribution of EI, yet it is important to underline some limitations of this study. First of all, the participants were a small group of Italian workers, and thus the results of the present study cannot be generalizable. It is also important to note that this is a cross-sectional correlational study such that causal inference cannot be assumed, with regard to the direction or causality of observed associations. Longitudinal research and intervention studies are needed to assess whether affect contributes to work meaning or vice versa and whether the relationship between negative affect and work meaning can be modified through EI intervention. It is also important to note that all measures were self-report and thus deserve corroboration with behavioral and observational measures, where possible. The participants were Italian workers from a multinational pharmaceutical company, and the data collected in the Italian context were not representative of the different organizational contexts. In future research it could be important to examine the associations among these variables also in other organizational contexts. It could be also interesting to replicate this research in other countries. Other models of emotional intelligence, as, for example, the Bar-On model (Bar-On, 1997) and the ability-based emotional intelligence model (Mayer & Salovey, 1997), could be also used in future research.

9.6 Conclusions

If the results of the present study are confirmed through future research, evidence to support the potentiality of trait emotional intelligence in relation to meaningful work will be strengthened and serve as a foundation for further research and intervention. In a strength-based prevention perspective (Di Fabio & Saklofske, 2021) as well as in primary prevention (Di Fabio & Kenny, 2016a, b; Hage et al., 2007; Kenny & Hage, 2009), trait emotional intelligence has emerged as a particularly promising resource since it can be increased through specific training in school and

work settings (Di Fabio & Kenny, 2011; Vesely et al., 2016) distinct from personality traits, which tend to be more stable (Costa & McCrae, 1992). Trait emotional intelligence training could thus be implemented to foster emotional strengths for promoting the well-being of workers, permitting them to acquire psychological resources and deepen their sense of work meaning to navigate in the complexity and turbulence of the twenty-first century and thus foster healthy organizations (Di Fabio, 2017a; Di Fabio et al., 2020). Individuals with higher emotional intelligence may prosper not only as individuals but also as workers who experience meaning in their work lives and contribute to healthier organizations (Di Fabio, 2017a; Di Fabio et al., 2020). Emotional intelligence could also be considered as a valuable individual resource in relation to the psychology of sustainability and sustainable development (Di Fabio, 2017a, b; Di Fabio & Rosen, 2018) and to the 17 UNESCO Sustainable Development Goals. In this regard, we highlight UNESCO goal number three, good health and well-being, which is under threat in the current postmodern scenario characterized by continuous change, transition, insecurity, and numerous challenges.

References

Arnold, K. A., Turner, N., Barling, J., Kelloway, E. K., & McKee, M. C. (2007). Transformational leadership and psychological well-being: The mediating role of meaningful work. *Journal of Occupational Health Psychology, 12*, 193–203. https://doi.org/10.1037/1076-8998.12.3.193

Austin, E. J., Saklofske, D. H., & Egan, V. (2005). Personality, well-being and health correlates of trait emotional intelligence. *Personality and Individual Differences, 38*(3), 547–558. https://doi.org/10.1016/j.paid.2004.05.009

Bar-On, R. (1997). *The Emotional Intelligence Inventory (EQ-i): Technical manual*. Multi-Health Systems.

Bauman, Z. (2000). *Liquid modernity*. Polity Press.

Blustein, D. L. (2006). *The psychology of working: A new perspective for counseling, career development, and public policy*. Lawrence Erlbaum Associates.

Blustein, D. L., Kenny, M. E., Di Fabio, A., & Guichard, J. (2019). Expanding the impact of the psychology of working: Engaging psychology in the struggle for decent work and human rights. *Journal of Career Assessment, 27*(1), 3–28. https://doi.org/10.1177/1069072718774002

Britt, T. W., Adler, A. B., & Bartone, P. T. (2001). Deriving benefits from stressful events: The role of engagement in meaningful work and hardiness. *Journal of Occupational Health Psychology, 6*, 53–63. https://doi.org/10.1037/1076-8998.6.1.53

Costa, P. T., & McCrae, R. (1992). *Revised NEO personality inventory (NEO-PI-R) and NEO five factor Model (NEO-FFI) professional manual*. Psychological Assessment Center.

Deci, E. L., & Ryan, R. M. (2000). The "what" and "why" of goal pursuits: Human needs and the self-determination of behavior. *Psychological Inquiry, 11*(4), 227–268. https://doi.org/10.1207/S15327965PLI1104_01

Di Fabio, A. (2017a). Positive healthy organizations: Promoting well-being, meaningfulness, and sustainability in organizations. *Frontiers in Psychology. Organizational Psychology, 8*, 1938. https://doi.org/10.3389/fpsyg.2017.01938

Di Fabio, A. (2017b). The psychology of sustainability and sustainable development for well-being in organizations. *Frontiers in Psychology. Organizational Psychology, 8*, 1534. https://doi.org/10.3389/fpsyg.2017.01534

Di Fabio, A. (2018). The work and meaning inventory (WAMI): Primo contributo alla validazione della versione italiana. [The work and meaning inventory (WAMI): First contribution to the validation of the Italian version]. *Counseling. Giornale Italiano di Ricerca e Applicazioni, 11*(1). https://doi.org/10.14605/CS1111808

Di Fabio, A., & Kenny, M. E. (2011). Promoting emotional intelligence and career decision making among Italian high school students. *Journal of Career Assessment, 19*, 21–34. https://doi.org/10.1177/1069072710382530

Di Fabio, A., & Kenny, E. M. (2016a). From decent work to decent lives: Positive Self and Relational Management (PS&RM) in the twenty-first century. *Frontiers in Psychology. Section Organizational Psychology, 7*, 361. https://doi.org/10.3389/fpsyg.2016.00361

Di Fabio, A., & Kenny, M. E. (2016b). Promoting well-being: The contribution of emotional intelligence. *Frontiers in Psychology. Organizational Psychology, 7*, 1182. https://doi.org/10.3389/fpsyg.2016.01182

Diener, E. D., Emmons, R. A., Larsen, R. J., & Griffin, S. (1985). The satisfaction with life scale. *Journal of Personality Assessment, 49*(1), 71–75. https://doi.org/10.1207/s15327752jpa4901_13

Di Fabio, A., Cheung, F., & Peiró, J. -M. (2020). Editorial Special Issue Personality and individual differences and healthy organizations. *Personality and Individual Differences, 166*. https://doi.org/10.1016/j.paid.2020.110196

Di Fabio, A., & Gori, A. (2021). Positive and Negative Affect Schedule (PANAS) at work: Psychometric properties. Counseling. *Giornale Italiano di Ricerca e Applicazioni, 14*(2). https://doi.org/10.14605/CS1422107

Di Fabio, A., & Kenny, M. E. (2016c). Promoting well-being: The contribution of emotional intelligence. *Frontiers in Psychology. Organizational Psychology, 7*, 1182. https://doi.org/10.3389/fpsyg.2016.01182

Di Fabio, A., & Kenny, M. E. (2019). Resources for enhancing employee and organizational well-being beyond personality traits: The promise of emotional intelligence and positive relational management. *Personality and Individual Differences* (Special Issue Personality, Individual Differences and Healthy Organizations), *151*, https://doi.org/10.1016/j.paid.2019.02.022

Di Fabio, A., & Palazzeschi, L. (2011). Proprietà psicometriche del Trait EI Questionnaire Short Form (TEIQue-SF) nel contesto italiano [Psychometric properties of the *Trait EI Questionnaire Short Form* (TEIQue-SF) in the Italian context]. *Counseling. Giornale Italiano di Ricerca e Applicazioni, 4*, 327–336.

Di Fabio, A., & Saklofske, D. H. (2014a). Comparing ability and self-report trait emotional intelligence, fluid intelligence, and personality traits in career decision. *Personality and Individual Differences, 64*, 174–178. https://doi.org/10.1016/j.paid.2014.02.024

Di Fabio, A., & Saklofske, D. H. (2014b). Promoting individual resources: The challenge of trait emotional intelligence. *Personality and Individual Differences, 65*, 19–23. https://doi.org/10.1016/j.paid.2014.01.026

Di Fabio, A., & Saklofske, D. H. (2021). The relationship of compassion and self-compassion with personality and emotional intelligence. PAID 40th anniversary special issue. *Personality and Individual Differences, 157*. https://doi.org/10.1016/j.paid.2020.110109, https://doi.org/10.1016/j.jvb.2003.10.005

Di Fabio, A. (2014). Meaningful Life Measure: Primo contributo alla validazione della versione italiana [Meaningful Life Measure: First contribution to the validation of the Italian version]. *Counseling Giornale Italiano di Ricerca e Applicazioni, 7*, 307–315.

Di Fabio, A., & Rosen, M. A. (2018). Opening the Black Box of psychological processes in the science of sustainable development: A new frontier. *European Journal of Sustainable Development Research, 2*(4), 47. https://doi.org/10.20897/ejosdr/3933

Duffy, R. D., Blustein, D. L., Diemer, M. A., & Autin, K. L. (2016). The psychology of working theory. *Journal of Counseling Psychology, 63*(2), 127–148. https://doi.org/10.1037/cou0000140

Fisher, C. D. (2014). Conceptualizing and measuring wellbeing at work. In P. Y. Chen & C. L. Cooper (Eds.), *Wellbeing: A complete reference guide* (Vol. 3, pp. 10–35). John Wiley & Sons.

Furnham, A., & Petrides, K. V. (2003). Trait emotional intelligence and happiness. *Social Behavior and Personality: An International Journal, 31*(8), 815–823. https://doi.org/10.2224/sbp.2003.31.8.815

Gagné, M., & Houlfort, N. (2007). Validation of the satisfaction with work scale. *International Journal of Psychological Studies, 8*, 26–39.

Hage, S. M., Romano, J. L., Conyne, R. K., Kenny, M., Matthews, C., Schwartz, J. P., & Waldo, M. (2007). Best practice guidelines on prevention practice, research, training, and social advocacy for psychologists. *The Counseling Psychologist, 35*, 493–566. https://doi.org/10.1177/0011000006291411

Hayes, A. F. (2018). *Introduction to mediation, moderation, and conditional process. Analysis: A regression-based approach* (2nd ed.). Guilford Press.

Johnson, S. J., Batey, M., & Holdsworth, L. (2009). Personality and health: The mediating role of trait emotional intelligence and work locus of control. *Personality and Individual Differences, 47*(5), 470–475.

Kashdan, T. B., Biswas-Diener, R., & King, L. A. (2008). Reconsidering happiness: The costs of distinguishing between hedonics and eudaimonia. *The Journal of Positive Psychology, 3*(4), 219–233. https://doi.org/10.1080/17439760802303044

Kenny, M. E., & Hage, S. M. (2009). The next frontier: Prevention as an instrument of social justice. *The Journal of Primary Prevention, 30*(1), 1–10. https://doi.org/10.1007/s10935-008-0163-7

King, L. A., Hicks, J. A., Krull, J. L., & Del Gaiso, A. K. (2006). Positive affect and the experience of meaning in life. *Journal of Personality and Social Psychology, 90*(1), 179–196. https://doi.org/10.1037/0022-3514.90.1.179

Mavroveli, S., Petrides, K. V., Rieffe, C., & Bakker, F. (2007). Trait emotional intelligence, psychological well-being and peer-rated social competence in adolescence. *British Journal of Developmental Psychology, 25*(2), 263–275. https://doi.org/10.1348/026151006X118577

May, D. R., Gilson, R. L., & Harter, L. M. (2004). The psychological conditions of meaningfulness, safety and availability and the engagement of the human spirit at work. *Journal of Occupational and Organizational Psychology, 77*, 11–37. https://doi.org/10.1348/096317904322915892

Mayer, J. D., & Salovey, P. (1997). What is emotional intelligence? In P. Salovey & D. Sluyter (Eds.), *Emotional development and emotional intelligence: Educational implications* (pp. 3–31). Basic Books.

Mayer, J. D., Salovey, P., & Caruso, D. (2002). *Mayer–Salovey–Caruso Emotional Intelligence Test (MSCEIT)*. MHS.

Mikolajczak, M., & Van Bellegem, S. (2017). Increasing emotional intelligence to decrease healthcare expenditures: How profitable would it be? *Personality and Individual Differences, 116*, 343–347. https://doi.org/10.1016/j.paid.2017.05.014

Morgan, J., & Farsides, T. (2009). Psychometric evaluation of the meaningful life measure. *Journal of Happiness Studies, 10*(3), 351–366. https://doi.org/10.1007/s10902-008-9093-6

Oscós-Sánchez, M. A., Oscós-Flores, L. D., & Burge, S. K. (2008). The Teen Medical Academy: Using academic enhancement and instructional enrichment to address ethnic disparities in the American healthcare workforce. *Journal of Adolescent Health, 42*, 284–293. https://doi.org/10.1016/j.jadohealth.2007.08.023

Petrides, K. V. (2009). Psychometric properties of the trait emotional intelligence questionnaire (TEIQue). In C. Stough, D. H. Saklofske, & J. D. A. Parker (Eds.), *Assessing emotional intelligence* (pp. 85–101). Springer.

Petrides, K. V., & Furnham, A. (2000). On the dimensional structure of emotional intelligence. *Personality and Individual Differences, 29*(2), 313–320.

Petrides, K. V., & Furnham, A. (2001). Trait emotional intelligence: Psychometric investigation with reference to established trait taxonomies. *European Journal of Personality, 15*, 425–428.

Petrides, K. V., Gómez, M. G., & Pérez-González, J. C. (2017). Pathways into psychopathology: Modeling the effects of trait emotional intelligence, mindfulness, and irrational beliefs in a clinical sample. *Clinical Psychology & Psychotherapy, 24*(5), 1130–1141. https://doi.org/10.1002/cpp.2079

Roberson, L. (1990). Functions of work meanings in organizations: Work meanings and work motivation. In A. Brief & W. Nord (Eds.), *Meanings of occupational work: A collection of essays* (pp. 107–134). Lexington Books.
Ryan, R. M., & Deci, E. L. (2001). To be happy or to be self-fulfilled: A review of research on hedonic and eudaimonic well-being. In S. Fiske (Ed.), *Annual review of psychology* (Vol. 52, pp. 141–166). Annual Reviews.
Ryff, C. D. (1995). Psychological well-being in adult life. *Current Directions in Psychological Science, 4*(4), 99–104. https://doi.org/10.1111/1467-8721.ep10772395
Schutte, N. S. (2014). The broaden and build process: Positive affect, ratio of positive to negative affect and general self-efficacy. *The Journal of Positive Psychology, 9*(1), 66–74.
Seligman, M. E. P. (2002). Positive psychology, positive prevention, and positive therapy. In C. R. Snyder & S. J. Lopez (Eds.), *Handbook of positive psychology* (pp. 3–9). Oxford University Press.
Sheldon, K. M., Ryan, R., & Reis, H. T. (1996). What makes for a good day? Competence and autonomy in the day and in the person. *Personality and Social Psychology Bulletin, 22*(12), 1270–1279. https://doi.org/10.1177/01461672962212007
Steger, M. F., Dik, B. J., & Duffy, R. D. (2012). Measuring meaningful work: The work and meaning inventory (WAMI). *Journal of Career Assessment, 20*(3), 322–337. https://doi.org/10.1177/1069072711436160
Steyer, M. F., Kashdan, T. B., & Oishi, S. (2008). Being good by doing good: Daily eudaimonic activity and well-being. *Journal of Research in Personality, 42*, 22–42. https://doi.org/10.1016/j.jrp.2007.03.004
Stough, C., Saklofske, D., & Parker, J. (2009). *Assessing emotional intelligence: Theory, research, and applications*. Springer.
Terraciano, A., McCrae, R. R., & Costa, P. T., Jr. (2003). Factorial and construct validity of the Italian positive and negative affect schedule (PANAS). *European journal of psychological assessment, 19*(2), 131–141. https://doi.org/10.1027/1015-5759.19.2.131
Vázquez, C., Hervás, G., & Ho, S. (2006). Intervenciones clínicas basadas en la psicología positiva: fundamentos y aplicaciones [Clinical interventions based on positive psychology: Fundamentals and applications]. *Psicología Conductual, 14*(3), 401–432.
Vesely, A. K., Saklofske, D. H., & Nordstokke, D. W. (2016). Program evaluation: Building teacher resilience through emotional intelligence training. *Personality and Individual Differences, 100*(101), 523–524. https://doi.org/10.1016/j.paid.2016.05.336
Waterman, A. S. (2008). Reconsidering happiness: A eudaimonist's perspective. *Journal of Positive Psychology, 3*, 234–252. https://doi.org/10.1080/17439760802303002
Waterman, A. S., Schwartz, S. J., Zamboanga, B. L., Ravert, R. D., Williams, M. K., Bede Agocha, V., et al. (2010). The questionnaire for Eudaimonic well-being: Psychometric properties, demographic comparisons, and evidence of validity. *Journal of Positive Psychology, 5*, 41–61. https://doi.org/10.1080/17439760903435208
Watson, D., Clark, L. A., & Tellegen, A. (1988). Development and validation of brief measures of positive and negative affect: The PANAS scales. *Journal of Personality and Social Psychology, 54*(6), 1063–1070. https://doi.org/10.1037//0022-3514.54.6.1063

Chapter 10
Social Identity in the Organizational Field: New Challenges for Sustainable Management in Today's Organizations

Samuel Fernández-Salinero and Gabriela Topa

10.1 Social Identity and Its Relevance to Organizational Management

The research area of the psychology of sustainability and sustainable development (Di Fabio, 2017a, b; Di Fabio & Rosen, 2018) highlighted the value of sustainable management (Di Fabio & Peiró, 2018).

Sustainability is a new and very necessary optic that has grown along with organizational management. Even recent studies have stated that sustainability in organizations may be at least as important as financial independence (Moldavanova & Geordel, 2017). Sustainability refers to the need to include future generations in the current procedures and approaching challenges through long-term responses (Osborne et al., 2014; Stazyk et al., 2016).

One way for building sustainable organizations is to look for worldwide healthy and long-term measures that may help the population to grow and develop. Some recent researches have focused on psychological, social, and interpersonal variables, showing the importance of its management in today's organizations (Di Fabio & Peiró, 2018; Di Fabio & Tsuda, 2018; Di Fabio & Saklofske, 2019).

Social identity theory demands its paper in this positive and sustainable organization's research. Recent studies have demonstrated, in the framework of positive psychology, that identity is a crucial factor related to some health variables such as addictions or mental health recovery and improvement (Best et al., 2016).

S. Fernández-Salinero (✉)
Psychology Department, King Juan Carlos University, Madrid, Spain
e-mail: samuel.fernandez@urjc.es

G. Topa
Social and Organizational Psychology Department, The National Distance Education University, Madrid, Spain

A. Di Fabio (ed.), *Cross-cultural Perspectives on Well-Being and Sustainability in Organizations*, https://doi.org/10.1007/978-3-030-86709-6_10

Managing policies are changing nowadays because organizations are changing as well. Adaptation is one of the most important challenges that society must consider. Due to this, sustainability is one of the most interesting purposes that current research is reflecting. Current organizations have gained in complexity.

Nowadays, organizations are bigger and composed of multiple workgroups which are, in turn, even composed of subgroups and subcultures (Yubero & Morales, 2006). Therefore, managers should be aware of this phenomenon, to achieve successful management strategies. Moreover, organizations are characterized by more complex phenomena. Managers should consider the relationship between variables and evaluate their impact in organizational field.

We share the optic that organizations are complex social structures, and, in this chapter, we will try to shed light upon the social identity concept and its history, strengths, and applications. Besides, we will try, through the exposition of some recent research, to explain the role of social identity as a direct or indirect variable. We will analyze its relationship with some important outcomes such as performance, conflict, or interactional justice. This analysis is yielded in our recent research in the organizational field. Hence, a wider optic of the organizational field is purposed, where not only linear relations are appropriate but also indirect relations are interesting as well.

Social identity has demonstrated to be an important variable in the understanding of mental health, relational intelligence, addictions, job performance, and some other valuable variables on the organizational field. Besides, it has been proven that mental health may be fostered through social learning, social control, and social contact. By last, recent meta-analyses have shown that there is a positive relationship between a workgroup and organizational identification with health. This is a piece of interesting evidence, but some recent researches have shown that there are some conditions where identifying with the workgroup may be harmful for the organization as a whole (Yubero & Morales, 2006). Based on this, we hypothesized that there are at least two ways of identifying with the workgroup with different outcomes which may be evaluated deeper in the future. In the line with these assumptions, recent research has verified that organizational identification enhances intergroup consensus, whereas group identification reduces it (Porck et al., 2020).

Delving a little bit more into the literature results, it was found that the relationship between social identity and health is stronger for the indicator of the presence of well-being than the absence of stress and for psychological than physical health (Steffens et al., 2021). Not only organizational identification works as an independent variable, but also it has demonstrated to work as an important modulator in previous research. For instance, it has showed its influence as moderator in the relationship between psychological contract and organizational citizenship behavior (Tufan & Wendt, 2020). Moreover, other research has shown that organizational identification plays an interesting role moderating the relationship between ethical leadership and turnover intention (Suifan et al., 2020). Having said this, we are going to initiate the chapter introducing the social identity theory and its most important concepts and thereupon presenting three of our experiments and the most relevant derived statements.

10.2 Social Identity and Self-Categorization Theory

Social identity theory (SIT from now on) has been one of the most influencing frameworks on social psychology in the last decades (Scandroglio et al., 2008). Since every single organization is composed of persons, managers must be aware their social identity. Incorrect management of social identity may lead to some undesirable outcomes like absenteeism, distress, burnout, or psychological contract break.

In the early 1970s, Tajfel (1970) and partners described the *minimal group paradigm*. This paradigm explains that the emergence of even arbitrary social categories leads to social competitiveness and intragroup favoritism. On the cognitive level, for achieving a positive social identity, the individual tends to maximize positive intragroup characteristics and exogroup differences. Tajfel and Brown (1978) coined a new term to group the new ideas rising within this topic; "Social identity theory" was born. The focus was on the cognitive bases underlying the categorization processes which shape and define identity.

The idea at the heart of social identity theory is that some of the aspects of the identity come from belonging to certain groups or social categories (Tajfel, 1981). People will strain for achieving a valuable social identity. The belongingness to a highly rated group will enhance the individual's self-esteem. In the classical framework, Tajfel thought that social behavior varies in function of two opposite poles: *intergroup* and *interpersonal*. However, this seemed a very simple explanation. Hereafter, self-categorization theory (SCT from now on) was purposed in order to expand the framework of SIT. SCT asserted that identity was organized according to a hierarchical structure (Turner et al., 1987), where, in a subject, there may be many identities activated in the same situation.

Relying on group interaction, a subject may be categorized in a superordinate way. If a situation activates the higher rank identity, a person can be categorized as a human being. On the contrary, if a subject is categorized in the subordinated way, he may act as a single person. All the processes mentioned before depend on the prominence of the activated category. Prominence, in turn, depends on a specific situation. For example, a worker may be acting as a member of his workgroup or of the whole organization depending on the nature of the situation. When a category is activated, cognitive, affective, and behavioral patterns encompassed by that category are activated as well.

To explain which aspects of the situation may be involved in activating identity, optimal distinctiveness theory (Brewer, 2001) was used. This theory stated that subjects will try to achieve an optimal balance between differentiation and similarity toward others. Besides, uncertainty reduction theory (Berger, 1988) postulates that people acquire their behavioral norms from group identification.

Numerous further developments have their roots on these first researches from Tajfel et al., and social identity theory has been used in many studies from varied organizational fields. Examples of these are some researches on cohesion (Hogg, 1992), conformity and influence, stereotyping (Turner et al., 1994), prejudice

(Brown, 1995; Fernández-Salinero & Topa, 2019), or organizational conflict (Ashmore et al., 2001) among others.

Additionally, when people are identified with a certain group, some identity processes such as depersonalization appear. Depersonalization refers to a condition where a subject idiosyncrasy disappears in favor of group attributes. A subject could act even in a way contrary to personal tendencies.

Due to this, to achieve successful managing processes in current organizations, we must be aware of these factors. Since the organizational world is composed of multiple workgroups within the same organization, we can assume that there are some groups that may imply a better social identity and others which may lead to an unfavorable social identity. Realistic conflict theory principles have offered a model that may explain that belongingness to a low rated group may trigger competitive processes in order to solve this situation and achieving a better social identity (Fernández-Salinero & Topa, 2019).

When a social comparison process generates negative outcomes, subjects tend to react in order to recover their positive social identity (Tajfel, 1981). Tajfel posed two kinds of possible comparison processes: on the one hand, secure comparisons, when the status structure is perceived as legitimate, and, on the other hand, insecure comparisons when that structure is considered as illegitimate.

When a subject has to face to these comparisons, different coping strategies may appear. For example, regarding social mobility beliefs, subjects may try to change their reference group and migrate to a higher rated group. When mobility is not possible, creativity or social competition may appear. These two last processes are activated when group barriers are not permeable, and subjects cannot change their reference groups. In our current organizations, which are composed of multiple groups that constitute even subcultures (Yubero & Morales, 2006), several pieces of research have found that some groups hold a higher status than others. This condition may trigger some competitive processes that may add complexities to the organizational field. To manage this, we should be aware of social identity processes because a conflict may remain latent and affect important organizational variables such as well-being or productivity.

10.3 Social Identity as Mediator of Motivational Orientations and Organizational Citizenship Behavior

Migratory processes are a very common strategy for improving the subject's quality of life. People looking for new opportunities may leave their native countries and try to start a new life in a foreign country. Work is one of the most important needs that everybody must face. On this basis, people differ in the way of addressing new situations. Where identity processes become more prominent such as in a migration restrictive policy, people who are looking for new opportunities may feel stressed (Fernández-Salinero & Topa, 2019). When a subject is treated differently in the function of some group characteristics, we may be talking of discrimination. Apparently, this phenomenon may lead to a harming of job satisfaction or to a deterioration of job performance.

Recent researches have proven that according to individual motivational orientations, the perception of being discriminated and the identification with the workgroup have different effects.

When people belong to a socially discriminated group, it is related to high levels of stress. This inevitably entails personal dissatisfaction, self-rejection, and low self-esteem. Moreover, global and stable categories, such as nationality or skin color, do more harm to well-being than perceived prejudice toward some particular feature (Branscombe et al., 1999). But, when subjects perceive their self-esteem threatened, they will tend to cope feeling and acting differently. Recent research has shown the critical negative effects that social isolation generates (Haslam et al., 2016). Concretely, there is evidence to affirm that social isolation and group disconnection have profound negative effects on mental health.

As noted above, SCT posed that subject may identify with simultaneous hierarchical different categories, even if they are contradictory. In a racial hostile country, people may try to gain a positive social identity by identifying their selves with their workgroups. Workgroup membership may serve to define oneself as a desirable citizen and may lead to behave as socially expected. People invest their efforts in becoming a desirable group member. Regarding this, individuals who have a negative racial identity and a positive workgroup identity may feel some contradictory situations that lead to wide variations in their identities.

Motivational orientations theory refers to the individual's guide to the interpretation of demands (Elliot & Trash, 2002). Basically, there are two kinds of motivational orientations: learning orientation and performance orientation. The first one refers to a subject desire to learn. The second one refers to the desire to gain positive social judgments or to avoid negative social judgments. Performance-oriented subjects are more focused on social signals, and it has been proved that they are more concerned to pay attention to their social identity (Fernández-Salinero & Topa, 2019).

In the organizational field, recent research has shown that learning-oriented people are more productive and showed more job satisfaction, no matter their social identification. However, learning-oriented people showed more organizational citizenship behaviors when identifying with their workgroups. Nevertheless, people who are more susceptible to social judgments were more vulnerable to signs of prejudice and behave differently according to the context.

When assessing subjects who were focused on gaining positive social judgments, we found that they were more motivated to perform organizational citizenship behaviors. Besides, when social discrimination was perceived, they tended to invest even more effort in being productive. Interestingly, they tended to identify with their workgroup less than learning-oriented people. It seems that social stable categorization such as nationality drives the subject's behavior to divest himself of the social stigma. Not always organizational citizenship behaviors may be carried out based on healthy principles.

Deepening into this, people oriented to avoid social negative judgments tended to show fewer organizational citizenship behaviors. Moreover, when perceived discrimination was low, these subjects tended to avoid organizational citizenship behaviors, trying to pass unnoticed. It may be necessary to evaluate, using longitudinal studies, the long-term impact that this way of coping may produce.

That said, social identification modulates the relation between personal characteristics and performance. This is a very important aspect in organizational management, not only because of the direct, single, and undeniable effect that organizational identification has on organizational performance but also because of the indirect effect. If managers do not pay attention to this, even potential good workers may be impaired because of the poor management of their social identity.

Notwithstanding, we must understand that not everybody responds to a negative social identity similarly. It seems clear that people oriented to evaluate other's judgments will tend to respond differently to contextual events, and human resources managers should be aware of this feature to better job design.

10.4 When Discrimination Occurs Inside the Same Organization: The Role of Organizational Identity

In the preceding epigraph, we have reflected when discrimination occurs based on nationality or racial stable attributes. In this epigraph, we are going to analyze discrimination within the same organization, specifically between workgroups, and the outcomes of this situation. As referred above, great inner complexity characterizes current organizations. Within these new organizations, subjects may identify stronger with their workgroups than with the organization as a whole (Hogg & Terry, 2000). This division may generate inner social circumstances which may lead to undesirable outcomes, such as conflictive intergroup situations (Bilbao & Dauder, 2004).

When there are different workgroups within the same organization, there may occur the perception that a subject is being treated in a different or unfair way according to their group membership (Sanchez & Brock, 1996). When this happens, and based on the SCT, subjects may be identifying themselves with their workgroup and perceiving the whole organization as an exogroup (Fernández-Salinero & Topa, 2019). Usually, the conflict appears when an individual or group perceives that the other part has frustrated or is about to frustrate some concern of his. Obviously, when there is the perception of a privileged group, social comparative processes may appear shaping the conflict. This may lead to some undesirable consequences such as a decrease in productivity or cohesion (Alcover et al., 2004).

Moreover, when inequality is perceived, compensatory processes may appear, and subjects may opt to adopt competitive patterns in order to achieve a better valued social identity. The classical resolution for this situation is trying to promote a higher level of self-categorization, such as identifying with the organization as a whole. Furthermore, when social mobility from the devalued to the desirable group is not possible, individuals must strive for achieving a successful social identity, trying to overpass the exogroup in some desirable characteristics. What is more, when in an organization the power distance is short, intergroup conflict may appear easily.

But, is there always a problem when the subject identifies with his workgroup instead of the whole organization? Recent research has shown that when discrimination is perceived, it leads to intergroup conflict easily. In general terms, organizational identification leads to a higher workgroup identification which drives to lower intergroup conflict values. But, when discrimination is perceived and organizational identification is high, group identification tends to decrease, decreasing intergroup conflict as well (Fernández-Salinero & Topa, 2019).

One possible explanation is that there may be numerous ways of identifying with workgroups. When division and devalued social identity is activated, it may drive to polarized group identification which may be harmful to organizational dynamics. Identifying only with the workgroup is not harmful for the organization; moreover, it may be helpful. The problem appears when conflict causes group division and compete process is activated. When this happens, the workgroup will tend to perceive the organization as an exogroup, activating cognitive biases as well, such as group homogeneity and intergroup favoritism.

10.5 Social Identification and Its Influence on the Relationship Between Perceived Conflict and Interactional Justice

As stated in the epigraph above, conflict is a constant in the organizational world. Due to this, organizations must pay attention to conflict dynamics and try to achieve a deep understanding of its complexity, to cope with it more effectively. When in the same organization, interactional conflict is triggered, it may elicit interpersonal and emotional responses such as loss of focus and hindering decision-making processes (Fernández-Salinero & Topa, 2019).

Organizational justice is a key variable of successful project delivery as well. When workers perceive justice, it has been related to more efficient problem-solving strategies, higher productivity, and respecting others leading to healthier social relationships. Following reciprocity theory, subjects make cognitive evaluations about their contributions to their organization and the outcomes they receive. Specifically, interactional justice refers to the way that workers perceive they are treated during the organization's decision-making process. This kind of justice has been referred to as the quality of interpersonal interaction between employees and employers.

As some previous research showed (Rani et al., 2012), when people are perceiving a conflict between the workgroups, perceived interactional justice will decrease, but, at this point, what would be the role of group identification? Will accentuate the polarity? Or maybe help to approach healthier problem-solving procedures?

Taking perceived conflict into account, we tested the effects of this variable over the perception of interactional justice, being this relationship mediated by group identification. To achieve a better understanding of these models considered the influence of job satisfaction between perceived conflict and group identification. As some research showed, a better emotional state, such as satisfaction, is related to healthier interactions with workmates (Orgambídez & Almeida, 2020).

Besides, the exclusive identification with workgroups has been related to reduced well-being and emotional exhaustion (Vogus et al., 2020). It is interesting to question how would be the interaction between workgroup identification and job satisfaction? To complete the complex optic, when perceived conflict is high, desirable categories are more salient, and competitive processes are triggered for achieving the best valued social identity (Tajfel, 1981).

Not surprisingly, perceived conflict was related to lower levels of interactional justice. However, group identification was related to higher levels of perceived interactional justice. Trying to go deeper into this, we see in our results that perceived conflict may lead to group identification, while job satisfaction may lead to another kind of group identification. Besides, interactional justice may be influenced by good relationships inside the workgroups. Despite this, it is obvious that there are different affective factors driving this group's social motivations.

Looking at the results, we see that when not taking conflict into account, job satisfaction and group identification were strongly and positively linked to the perception of interactional justice. On the other hand, when introducing perceived conflict in the model, group identification was related to lower levels of interactional justice. When conflict appears, people are motivated to keep their positive social identity, but when identity is in trouble, subjects will strive and fight in order to overpass the exogroup. We can imagine the change and the negative outcomes over the interactional processes and over the whole organization if this happens. Having said this, the conflict must be tackled from the very beginning, avoiding that way future difficulties.

10.6 Conclusions

Social identity theory is a very influencing theory in the organizational field. Even though it is a theory with more than 50 years, it still stimulates new developments. Nowadays, for approaching organizations, we shall adopt complex optics where variables interact between them and react differently as a function of the context. As mentioned above, organizational identification can be considered as an independent variable related to interesting organizational outcomes such as performance (Riketta, 2005; Van Knippenberg, 2000; van Knippenberg & van Schie, 2000. Cited in He & Brown, 2013). On the other hand, organizational identification has been widely applied as a modulator variable in different organizational models (Peña-Gonzalez et al., 2021).

Social identity theory provides an interesting optic for managers, and recent publications state that it covers a huge angle of organizational psychological and behavioral phenomena associated with group membership, status, and individual motivation (van Knippenberg, 2020). As discussed in the chapter, it is related to some desirable and undesirable outcomes in the organizational field. Besides, we have discussed and concluded the great importance of not forgetting to manage it and to avoid emerging negative stereotypes and enhance healthier organizations (Zagencyk et al., 2020).

In this chapter, we have explained social identity theory and presented it as a very valuable modulator factor and an inestimable help in promoting healthy organizations. Today, managers are more concerned with promoting healthier and sustainable organizations, and managing social identity processes has become a need to this end. However, as recent research shows, organizational identification should show a dark side related to unethical pro-organizational behavior and pro-social rule breaking (Irshad & Bashir, 2020). This statement shows the necessity of more research that contributes to clarify the conditions that contribute to healthy or damaging organizations.

As current organizations are changing, managing policies must change as well; in this line, social identity theory may contribute to develop organizational applied policies. The psychology of sustainability and sustainable development (Di Fabio, 2017a, b; Di Fabio & Rosen, 2018) represents the framework for a very interesting alternative to pay attention to future generations and long-term organizational measures. The research area of the psychology of sustainability and sustainable development underlines the value of sustainable management approaches (Di Fabio & Peiró, 2018) to promote flourishing workers and healthy organizations (Di Fabio et al., 2020). Furthermore, we should invest efforts in this perspective developing cross-cultural human resources practices oriented to promote health and sustainability in organizations all around the world.

We must think about building the next healthy organizations today. Social identity is a very important variable that underlies lots of organizational dynamics, as we have seen in this chapter. As recent research has stated, we must think in new generation organizations within the optic of emotional intelligence and positive relational improvements (Di Fabio & Saklofske, 2019). In short, we should develop new comprehensive and complex models that try to shed light on this complexity and defiant social reality in the perspective of the psychology of sustainability and sustainable development (Di Fabio, 2017a, b; Di Fabio & Rosen, 2018).

References

Alcover, C. M., Iñigo, D., Rodríguez Mazo, F., & Bilbao, R. (2004). *Introducción a la Psicología del trabajo* (Introduction to work psychology). McGraw-Hill.

Ashmore, R. D., Jussim, L., Wilder, D., & Heppen, J. (2001). Toward a social identity framework for intergroup conflict. Social identity, intergroup conflict, and conflict reduction, 213–249.

Berger, C. R. (1988). Uncertainty and information exchange in developing relationships. In S. Duck, D. F. Hay, S. E. Hobfoll, W. Ickes, & B. M. Montgomery (Eds.), Handbook of personal relationships: Theory, research and interventions (pp. 239–255). John Wiley & Sons.

Best, D., Beckwith, M., Haslam, C., Alexander Haslam, S., Jetten, J., Mawson, E., & Lubman, D. I. (2016). Overcoming alcohol and other drug addiction as a process of social identity transition: The social identity model of recovery (SIMOR). *Addiction Research & Theory, 24*(2), 111–123.

Bilbao, R., & Dauder, S. (2004). *Introducción a la teoría del conflicto en las organizaciones* (Introduction to organizational conflict theory). Universidad Rey Juan Carlos.

Branscombe, N. R., Ellemers, N., Spears, R., & Doosje, B. (1999). The context and content of social identity threat. In N. Ellemers, R. Spears, & B. Doosje (Eds.), *Social identity: Context, commitment, content*. Blackwell.

Brewer, M. B. (2001). Ingroup identification and intergroup conflict: When does ingroup love become outgroup hate? In R. D. Ashmore, L. Jussim, & D. Wilder (Eds.), *Social identity, intergroup conflict and conflict reduction* (pp. 17–41). Oxford University Press.

Brown, R. J. (1995). *Prejudice in Social Psychology*. Blackwell.

Di Fabio, A. (2017a). Positive Healthy Organizations: Promoting well-being, meaningfulness, and sustainability in organizations. Research Topic in *Frontiers in Psychology. Organizational Psychology, 8*, 1938.

Di Fabio, A. (2017b). The psychology of sustainability and sustainable development for well-being in organizations. Research Topic in *Frontiers in Psychology. Organizational Psychology, 8, 1534*.

Di Fabio, A., & Peiró, J. M. (2018). Human Capital Sustainability Leadership to promote sustainable development and healthy organizations: A new scale. In A. Di Fabio (Ed.) Psychology of Sustainability and Sustainable Development. Special Issue in *Sustainability* MDPI, *10*(7), 2413.

Di Fabio, A., & Rosen, M. A. (2018). Opening the black box of psychological processes in the science of sustainable development: A new frontier. *European Journal of Sustainable Development Research, 2*(4), 47.

Di Fabio, A., & Saklofske, D. H. (2019). Positive relational management for sustainable development: Beyond personality traits—The contribution of emotional intelligence. *Sustainability, 11*(2), 330.

Di Fabio, A., & Tsuda, A. (2018). The psychology of harmony and harmonization: Advancing the perspectives for the psychology of sustainability and sustainable development. *Sustainability, 10*(12), 4726.

Di Fabio, A., Cheung, F., & Peiró, J. M. (2020). Editorial Special Issue Personality and individual differences and healthy organizations. *Personality and Individual Differences, 166*.

Elliot, A., & Trash, T. (2002). Approach-avoidance motivation in personality: Approach and avoidance temperaments and goals. *Journal of Personality and Social Psychology, 82*(5), 804–818.

Fernández-Salinero San Martín, S., & Topa, G. (2019). Motivational orientations and organizational citizenship behaviors: The moderator role of perceived discrimination in the Brexit context. *Behavioral Sciences, 9*(3), 31.

Haslam, C., Cruwys, T., Haslam, S. A., Dingle, G., & Chang, M. X. L. (2016). Groups 4 Health: Evidence that a social-identity intervention that builds and strengthens social group membership improves mental health. *Journal of affective disorders, 194*, 188–195.

He, H., & Brown, A. D. (2013). Organizational identity and organizational identification: A review of the literature and suggestions for future research. *Group & Organization Management, 38*(1), 3–35.

Hogg, M. A. (1992). *The social psychology of group cohesiveness: From attraction to social identity*. Harvester Wheatsheaf and New York University Press.

Hogg, Y, M., & Terry, D. J. (2000). Social identity and self-categorization processes in organizational context. Academy of Management Review, 25(1).

Irshad, M., & Bashir, S. (2020). The dark side of organizational identification: A multi-study investigation of negative outcomes. *Frontiers in Psychology, 11*, 2521.

Moldavanova, A., & Goerdel, H. (2017). Understanding the puzzle of organizational sustainability: Toward a conceptual framework of organizational social connectedness and sustainability. *Public Management Review, 20.*

Orgambídez, A., & Almeida, H. (2020). Social support, role clarity and job satisfaction: a successful combination for nurses. *International Nursing Review, 67*(3), 380–386.

Osborne, S. P., Radnor, Z., Vidal, I., & Kinder, T. (2014). A sustainable business model for public service organizations? *British Journal of Management, 0*, 1–15.

Peña-González, K., Nazar, G., & Alcover, C. M. (2021). The mediating role of organizational identification in the relation between organizational social capital, perceived organizational prestige, perceived employability and career satisfaction. *The Spanish Journal of Psychology, 24.*

Porck, J. P., van Knippenberg, D., Tarakci, M., Ateş, N. Y., Groenen, P. J., & de Haas, M. (2020). Do group and organizational identification help or hurt intergroup strategic consensus? *Journal of Management, 46*(2), 234–260.

Rani, R., Kumar, N., Rastogi, R., & Garg, P. (2012). Quality of work life: Predictor of psychological wellbeing of police employees. *Indian Journal of Positive Psychology, 3*(4), 356–364.

Riketta, M. (2005). Organizational identification: A meta-analysis. *Journal of Vocational Behavior, 66*(2), 358–384.

Sanchez, J., & Brock, P. (1996). Outcomes of perceived discrimination among Hispanic employees: Is diversity management a luxury or a necessity? *Academy of Management Journal, 39*, 704–719.

Scandroglio, B., López Martínez, J. S., & San José Sebastián, M. (2008). La teoría de la Identidad Social. Una síntesis crítica de sus fundamentos, evidencias y controversias. [Social Identity Theory, a critical synthesis of its foundations, evidence and controversies]. *Psicothema, 20*(1), 80–89.

Stazyk, E. C., Moldavanova, A., & Frederickson, H. G. (2016). Sustainability, intergenerational social equity, and the socially responsible organization. *Administration & Society, 48*(6), 655–682.

Steffens, N. K., LaRue, C. J., Haslam, C., Walter, Z. C., Cruwys, T., Munt, K. A., Haslam, S. A., Jetten, J., & Tarrant, M. (2021). Social identification-building interventions to improve health: A systematic review and meta-analysis. *Health psychology review, 15*(1), 85–112.

Suifan, T. S., Diab, H., Alhyari, S., & Sweis, R. J. (2020). Does ethical leadership reduce turnover intention? The mediating effects of psychological empowerment and organizational identification. *Journal of Human Behavior in the Social Environment, 30*(4), 410–428.

Tajfel, H. (1970). Experiments in intergroup discrimination. *Scientific American, 223*, 96–102.

Tajfel, H. (1981). *Grupos humanos y categorías Sociales*. [Human Groups and Social Categories]. Herder.

Tufan, P., & Wendt, H. (2020). Organizational identification as a mediator for the effects of psychological contract breaches on organizational citizenship behavior: Insights from the perspective of ethnic minority employees. *European Management Journal, 38*(1), 179–190.

Turner, J. C., Hogg, M. A., Oakes, P. J., Reicher, S. D., & Wetherell, M. S. (1987). *Rediscovering the social group: A self-categorization theory*. Basil Blackwell.

Turner, J. C., & Brown, R. (1978). Social status, cognitive alternatives, and intergroup relations. In H. Tajfel (Ed.), *Differentiation between social groups: Studies in the social psychology of intergroup relations*. Academic Press.

Turner, J. C., Oakes, P. J., Haslam, S. A., & McGarty, C. (1994). Self and Collective: Cognition and Social Context. Personality and Social Psychology Bulletin, 20(5), 454–463. https://doi.org/10.1177/0146167294205002

Van Knippenberg, A. (2020). Strategies of identity management. In *Ethnic minorities* (pp. 59–76). Garland Science.

van Knippenberg, D. (2000). Work motivation and performance: A social identity perspective. *Applied Psychology, 49*(3), 357–371.

van Knippenberg, D., & van Schie, E. C. M. (2000). Foci and correlates of organizational identification. *Journal of Occupational and Organizational Psychology, 73*, 137–147.

Vogus, T. J., Ramanujam, R., Novikov, Z., Venkataramani, V., & Tangirala, S. (2020). Adverse events and burnout: The moderating effects of workgroup identification and safety climate. *Medical care, 58*(7), 594–600.

Yubero, S., & Morales, J. F. (2006). La identificación en las organizaciones y su relación con la prototipicidad y el conflicto entre grupos profesionales. [Identification within organizations and its relationships with prototyping]. *Psicothema, 18*(3), 400–406.

Zagenczyk, T. J., Purvis, R. L., Cruz, K. S., Thoroughgood, C. N., & Sawyer, K. B. (2020). Context and social exchange: Perceived ethical climate strengthens the relationships between perceived organizational support and organizational identification and commitment. *The International Journal of Human Resource Management*, 1–20.

Chapter 11
Actual Situation of Primipara's Sleep Acquired by the Mobile Sleep Diary for One Month After Giving Birth in Japan: Contributions to Sustainable Parenting

Nami Fujoka, Chikako Hata, and Akira Tsuda

11.1 Introduction

The Psychology of Sustainability and Sustainable Development (Di Fabio, 2017a, b; Di Fabio and Rosen, 2018) aims at opening the black box of psychological processes, enlarging the concept of sustainability and sustainable development. In this framework, the Psychology of Harmonization (Di Fabio & Tsuda, 2018) constitutes a fundamental theoretical point of reference. The Psychology of Harmonization (Di Fabio & Tsuda, 2018) recognized both geographical and temporal directions, including meaningful construction processes from the past, to the present, toward the future, using reflexivity processes at individual, group, intergroup, social, community, organization, interorganization, national, and cross-national levels. The Psychology of Harmonization also underlines the value of taking care of a harmonic recomposition on many levels of internal and external complexity to promote health and well-being in different environments. The study is situated in this framework focusing on the issue of primipara's sleep for health and well-being.

In Japan, population of aged primipara has been increasing due to late marriage and overdue birth and moreover by development of assisted reproductive technology, and mothers' average ages at the time of the first childbirth have exceed 30 years old and been in uptrend (Ministry of Health Labor and Welfare, 2016). Women with late childbearing bear their children under various risks from the pregnant period with aging as a factor (Biro et al., 2012). After childbearing, they start

N. Fujoka (✉) · C. Hata
Kwassui Women's University, Nagasaki, Japan
e-mail: fujioka73@kwassui.ac.jp

A. Tsuda
Kurume University, Kurume, Japan

A. Di Fabio (ed.), *Cross-cultural Perspectives on Well-Being and Sustainability in Organizations*, https://doi.org/10.1007/978-3-030-86709-6_11

learning parenting technique such as breastfeeding before they recover from fatigue. Moreover, mothers are forced to divide their sleep for parenting.

In general, there is an established correlation between sleep and mental health, and poor sleep is correlated with poor mental health (Goyal et al., 2007). It was reported that sleep time in the early puerperium was shorter than that at the end of pregnancy and arousal time after falling sleep increased (Shinkoda et al., 2001). Disturbance of sleep pattern and shortening of sleep time are outstanding particularly in primiparas. The maternity blue symptom strongly develops after discharge, and it may shift to mental disorders (Matsumoto et al., 2011). Results of a large-scale survey in Japan revealed that most of postpartum women received support for housekeeping, while on the other hand 73% of primiparas and 56% of parous women complained about fatigue caused by lack of sleep (Shimada et al., 2001).

In particular, all is the first experience for primiparas. Incidence of depression after childbirth due to "alteration of sleep," "anxiety" (Mizukami et al., 1994), and "accumulation of fatigue and stress" (Dennis & McQueen 2009; Bozoky & Corwin, 2002) and parenting stress have become considered problematic, indicating that mental health care after childbearing is needed.

However, there have been only few studies that focused on actual situation of postpartum women's sleep secured by dividing it and on factors to inhibit and promote their sleep.

In the case of elderly puerperal women, they cause sleep disorder when getting closer to menopause. Women around the age of 40 experience mood and sleep problems due to a decline in estrogen production. Moreover, they cannot sleep deeply due to "rush of blood to the head" and "diaphoresis/palpitation" (Shibui, 2005). It was reported that 42% of the climacteric women were insomniac (Owens & Matthews, 1998). Based on previous researches, it is important to investigate the actual state of sleep and mental health of elderly first-time mothers and to consider countermeasures.

Based on the above, we investigated the actual sleep status and mental health status of elderly first-time mothers. It is important to use the results of this study as an aid to countermeasures.

Therefore, the authors have developed a mobile sleep diary (E-Mom Sleep Diary: EMSD) that allows timely survey with a smartphone. This paper reports actual situation of over 35 postpartum women's sleep during 1 month after giving birth clarified by the diary developed this time.

11.2 Methods

Participants

Participants of this study were primiparas who were 35 years or older and agreed to participate in the study. They started brooding behavior at the next day of childbearing. And, they had smartphones and were able to participate in EMSD online.

Measures

We developed the sleep diary (EMSD) to grasp the hypnic actual situations of women's sleep 1 month after giving birth. The EMSD was constructed based on references and previous studies. Participants filled in the EMSD using their smartphones. Before participating in the EMSD, we surveyed these women on the following topics. The survey included age, marital status, occupation, fertility status, mode of delivery, and total daily sleep time during pregnancy and after delivery. These surveys were conducted using the questionnaire.

EMSD was constituted so that it was introduced 5 days after giving birth by QR code, and the participants filled in the sleep diary 1 week, 2 weeks, 3 weeks, and 1 month after giving birth. Furthermore, in order to grasp their thought after giving birth, a communication tool useful for the postpartum women was set.

From 5 days to 1 month after giving birth, the participants' actual sleep situation (total sleep time of the day, main sleep time, feeling of sound sleep) were investigated by EMSD using their smartphones. Participants will be discharged after participating in EMDS on the 5th postpartum day. Since then, they have used EMSD at home.

The feeling of sound sleep was rated by 4-point Likert scale (1 pt, no feeling of sound sleep at all; 2 pts, not much feeling of sound sleep; 3 pts, fairly good feeling of sound sleep; 4 pts, very good feeling of sound sleep). Furthermore, promotional factors and constraints for sleep were surveyed by free description.

Procedure

First, we obtained consent for research cooperation from the head of the obstetrics and gynecology facility to conduct this study. Next, first-time mothers aged 35 years or older who gave birth at a hospital consented to the research collaboration. We invited participants who were 5 days postpartum to participate in the EMSD.

The survey period was from June 2017 to December 2018.

They are requested by email to record on EMSD 5 days, 1 week, 2 weeks, 3 weeks, and 1 month after the delivery. At the same time, from the first day after birth to the first month, the participant was wearing an activity meter which measured sleep-wake rhythms.

Data Analysis

Descriptive statistics values were calculated for basic attributes. Friedman test was performed for temporal changes in sleep time and feeling of sound sleep, and then subsequent tests were performed. The free descriptions about promotional factors and constraints for sleep in each period were encoded, and their contents were analyzed temporally according to similarity of the meaning. The statistics software SPSS ver.23 was used for the analysis with the significant level of 5%.

The free descriptions about promotional factors and constraints for sleep in each period were encoded, and their contents were analyzed temporally according to similarity of the meaning.

Ethical Consideration

This study underwent ethical review by the organization that the researchers belong to and a research agency and was approved by them (Approval number 16–018) prior to its execution. The purpose of the study was explained to the participants orally and in document, and they were requested to participate in the study based on free will. Their consent was obtained by signature. There is no conflict of interests.

11.3 Results

Background and Actual of the Sleep Characteristics in Maternity Cycle

The participants of this study were 80 primiparas who were 35 years or older, with average age of 36.7+−2.0 years old and age of marriage of 34.4 +− 2.6 years old. The 52 (65.0%) participants had occupation. 24 participants (30.0%) became pregnant by infertility treatment, and 44 (55.0%), more than a half of them, gave birth by Caesarean section (Table 11.1). The participant had been hospitalized until 5 days after delivery and had been at home since then. The transition of the participant's total sleep time and main sleep time during 1 month after childbirth is shown in Fig. 11.1. The average total sleep time were 6.7+−0.9 h at non-pregnant state, 7.0+−1.6 h during pregnancy, 5.5+−2.1 h 1 day after giving birth, 5.1+−1.6 h 3 days after giving birth, 4.8+−1.2 h 5 days after giving birth, 5.2+−1.5 h 1 week after giving birth, 5.7+−1.5 h 2 weeks after giving birth, 5.7+−1.5 h 3 weeks after giving birth, and 5.9+−1.5 h 1 month after giving birth, respectively. Significant difference was recognized in temporal change (p<.001), at non-pregnant state and all periods after giving birth (p<.001 for all), and total sleep time were shorter than those of non-pregnant state.

The participants with a total sleep time of less than 5 h on each survey day were as follows: 4 (5.0%) of non-pregnant, 26 (32.5%) of participants on the 1st postpartum, and 35 (43.8%) of participants on the 3rd postpartum. Among them, 31 (38.8%), 27 (33.8%) of the participants in the 1 week postpartum, 16 (20.0%) in the 2 weeks postpartum, and 17 (21.3%) in the 3 weeks postpartum and 16 (20.0%) of

Table 11.1 Basic attributes

	Mean	SD	Max
Marriage age	34.4	2.6	40
Age	36.7	2	43
	Yes	No	
Occupation	52(65%)	28(35%)	
Fertility treatment	24(30%)	56(70%)	
Delivery style	Vaginal delivery	Caesarean section	
	36(45%)	44(55%)	

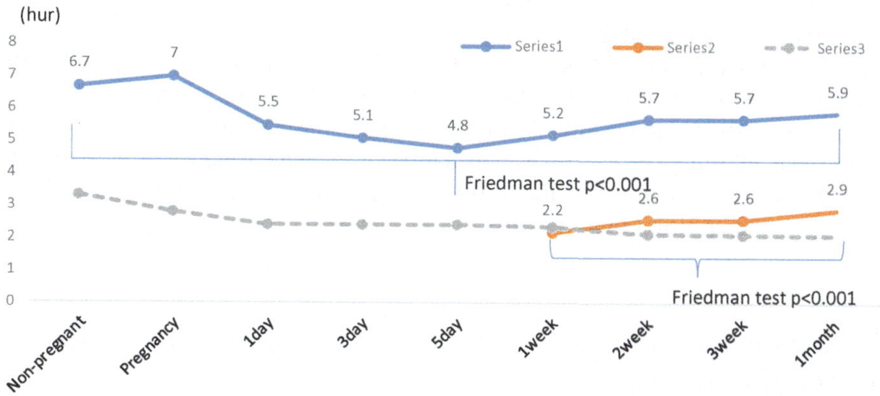

Fig. 11.1 Sleeping time from non-pregnancy to one month after childbirth

the participants one month after childbirth. Those who had less than 5 h of total sleep were most likely on the third day after birth (43.8%).

Comparison between the pregnancy period and all periods after giving birth revealed significant difference (all $p<.001$), indicating total sleep time were shortened after giving birth. In particular, the total sleep time on postpartum day 5 was shorter than on any of the measurement days ($p<.001$). And after 5 days of postpartum, the total amount of sleep became gradually longer. In addition, the total amount of sleep time 3 days after giving birth was shorter than 1 month after giving birth. The total sleep time for 1 week after delivery was shorter than that for 1 month after delivery ($p<.001$). Significant difference was seen in main sleep time from 1 week to 1 month after giving birth ($p<.001$). In particular, the main sleep time for 1 week after childbirth (2.2±0.9 h) was shorter than at any other measurement time (2 weeks: 2.6±0.9 h, $p<.05$; 3 weeks: 2.6±1.0 h, $p<.001$; 1 month: 2.9±0.9 h, $p<.001$) (Fig. 11.1).

The feeling of a good "sound sleep score" was significantly changed ($p<0.001$). The mean scores for the sense of sound sleep were 3.3±0.6 during non-fertility, 2.8 ± 0.6 during pregnancy, 2.4 ± 0.8 on postpartum day 1, 2.4 ± 0.7 on day 3, 2.4 ± 0.7 on day 5, 2.4 ± 0.7 on week 1, 2.1 ± 0.7 on week 2, 2.1 ± 0.7 on week 3, and 2.1 ± 0.7 on month 1. The feeling of a sound sleep was highest during non-pregnancy. Compared to non-pregnancy, the feeling of sound sleep was lower at all periods after delivery. The correlation coefficients between total sleep time and feeling of sound sleep were $r = .79$ ($p<.001$), $r = .71$ ($p<.001$), $r = .09$ ($p < .05$), $r = -.65$ ($p<.001$), $r = -.58$ ($p<.001$), r = −.61 ($p<.001$), and $r = -.59$ ($p < .001$) for 1 day, 3 days, 5 days, 1 week, 2 weeks, 3 weeks, and 1 month after giving birth. Significant positive correlations were seen between total sleep time and feeling of sound sleep until 5 days after giving birth when the participants were hospitalized. On the other hand, there was a significant negative correlation between total sleep time and sound sleep for 1 month after discharge.

In addition, correlation coefficients between the main sleep time and feeling of sound sleep after 1 week revealed significantly negative correlations in all periods ($p<.001$), and correlation coefficients were $r = -.37$, $r = -.53$, $r = -.55$, and $r = -.49$ for 1 week, 2 weeks, 3 weeks, and 1 month after giving birth, respectively. We calculated the correlation coefficients between main sleep time and total sleep time from 1 week to 1 month after birth. As a result, the positive correlations were observed at all times ($p<.001$). The correlation coefficients were $r = .44$ for 1 week after birth, $r = .57$ for 2 weeks after birth, $r = .53$ for 3 weeks after birth, and $r = .51$ for 1 month after birth (Table 11.2).

Promotional Factor and Constraints for Sleep from 1 Week to 1 Month After Giving Birth

As a result of encoding the free descriptions stated in the EMSD, promotional factor and constraints were categorized into four (promotional factor and constraints due to breastfeeding, those due to the first parenting, those due to mental and physical conditions, and those due to environment). Codes constituting each factor and the number of codes are shown below.

1. *Promotional Factor of Sleep from 1 Week to 1 Month After Giving Birth*

The promotional factors due to breastfeeding are 31 codes. One week after giving birth, they consisted of "Rest and sleep in accordance with nursing rhythm (6)," "Habituation of breastfeeding (1)," and "The child sucks well (1)." Two weeks after giving birth, they consisted of "Rest and sleep in accordance with breastfeeding rhythm (4)," "Secure the amount by additional milking and milk (2)," and "Increase in quantity of milk (2)." Moreover, they consisted of "Secure the amount by additional milking and milk (2)" and "Extension of breastfeeding interval (4)" 3 weeks after giving birth and "Extension of breastfeeding interval (9)" 1 month after giving birth.

The promotional factors due to the first parenting were 41 codes, and they consisted of "The child sleeps well (4)," "Sleep with the child (3)," and so on 1 week after giving birth and of the same 2 weeks after giving birth. In particular, the number of the codes for "The child sleeps well" increased chronologically.

Table 11.2 The correlation coefficients between sleep time and feeling of sound sleep

Score for sound sleep	Total sleep time – Feeling of sound sleep		Main sleep time – Feeling of sound sleep		Friedman test-Scheffe test
After birth	Correlation coefficient(r)	*P* value	Correlation coefficient®	P value	
1 day(2.4±0.8)	0.79	*p*<0.001			↑
3 days (2.4±0.7)	0.71	*p*<0.001			
5 days (2.4±0.7)	0.09	*p*<0.05			
1 week (2.4±0.7)	−0.65	*p*<0.001	−0.37	*p*<0.001	
2 weeks (2.1±0.7)	−0.58	*p*<0.001	−0.53	*p*<0.001	
3 weeks (2.1±0.7)	−0.61	*p*<0.001	−0.55	*p*<0.001	
1 months (2.1±0.7)	−0.59	*p*<0.001	−0.49	*p*<0.001	↓ *P*<0.01
Not pregnant (3.3±0.6)					
Pregnant (2.8±0.6)					

The promotion factors due to mental and physical conditions were 24 codes. They consisted of "Have feeling of sound sleep (1)" and so on 1 week after giving birth. "Have feeling of sound sleep" increased chronologically after 2 weeks after giving birth, and "prolongation of sleep time (8)" was extracted from the period 3 weeks after giving birth.

The promotional factors due to environmental factors are 11 codes, and "Support of family" was extracted from the period 2 weeks after giving birth. Total 110 codes constituted the promotional factors and were conspicuously fewer than the number of codes that constituted the constraints (Table 11.3).

2. *Constraints of Sleep from 1 Week to 1 Month After Giving Birth*

Constraints due to breastfeeding are 158 codes, which consisted of "Frequent breastfeeding (31)" which was extracted most, followed by "Night breastfeeding (12)," "Inexperienced breastfeeding technique (9)," and so on 1 week after giving birth. "Frequent breastfeeding" was extracted most during 1 month after giving birth, and "Night breastfeeding" was extracted second most. However, the code that sleep was disturbed by "Inexperienced breastfeeding technique" decreased after 2 weeks after giving birth and was not extracted 1 month after giving birth. Moreover, the code "Anxious to breast milk output" was extracted from the beginning to the end. Constraints due to the first parenting were 161 codes. "The child does not fall asleep (14)" was extracted most from 1 week after giving birth, and sleep disturbance for this reason continued. Moreover, the code "Parenting takes time (7)" extracted from 1 week after giving birth disappeared until 2 weeks after giving birth (9 codes during 2 weeks after giving birth). For the code "The child does not stop

Table 11.3 The promotional factors of sleep

1w		2w		3w		1m	
① Due to breastfeeding							
Rest and sleep in accordance with nursing rhythm	6	Rest and sleep in accordance with breastfeeding rhythm	4	Secure the amount by additional milking and milk	2	Extension of breastfeeding interval	9
Habituation of breastfeeding	1	Secure the amount by additional milking and milk	2	Extension of breastfeeding interval	4		
The child sucks well	1	Increase in quantity of milk	2				
② Due to the first parenting							
The child sleeps well	4	The child sleeps well	4	The child sleeps well	11	The child sleeps well	11
Sleep with the child	3	Sleep with the child	1	Sleep with the child	1	Sleep with the child	2
Leave the child	2			Decrease been fussy	1	Modulation of life through childcare	1
③ Due to mental and physical conditions							
Have feeling of sound sleep	1	Have feeling of sound sleep	2	Have feeling of sound sleep	6	Have feeling of sound sleep	5
Reduces headache	1	Decrease fluffy feeling	1	Prolongation of sleep time	8		
④ Due to environmental factors							
		Support of family	5	Support of family	4	Support of family	2

crying," 8 codes were extracted 1 week after giving birth though they gradually decreased, and only 2 codes were extracted 1 month after giving birth.

Although the number of codes was small, the code "There is always something I'm anxious about (parenting)" was extracted through 1 month (1 code after 1 week, 3 codes after 2 weeks, 3 codes after 3 weeks, 2 codes after 1 month), indicating that the participants' sleep was disturbed (Table 11.4).

Constraints due to mental and physical conditions were 94 codes. "Sleep is divided (9)" was extracted most from 1 week after giving birth and was extracted through 1 month after giving birth. Moreover, "Life rhythm was disturbed" was extracted from 3 weeks after giving birth, and "Have problem with falling asleep" was extracted continuously 1 month after giving birth. Further, "Don't have time for my own (7)" was newly extracted 1 month after giving birth.

Constraints due to environmental factors were 18 codes, which were extracted most from 1 week after giving birth, consisting of "I am not get used to the environment (3)," "Feel anxious thinking about life after discharge (3)," "Continuous meeting (1)," and so on. "Feel anxious thinking about future life (1)" and "Cannot get used to the environment (1)" were extracted 2 weeks after giving birth. "My

Table 11.4 Constraint factors of sleep

1w		2w		3w		1m	
① Due to breastfeeding							
Frequent breastfeeding	31	Frequent breastfeeding	18	Frequent breastfeeding	17	Frequent breastfeeding	14
Night breastfeeding	12	Night breastfeeding	13	Night breastfeeding	7	Night breastfeeding	5
Inexperienced breastfeeding technique	9	Anxious to breast milk output	5	Anxious to breast milk output	6	Anxious to breast milk output	7
Anxious to breast milk output	5	Inexperienced breastfeeding technique	3	Breast trouble (mastitis, etc.)	1	Breast trouble (mastitis, etc.)	2
Breast trouble (mastitis, etc.)	1	Breast trouble (mastitis, etc.)	1	Inexperienced breastfeeding technique	1		
② Due to the first parenting							
The child does not fall asleep	14	The child does not fall asleep	13	The child does not fall asleep	14	The child does not fall asleep	12
The child does not stop crying	8	Parenting takes time	9	The child is fussing	11	The child is fussing	7
Parenting takes time	7	The child does not stop crying	7	Children's day and night reversal	6	Baby crying at night	6
The child sleeps on the way	3	The child is fussing	5	The child does not stop crying	5	The child does not stop crying	3
Body weight of a child does not increase	1	Diaper exchange frequently	4	Baby crying at night	4	Diaper exchange frequently	2
Always anxious about (parenting)	1	Always anxious about (parenting)	3	Diaper exchange frequently	4	Always anxious about (parenting)	2
		Children's day and night reversal	2	Always anxious about (parenting)	3		
		The child sleeps on the way	2	The child sleeps on the way	2		
		Body weight of a child does not increase	1				
③ Due to mental and physical conditions							
Sleep is divided	9	Sleep is divided	7	Sleep is divided	11	Sleep is divided	7
Have problem with falling asleep	3	Poor physical condition	5	Pain	3	Don't have time for my own	7
Defatigation	2	Have problem with falling asleep	2	Housework	3	Life rhythm was disturbed	5
Itching (wounded area)	2	Housework	2	Have problem with falling asleep	2	Defatigation	5
Pain	1	Pain	1	Life rhythm was disturbed	2	Have problem with falling asleep	4

(continued)

Table 11.4 (continued)

1w		2w		3w		1m	
The little things make me cry.	1	Defatigation	1	Poor physical condition	1	Poor physical condition	3
						Housework	3
						Pain	2
④ Disincentive caused by environmental factors							
Cannot get used to the environment	3	Feel anxious thinking about future life	1	My husband does not support	1	My biological mother's fatigue	3
Feel anxious thinking about life after discharge	3	It does not got professional advice	1			My husband does not support	2
Professional advice is not provided usefully	1	I am not get used to environment	1			Quarrel with my biological mother	1
Continuous meeting	1	Advice to childcare of biological parents	1				

husband does not support (1)" was extracted 3 weeks, and "My biological mother's fatigue (1)," "Quarrel with my biological mother (1)," and "My husband does not support (2)" were extracted from 1 month after giving birth due to dissatisfaction with family and worries.

11.4 Discussion

Sleep Characteristics in Maternity Cycle of over 35 Primipara

Elderly primiparas in this study had an average marriage age as late as 34 years. Also, most of them were working and have a social role. The popularization of higher education for Japanese women and their advance into society are accelerating late marriage and overdue birth, and the percentage of unmarried young people in their 20–30s is increasing in particular, and the birth rate of women in their 20s, who have high ability for childbearing, is decreasing coupled with that (National Institute of Population and Social Security Research, 2019). The mean age of marriage of these participants was as high as 34 years old, like the trend of Japan, and it took them approximately 2 years to have their first birth. In addition, 30% of the participants were conceived through assisted reproductive health care. This result is very high than the 5.5% rate of pregnancy due to assisted reproductive health care in Japan (Ministry of Health, Labor and Welfare, 2018). And 55%, more than a half of them, were not able to deliver vaginally and required a Caesarean section. We considered elderly first-time mothers to be at risk for reduced fertility and abnormal deliveries. The participant's total sleep time during the maternity cycle was found to be longest during pregnancy. Sleep quality during pregnancy was worse than sleep quality during non-pregnancy due to influence such as hormonal dynamics (Fast et al., 1987). Most of pregnant women experience sleep alterations. Since women in early gestation sleep excessively and they feel sleepy night and day, sleep time of 1 day is extended.

Sleep time at mid-pregnancy is relatively stable, and arousal at night and sleepiness in daytime at the end of pregnancy increase (Shibui, 2005). The total sleep time of the participants of this study was longest in the pregnant period, and it was caused by characteristics of pregnant women mentioned above, we guess.

On the other hand, average total sleep time did not reach 5 h particularly 5 days after giving birth. Additionally, those who had less than 5 h of total sleep were most likely on the third day after birth (43.8%), followed by 5 days after birth (38.8%). In other words, it was found that the early puerperium sleep time was noticeably short. According to the reports (Ando et al., 2017), more than half of their participants have lack of sleep at an early stage after discharge from their obstetrics medical institution, and they did houseworks and parenting while having pain at their external genitalia and that of the wound from Caesarean section. Therefore, most of their participants wished some kinds of care while holding anxiety, and furthermore, more than 85% of them wished "relaxation" after giving birth. These results suggest that the participants of this study also wanted to receive care due to their anxiety after discharge from the hospital. It was suggested that the inability to satisfy the needs of the postpartum women was one of the causes of sleep deprivation.

In Japan, a nationwide postpartum care model program aiming at continuous support from pregnancy to birth and childrearing periods corresponding to each local characteristic has been enforced since 2014 (Ministry of Health, Labour and Welfare, 2015). "Mental and physical rest of mothers themselves" is highest in care needs after giving birth. It is suggested that the target population similarly needs to be provided with care that promotes postpartum sleep. The Japan Medical Association described it as follows (Japan Medical association, 2019). A general sleep cycle repeats non-REM sleep for 90 minutes and REM sleep for 90 minutes. A healthy period of sleep consists of four rounds of this cycle. In other words, we can get enough sleep in 6 h. First of all, women's body is affected by hormones. Throughout her life stages (menorrhea, the pregnancy, delivery, childcare terms), it secretes hormones, such as estrogen, progesterone, oxytocin, prolactin, and so on. These hormones influence their psychosomatic conditions. So, the women had high insomniac prevalence compared to men (Kim et al., 2003). It was thought that women had certain characteristics that affect their sleep quality.

According to the survey by the Ministry of Health, Labour and Welfare, the most observed average sleep time of 30–39-year-old women was 6–7 h, accounting for 34.2% of them. Furthermore, among this generation of women, 71.9% of women had more than 6 h of sleep. On the other hand, only 5.5% of the respondents slept less than 5 h (Ministry of Health Labor and Welfare, 2017). The percentage of those who slept less than 5 h per day after childbirth was higher than that of Japanese women of the same age (5.5%). Furthermore, their survey revealed that approximately 70% of the women whose sleep time was within 5 h were not satisfied with the quality of their sleep. Therefore, it could be assumed that the majority of our study participants were also not satisfied with the quality of their sleep.

Most of the postpartum women sleep between childcare. If the puerperium takes a long time to care for her baby, it will prevent the puerperal woman from falling asleep. The longer she falls asleep, the shorter her sleep time. The total sleep time of the participants was longer during pregnancy than during non-pregnancy. However, postpartum sleep hours varied very shortly, averaging 4–5 h per day.

The sound sleep is affected by the environment, sleeping time, and depth of sleep. In other words, we inferred the following: the participant may be at risk of falling asleep or insomnia due to the child's irregular sleep-wake rhythm, which may cause the sleep-wake rhythm to be difficult to modulate.

Positive correlations were seen between total sleep time and feeling of sound sleep in early puerperium while they were hospitalized.

On the other hand, after 1 week postpartum (after discharge from the hospital), total sleep time was negatively correlated with the feeling of sound sleep. In addition, positive correlations were seen between total sleep time and main sleep time of the participants during their puerperal period. While in the hospital, participants can receive support from specialists for physical recovery, breastfeeding establishment, and childcare support. In addition, the participant can always consult with the specialist if they feel anxious or worried. Thus, the participant may be relaxed and able to sleep soundly during their hospitalization. During the hospitalization, it was suggested that the participants were well cared for by professionals to help them get a sound sleep.

A factual survey on sleep of family members taking care of their child revealed that mothers having family member's replacement for attendance had better feeling of sound sleep than those not having such people (Tsuruwaka et al., 2009). In other words, we suggested that when the participant was aware of the shift's presence, the participant could feel more comfortable, which could lead to better sleep quality. It is necessary to coordinate environment for over 35 primiparas in the early puerperium in which family members can take care of the child so that they do not need to worry themselves. In addition, practical support is needed for post-discharge women to take care of their bodies and minds at home without anxiety. In order to secure this practical support, it is necessary to "instruct their supporters on the specific content of support just before discharge" and "connect them to community specialists in an integrated and seamless manner." For example, we introduced a "postpartum care service" where postpartum mothers could receive support from a specialist at home to give them peace of mind. It is expected that the postpartum women will be able to get a good sound sleep.

In our study, the average of main sleep time which indicates longest sleep time during a day was 2.2 h, the shortest week after birth. It was speculated that childcare divided the participant's sleep time. Therefore, the quality of their sleep was very low, we assume. One of the precedent studies reported that mothers 1 month after giving birth do not have feeling of sound sleep physically, being anxious about the amount of milk production while they are inexperienced for parenting and feeling it is hard (Isoyama et al., 2017), and there are many other studies that show similar results (Nowakowski et al., 2013). The midwives assess the participant's lactation and provide care for it. This can reduce the participant's anxiety about lack of milk production. It is expected that appropriate milk addition secures not only the nutrition of the newborn baby but also rest of the postpartum woman. In particular, milk production in the early puerperium is absolutely fewer than that required for a newborn baby, and it is outstanding in primiparas. They breastfeed every 3 h until the child's weight gain is consistent. Therefore, breastfeeding is closely related to sleep. When breastfeeding takes longer, sleep is gained for longer.

However, if breastfeeding takes a long time, the participants do not gain sleep. Also, the unconsidered milk addition reduces the number of times of direct breastfeeding and nipple stimulation in breastfeeding, becoming a factor of reduction of milk quantity (Narita et al., 2017).

It is desirable that the newborn be given an appropriate supply of milk by the breast. It is exceedingly difficult for a first-time mother to determine the balance between proper follow-up milk and unintentional milk additions. We assumed that the reasons for the positive correlation between total sleep time and sound sleep during hospitalization was due to the support of midwives and other professionals who supported them with regard to breastfeeding. On the other hand, there was a negative correlation between total sleep time and the feeling of sound sleep after discharge from the hospital. This means that after discharge from the hospital, the professionals are not able to be close enough to assess and care for the situation of the participant who is breastfeeding and parenting the child.

After discharge from the hospital, a negative correlation was found between the participants' sleep time and their sense of sound sleep. The reason why the participants were unable to sleep soundly due to anxiety and tried to sleep for longer periods of time was thought to be due to their anxiety. Therefore, the conduct of professional breastfeeding assessment and care in the early post-discharge period, especially in the case of first-time mothers, may have an effect on the feeling of sound sleep.

However, the participants did not feel sound asleep at any time in the postpartum period compared to during pregnancy. In addition, we found that the feeling of a sound sleep was lower after discharge from the hospital than during hospitalization. And, the relationship between the main sleep time and the feeling of deep sleep from 1 week to 1 month after birth was the same as the relationship between the total sleep time and the feeling of deep sleep. After discharge from the hospital, it was speculated that the participant was trying to secure a longer sleeping time at home because of the feeling of deep sleep. If they feel frustrated that they can't fall asleep, their commitment to trying to sleep grows stronger. And, they have an increased risk of not falling asleep (Uchiyama, 2012).

The facts that the participants did not feel sound sleep after being discharged from the hospital suggests that the participants had become anxious about sleep and was trying to sleep longer. The elderly primiparas in our study participants are all new experiences because they are "first-time mothers" and have more anxiety and worry than the multiparous women. In addition, because the baby was born at an advanced age, the postpartum recovery was slower than that of postpartum mothers of adjusted age. We guess that these facts are a risk when considering sleep health. "Mothers who are raising children for the first time" or "high-risk women with elderly babies" may need to be emphasized so that participants can get a good night's sleep. It is natural that the life changes to the other with the child as the main focus. However, at the same time as the process of forming a "new family" with the addition of a newborn baby, the participants are forced to recover their own physical condition and learn childcare skills.

Improving sleep deprivation can help reduce anxiety (Okajima & Chen, 2017). It is necessary to establish the support system for the participants after discharge from the hospital in order to get feelings of a good sleep and to get time to sleep. It is also desirable to provide support that respects the wishes and intentions of the elderly primiparas for childcare.

Need of Comprehensive Parenting Support That Utilizes Sleep Promotional Factors and Reduces Constraints

The participants' sleep promotional factors were considerably fewer than the codes that constituted constraints. However, it is expected that appropriate milk addition secures not only the nutrition of the newborn baby but also rest of the postpartum woman. "Child sleeping well" and "Sleep with the child" were extracted most, and securing of a newborn baby's sleep is indispensable for securing over 35 primiparas' sleep after giving birth. In other words, support for appropriate provision of milk quantity sufficient for a child is required, we presume. Moreover, support by family was also mentioned as a sleep promotional factor. In particular, they recognized that support by family 2 and 3 weeks after giving birth just after discharge promoted their sleep. We believe that it is necessary to coordinate environment for over 35 primiparas in the early puerperium in which family members can take care of the child so that they do not need to worry themselves. In addition, it was reported that self-efficacy is low so that age is high in the investigations 1 month after giving birth (Bryanton et al., 2008). Therefore, we guess that the self-efficacy of the over 35 primiparas influences the feeling of their sound sleep

On the other hand, the number of constraints for sleep was ten times as great as that of codes extracted in promotional factors, implying that they had a lot of factors. In particular, they recognized that frequent breastfeeding interrupts sleep for 1 month, while they believed that night breastfeeding is also a constraint. We have obtained similar results with respect to mothers 1 month after giving birth, while "Inexperience in breastfeeding" disappeared 2 weeks after giving birth. Moreover, as for sleep obstruction caused by frequent breastfeeding, the number of codes extracted 1 week after giving birth is double than those obtained 1 month after giving birth, indicating that it decreased to a half in 1 month.

It seems that contents of such sleep constraints caused by breastfeeding have varied temporally.

Therefore, it is necessary to respond to the needs in accordance with the course of puerperium so as to reduce sleep constraints due to breastfeeding of postpartum women. For constraints due to mental and physical conditions of postpartum women, since over a half of participants underwent Caesarean section, they suffered pains and itching at wound sites during 1 month of puerperium, which interrupted sleep. Moreover, codes indicating that they recognized divided sleep as a constraint were extracted for mental and physical conditions of over 35 postpartum women, which makes us infer that they had a strong desire for acquiring sleep.

It is natural that the life changes to the other with the child as a main focus. However, in order to adapt to the environment, it is important to find a method by which mothers are satisfied even if they have to divide their sleep. There is an urgent need to set an approach for recognition that the child is not satisfied by breastfeeding such as "frequent breastfeeding," "lack of the amount of milk production," and "the child does not fall asleep." The participants' sleep was less than 6 h. In particular, postpartum sleep is shortened on the fifth day. There was a positive correlation between total sleep time and main sleep time. And, there was a positive correlation between sleep time during hospitalization and feeling of deep sleep. However, after discharge, the correlation turned negative. In other words, during the hospitalization, care is provided to ensure that the participant feels a good sleep. However, after discharge from the hospital, due to the lack of care, the participants were unable to acquire a sound sleep and had a poor sleep environment.

Therefore, sleep time and feeling of sound sleep by practical support after discharge are required.

The Limitation of This Study

The participants of this study were primiparous over 35. Therefore, we have not made comparisons with under 35-year-old primiparous women. It is not necessarily a characteristic of older first-time mothers only. We also assume that the results of the survey are limited to a certain area, and thus bias is also apparent.

11.5 Conclusions

The participants' sleep was less than 6 h. In particular, postpartum sleep is shortened on the fifth day.

There was a positive correlation between total sleep time and main sleep time. In addition, there was a positive correlation between sleep time during hospitalization and feeling of deep sleep. However, after discharge, the correlation turned negative.

We speculated as follows. During hospitalization, the subject can sleep soundly with the help of the specialists. However, after discharge from the hospital, they cannot receive easy assistance from specialists and cannot sleep soundly because of childcare.

Therefore, sleep time and feeling of sound sleep by practical support after discharge are required. It is necessary to coordinate environment for over 35 primiparas in the early puerperium in which family members can take care of the child so that they do not need to worry themselves.

In the framework of the Psychology of Sustainability and Sustainable Development (Di Fabio, 2017a, b; Di Fabio & Rosen, 2018) and in relation to personal environment and individual well-being, the results of the present study offer promising perspectives for sustainable parenting.

References

Ando, F., Nakanishi, N., Kuratomi, J., Takaushima, Y., Nagao, M., Hashiba, H., et al. (2017). Investigation study for need of Postpartum Program. *Journal of Osaka Society of Maternal Health, 53*(1), 49–54. (in Japanese).

Biro, M. A., Davey, M. A., Carolan, M., & Kealy, M. (2012). Advanced maternal age and obstetric morbidity for women giving birth in Victoria Australia: A population-based study. *Australian and New Zealand Journal of Obstetrics and Gynaecology, 52*(3), 229–234.

Bozoky, I., & Corwin, E. J. (2002). Fatigue as a predictor of postpartum depression. *Journal of Obstetric, Gynecologic, and Neonatal Nursing, 31*(4), 436–443.

Bryanton, J., Gagnon, A. J., Hatem, M., & Johnston, C. (2008). Predictors of early parenting self-efficacy: Results of a prospective cohort study. *Nursing Research, 57*(4), 252–259.

Dennis, C. L., & McQueen, K. (2009). The relationship between infant-feeding outcomes and postpartum depression, a qualitative systematic review. *Pediatrics, 123*(4), 736–751.

Di Fabio, A. (2017a). Positive Healthy Organizations: Promoting well-being, meaningfulness, and sustainability in organizations. In G. Arcangeli, G. Giorgi, N. Mucci, J.-L. Bernaud, & A. Di Fabio (Eds.), Emerging and re-emerging organizational features, work transitions and occupational risk factors: The good, the bad, the right. An interdisciplinary perspective. Research Topic in *Frontiers in Psychology. Organizational Psychology, 8*, 1938. https://doi.org/10.3389/fpsyg.2017.01938

Di Fabio, A. (2017b). The psychology of sustainability and sustainable development for well-being in organizations. In G. Arcangeli, G. Giorgi, N. Mucci, J.-L. Bernaud, & A. Di Fabio (Eds.), Emerging and re-emerging organizational features, work transitions and occupational risk factors: The good, the bad, the right. An interdisciplinary perspective. Research Topic in *Frontiers in Psychology. Organizational Psychology, 8*, 1534. https://doi.org/10.3389/fpsyg.2017.01534

Di Fabio, A., & Rosen, M. A. (2018). Opening the black box of psychological processes in the science of sustainable development: A new frontier. *European Journal of Sustainable Development Research, 2*(4), 47. https://doi.org/10.20897/ejosdr/3933

Di Fabio, A., & Tsuda, A. (2018). The psychology of harmony and harmonization: Advancing the perspectives for the psychology of sustainability and sustainable development. *Sustainability MDPI, 10*(12), 4726. https://doi.org/10.3390/su10124726

Fast, A., Shapiro, D., Ducommun, E. J., Friedmann, L. W., Bouklas, T. B. S., & Floman, Y. (1987). Low-back pain in pregnancy. *Spine, 12*(4), 368–371.

Goyal, D., Gay, C. L., & Lee, K. A. (2007). Patterns of sleep disruption and depressive symptoms in new mothers. *The Journal of Perinatal & Neonatal Nursing, 21*(2), 123–129.

Isoyama, A., Kudo, T., Watabiki, H., Hatakayama, M., & Fujimoto, F. (2017). Physical and psychologic conditions of the mothers who gave birth after one month, and their care needs investigation. *The Journal of Japanese Midwives' Association, 71*(2), 78–82. (in Japanese).

Japan Medical Association. (2019). Science of sleeping (in Japanese). https://www.med.or.jp/forest/health/live/05.html. Accessed 8 June 2019.

Kim, K., Uchiyama, M., Okawa, M., Lix, X., & Ogihara, R. (2003). An epidemiological study of insomnia among the Japanese general population. *Sleep, 23*(1), 41–47. (in Japanese).

Matsumoto, K., Tsuchiya, K., Itoh, H., Kanayama, N., Suda, S., Matsuzaki, H., et al. (2011). Age-specific 3-month cumulative incidence of postpartum depression. *Journal of Affective Disorders, 133*(3), 607–610. (in Japanese).

Ministry of Health Labor and Welfare. (2015). The case report of the pregnancy and delivery Comprehension support (in Japanese). http://www.mhlw.go.jp/file/06-Seisakujouhou-11900000–.pdf. Accessed 8 June 2019.

Ministry of Health Labor and Welfare. (2016). The monthly and the annual reports of Vital statistics. (in Japanese). https://www.mhlw.go.jp/toukei/saikin/hw/jinkou/geppo/nengai16/dl/h3-4.pdf. Accessed 8 June 2019.

Ministry of Health Labor and Welfare. (2017). health/nutrition survey in Japan, An average sleep hour per a day. Figure 35. (in Japanese). https://www.mhlw.go.jp/stf/houdou. Accessed 17 June 2020.

Ministry of Health Labor and Welfare. (2018). The Current State of Assisted Reproductive Medicine. https://www.shojihomu.or.jp/documents/10448/7388932/1128sankou-siryou2-5.pdf. Accessed 17 June 2020.

Mizukami, A., Taniguchi, M., & Baba, N. (1994). The infant care concern of mother after childbirth. *Memoirs of the Faculty of Education, Kumamoto University, 197*(43), 89–97. (in Japanese).

Narita, E., Yoshida, Y., Asakawa, Y., & Nishijima, A. (2017). Efforts to improve breastfeeding rates-approach to early breastfeeding and frequent breastfeeding. *Journal of Sunagawa City Medical Center, 30*(1), 47–50. (in Japanese).

National Institute of Population and Social Security Research. (2019). The 2019 Collections of references for Population statistics, The women age-specific fertility rates (Table 4) (in Japanese). http://www.ipss.go.jp/syoushika/tohkei/Popular/Popular2019. Accessed 8 June 2019.

Nowakowski, S., Meers, J., & Heimbach, E. (2013). Sleep and women's health. *Sleep Medicine Research, 4*(1), 1–22.

Okajima, I., & Chen, J. (2017). The effect of insomnia on changes in anxiety, depression, and social function after a transdiagnostic treatment targeting excessive worry. *Sleep Biological Rhythm, 15*, 243–249.

Owens, J. F., & Matthews, K. A. (1998). Sleep disturbance in healthy middle-aged women. *Maturitas, 30*, 41–50.

Shibui, K. (2005). Women's sleep and hormones. *Journal of Society of Biomechanisms Japan, 29*(4), 205–209. (in Japanese).

Shimada, M., Sugimoto, M., Agata, T., Nitta, N., Seki, K., Ohashi, K., et al. (2001). Nationwide survey on maternal anxiety in one-month postpartum, needs for childrearing supports and environment at five years after healthy parents and children 21 program. *Pediatric Health Research, 60*(5), 671–679.

Tsuruwaka, M., Sasagawa, S., & Sumiyoshi, T. (2009). Study about the hypnic actual situations – The factors to influence feelings of sound sleep that the family who attends children have. *Journal of Japanese Nursing Society, Childcare, 39*, 107–109. (in Japanese).

Uchiyama, M. (2012). Sleep disorders. *Journal of Nihon Medical Association, 71*(6), 396–400. (in Japanese).

Chapter 12
Resolving Career Indecision: The Power of Integrating Conscious and Subconscious Advice Through Career Construction Counselling

Jacobus G. (Kobus) Maree

12.1 Chapter Structure

The first part of the chapter briefly discusses the psychology of career choice decision-making as well as key characteristics of career construction theory. The second part covers the rationale for and goals of the chapter, while the third part contextualises the study. The fourth part discusses and includes a brief discussion of the career counselling intervention.

12.2 Introduction

Most learners and workers in the twenty-first century are experiencing increased career decision-making pressure because of the many changes in the world of work—changes that are making them rethink their career-lives, including whether jobs will disappear or be changed to such an extent by Work 4.0 developments that they will no longer "fit" into these careers or meet the new demands posed by changing work environments (Urbanaviciute et al., 2019). We as career counselling theorists, researchers, and practitioners need to respond to these changes in a timely and

This chapter is based on an article (Maree, 2020) that has been previously published in the *South African Journal of Education* (Maree, 2016). Permission was requested and granted by the Editor, Prof. Ronél Ferreira, to rework the article into a book chapter for publication in this volume.

J. G. (Kobus) Maree (✉)
Department of Educational Psychology, Faculty of Education, University of Pretoria, Pretoria, South Africa
e-mail: kobus.maree@up.ac.za

A. Di Fabio (ed.), *Cross-cultural Perspectives on Well-Being and Sustainability in Organizations*, https://doi.org/10.1007/978-3-030-86709-6_12

appropriate manner to ensure that what we do remains relevant and continues to meet people's career counselling needs.

For a variety of reasons, education and training issues such as changing schools, dropping subjects in favour of other subjects, and changing fields of study are common in people's academic lives. Many learners, for instance, receive career counselling only to discover that they are in an "inappropriate" field of study. Others enter careers that leave them frustrated and unfulfilled, causing them to seek career counselling to help them decide on more appropriate careers.

Gati et al. (2011), Kelly and Lee (2002), and Maree (2016a, b, c) agree that career decision difficulties should be addressed and dealt with as early as possible in the career decision-making process. Gati et al. (2011) maintain that "[m]ost young adults report facing difficulties in making career decisions; indeed, many approach career counselors to get help in making these decisions" (p. 4). They add that "[s]ome of these deliberating individuals have only temporary or developmental career decision-making difficulties, considered a normative phase in vocational development; these individuals are typically labelled as undecided" (p. 4). I agree with this view, which is the lens through which this chapter should be read and interpreted.

Psychology of Career Choice Decision-Making

Points of Transition and Changing Career Decisions

The reasons for changing career decisions can be grouped into five broad categories.

First, many learners experience career decision-making insecurity at "natural" points of transition or crossroads during their academic careers. This includes transitioning from one school to another (where they may not be able to take all their chosen subjects), from one school phase to another (and having to choose between subjects), at the point of completing their school years (and having to decide on a future career and field of study), and at the point of completing their tertiary studies (and having to find employment).

A second group of learners receives or stumbles on (serendipitously or through "happenstance", cf. Krumboltz, 2003) additional information about the world of work and begin to question their chosen subjects, schools, fields of study, and tertiary training institutions.

A third group of learners discovers more about themselves and begin to doubt whether they will "fit" into their chosen field of study.

A fourth group of learners develops ("matures") naturally, becomes more motivated to study and achieve better marks, and discovers that they are capable of "constructing" themselves and their careers at higher levels.

A fifth group of learners is helped by skilled career construction counsellors to uncover key life themes (e.g. a desire to heal sick animals or to devote their lives to educate and facilitate opportunities for learners from resource-constrained environments) and to consider entering different fields of study that may be better suited to fulfilling these key life themes.

Different Kinds of Career Indecision (or Undecidedness) Difficulties

A brief overview of earlier research on career decision-making difficulties suggests that a distinction can be drawn between the following five kinds of such difficulties (Career indecision, n.d.; Gati & Asher, 2001; Kelly & Lee, 2002): (i) lack of and need for career choice information; (ii) identity diffusion (an inadequate sense of self); (iii) trait indecision (inability to make career choice decisions); (iv) choice anxiety (anxiety regarding career choices); and (v) disagreement with others (observing objectively and/or subjectively that significant others disagree with or disapprove of their career choices).

These authors argue also that two different kinds of career indecision (each requiring its own kind of intervention) can be distinguished:

First, people with "developmental" indecision who need basic information (such as the differences between a university and a university of technology) as well as psychoeducational and psychosocial information. For this group, brief career counselling (when specific career choice information is provided) should be sufficient.

Second, people with long-lasting indecisiveness who often exhibit emotional problems relating to, among other things, anxiety, insecurity, inadequate self-efficacy, or a poor (underdeveloped) sense of self and identity. These people require career counselling embedded in life design (Savickas et al. 2009).

Distinction Between Career Indecisiveness and Career Indecision

Some authors, including Di Fabio et al. (2013), argue in favour of a further distinction between career indecisiveness and career indecision. They believe that career indecisiveness is indicative of general indecisiveness regarding anxiety and problem-solving challenges and that career indecision is negatively correlated with people's emotional-social intelligence (revealed by, for instance, their ability to be in touch with their emotions and the emotions of others, to manage stress, and to be optimistic generally). These authors maintain that different interventions are needed to deal with the phenomena of career indecision and career indecisiveness. Gianakos' (1999) research, for example, revealed that career decision self-efficacy improved his student participants' decision-making ability and, simultaneously, decreased their career indecision.

In similar vein, Guay et al. (2003) contend, on the basis of career indecision as defined in self-determination theory, that sound interpersonal relationships between college students and their parents and friends can influence their career indecision positively and vice versa. The authors found that the behaviour of parents who displayed controlling behaviour correlated strongly negatively with their children's self-efficacy and decision-making independence. Likewise, Garcia et al. (2012) found that the desire to become highly skilful correlated strongly negatively with measured career indecision. In summary, factors such as adequate self-efficacy levels, emotional-social intelligence, a positive attitude, intentionality (Maree, 2016a, b, c), and self-sufficiency seemingly correlate strongly positively with decreased career indecision. In seeking ways to deal with career indecision, these and other related factors should always be borne in mind. The above findings have theoretical as well as counselling implications in that they can enhance our attempts to deal with people's career decision-making challenges by helping us come up with the best coping strategies for them.

In this chapter, I examine the participant's career decision-making needs through the lens of career construction theory (Savickas, 2019). I will here discuss briefly only the key characteristics of this theory.

Career Construction Theory

Career construction theory (CCT) (Hartung, 2011, 2013; Savickas, 2015a, b, 2019) brings together seemingly divergent theoretical career counselling traditions. First, CCT draws on the differential approach (Holland, 1997) by using tests that help counsellors and clients identify key differences between people as well as key similarities between their individual profiles and the characteristics of certain jobs. The aim is to find an "optimal" match between people's personality traits and particular work environments. Second, CCT brings the developmental tradition (Super, 1957) into play by applying educational strategies to help people adapt. This approach emphasises the numerous roles people play in their lives and also the different life stages they go through. According to Savickas (2019, p. 10), "to complement the matching model's focus on the content of career choices Super (1957) applied the life-cycle paradigm to psychosocial education about the process of coping with a progressive sequence of career stages and developmental tasks". Super (1957) stresses the importance of assisting people to choose and adapt to careers instead of focusing merely on possible work-related choices.

Third, the CCT draws on the psychodynamic (narrative or storied tradition) (Freud, 1963). A key component of this approach is the eliciting of people's numerous micro-stories (thereby promoting their narratability) and then helping them reflect on these stories (their autobiographies) to identify their key life themes. Eliciting such a theme is possible when counsellors adopt a not-knowing approach that enables them to listen for people's stories instead of listening to their stories (Savickas, 2016; Welty, 1998). Events during people's early lives often become part of their subconscious and can powerfully affect their career-life stories and influence their conscious cognitions and actions (including their career-related motivations and choices and the executions of these choices). A carefully selected number of the micro-, meso-, and macro-stories that collectively constitute people's career-life stories therefore need to be elicited. People can then be helped by career counsellors skilled in the narrative approach to "make meaning" of these stories. For this to happen, it is important to draw on what is referred to as (auto)biographical reasoning. This kind of reasoning starts with the examination of early career-life stories and locating them in educational-social contexts. Thus, conscious knowledge (what people know consciously) is connected with subconscious advice through career construction counselling to resolve career choice indecision (Maree, 2019, 2020). The result is that their sense of self and identity is bolstered.

Rationale for the Study

Narrative, subjective, or qualitative career counselling approaches (and associated techniques) have received global recognition over the past few decades as equally (if not more) effective and useful than "traditional" quantitative approaches (Maree, 2018). Research conducted globally confirms the effectiveness of career construction counselling intervention in resolving career indecision generally. More specifically, it has been shown that enhanced career decision skills often correlate positively with enhanced "meaning making" in people's careers and with people acquiring a sense of purpose and hope in their work-lives (Maree, 2017b). However, research on the influence of career construction intervention on the career indecision of disadvantaged students (black students in particular) has received scant attention. The situation in Africa especially is a matter of concern to those of us who conduct career counselling research and practise career counselling here and have witnessed its value in other contexts. I believe that this case study on the implementation of career construction counselling with a disadvantaged black student experiencing career indecision can shed light on the value of this kind of approach for people from different cultures and disadvantaged people especially and can contribute to the debate on the topic.

Goals of the Study

The goal of this study was to establish the influence of career construction counselling (based on co-constructive conversations) on the career choice indecision of a master's degree student in psychology. The following specific research questions guided the research.

(a) What was the influence of the intervention on the participant's career indecision?
(b) What changes did the intervention bring about in the participant?
(c) What factors prompted these changes?

12.3 Method

Participant and Context

The participant was a conveniently selected 27-year-old Zulu-speaking man who had received his education in English and who had requested career construction counselling (one of the selection criteria). Together with some other participants, Lionel (pseudonym; at the time of the study, a master's degree student in counselling psychology) had volunteered to play the role of "client" in a live demonstration of career construction counselling. The demonstration comprised part of a course in narrative career counselling in 2018. Prior to the demonstration, the students were requested to write their names on pieces of paper, which were then put into a hat, and then one name was randomly drawn. This sampling style can be described as convenience sampling with an element of randomness. After the demonstration, three of the attendees were invited to thank the participant and say what her or his story had meant to them. All the students were reminded about the ethical aspects of such a demonstration, for example, not a word should be said about the demonstration outside the course venue (the principle of privacy) (Sieck, 2012).

Procedure

The intervention took place in Pretoria. During the first ("elicitation") phase, the *Career Construction Interview* (*CCI*) (Savickas, 2019) was administered (lasting about 65 min). However, since the information obtained from the *CCI* did not yield sufficient information to proceed with Phase 2 of the intervention, the *CCI* administration was followed by the administration of the *Career Interest Profile* (*CIP*) (Maree, 2017a) (lasting about 70 min). The following morning, during the "authorisation" phase, Lionel and I together analysed the outcomes of the intervention.

After a 30-min break, we entered into the third phase (devising action steps to facilitate forward movement) and delved into Lionel's reason for consulting me (this phase lasted about 120 min). We commenced by examining the outlines of his career-life story and connecting the dots that emerged. This enabled us to identify ways to deal with his career choice indecision by investigating the gap between what he knew and understood consciously and what he could access with subconsciously (advice), devise possible strategies, and draw up a "road map" or course of action to help him move forward.

Mode of Inquiry

An explorative, descriptive, intrinsic, intervention individual case study design was implemented in a one-on-one research setting. The data generation instruments (see below) were purposively selected to promote critical self-reflection and reflexivity to uncover more psychologically healthy ways of self- (Guichard, 2009) and career construction (Savickas, 2015a, b) by linking the past, the present, and the future prospectively. Crystallization (Richardson, 2020) and triangulation (Stewart et al., 2017) were used to promote the rigor of the study.

Data-Generating Instruments

The Career Construction Interview (CCI) (Savickas, 2019)

The *Career Interest Profile* (*CIP*, Version 6) (Maree, 2017a)

The *CIP* is used to elicit career story information qualitatively. The *CIP* is grounded in career construction theory (Savickas, 2019), self-construction theory (Guichard, 2009), career style assessment and counselling (Adler, 1958), narrative career counselling theory and practice (Cochran, 1997), and Erikson's (1968) self-concept and psychosocial development theory. The *CIP* questions are used to elicit information that relates to the differential, developmental, and psychodynamic (storied) traditions. The small quantitative part of the *CIP* (Part 2) has demonstrated satisfactory test-retest reliability and content and criterion-related validity (Maree & Sommerville, 2008). It can be administered individually and in groups. Carefully and purposefully structured, the *CIP* questions promote reflection and, ultimately, reflexivity to develop the processes of narratability and biographicity and to facilitate the clarification and construction of career-life stories (Maree, 2013).

Data Analysis

Qualitative Data

I used Savickas' (2019) guidelines to analyse the kind of data generated by the intervention (see below) and also considered Merrill and West's (2009) biographical research data analysis approach. First, I asked the participant repeatedly to reflect critically on and interpret his own reflections (his responses to questions). Meta-reflection was thus promoted. Second, repeated words, expressions, and phrases were shared with Lionel, and we jointly attempted to identify patterns from them. This bolstered reflexivity, which, in turn, advanced narratability and biographicity and promoted clarification and co-construction of Lionel's career-life stories (Maree, 2013; ML Savickas, pers. comm.). Third, I read Lionel's words and expressions out to him verbatim so that he could listen to and hear himself and feel in control of the assessment situation. Fourth, we began by discussing his three earliest recollections and his responses to the three hurtful experiences to uncover his key career-life themes. Fifth, we carefully worked our way through the rest of the questions (in the order presented in the Outcomes section). Sixth, although Lionel and I reflected on his answers together, I took care to highlight certain key themes that emerged from the answers. Seventh, we attempted to find a link between (a) Lionel's integrated elicited themes (subconscious motivations and strivings) and (b) his responses to Part 2 of the *CIP* (his conscious insights) in relation to his reason for seeking career counselling.

Outcomes

Outcomes of the "Quantitative" Part of the Assessment

According to Part 2 of the *CIP*, Lionel's preferred careers were professional athlete, coach, team manager, and business manager.

Outcomes of the "Qualitative" Part of the Assessment

Before commencing the intervention, I explained briefly to Lionel that we would be eliciting a number of his micro- (smaller) career-life stories that would be analysed afterwards in search of patterns that might indicate (advice to himself) his reason for seeking career counselling. I also said that, theoretically, his stories could contain guidance regarding the decisions he needed to take (Reid, 2018). Stated differently, we would be examining his subconscious preoccupations, that is, what he had been rehearsing throughout his life (Savickas, 2019), in an effort to arrive at the root of his indecision, find an answer to his reason for consulting me, and establish connections between his conscious cognitions and subconscious insights.

Below, I list Lionel's answers to selected questions in Part 4 of the *CIP* (because of space constraints, only a few selected responses are given). Lionel's responses are edited only lightly to preserve their authenticity.

"How can I be of use, help, or value to you? What are you hoping to gain from this session?"

"I think my interest in taking part is to validate my thoughts in terms of my career path. Get validation that I am heading in the right path in terms of my career".

(i) "What are your

(a) greatest strengths?"

"I get along well with people from different backgrounds, I am a keen communicator and listener, I can speak many African languages, I am resilient, humble, serious, disciplined, hard working. I like making witty, cheeky jokes, I am open-minded, and I am curious to learn many things".

(b) areas for growth?"

"I can read more. I sometimes procrastinate. I am unsure of some things".

(ii) "Whom did you admire or who were your role models when you were young and why?" (His sense of who he wants to be or become.)

> "*Way, way back. Groups representing same traits: cartoon characters: Goku, Shakes Mokwena, Dexter, Johnny Bravo: funny, laughed readily, strong, leaders to an extent in their own field*".
>
> "*Lionel Messi* [Barcelona's captain] *I love watching him play. He plays his game in a natural and seemingly effortless manner, yet so unstoppable. He epitomises self-expression and joy. He is described to be the most humble and down-to-earth person*".
>
> "*Sachin Tendulkar* [Indian cricketer]. *At the time I was playing cricket and developing my skills as a cricket player: left-handed (just like me), classy; the way he played the sport; he was short like me*". (SMILES)

(iii) (Lionel's advice to himself. His explanations of what these quotations mean to him appear between brackets.)

> "*If I get a cow, I am going to milk it*". ("*Make the most of that opportunity*".)
>
> "*Eendag sal die poppe dans*" (Afrikaans) ["*The dolls will dance*"]. ("*To me, it says that one day hard work and sacrifice will be rewarded*".) (Here, too, Lionel's explanation differed from the generally understood meaning of the expression, namely, that someday there is going to be trouble. This indicates the importance of making sure his understanding of any matter is correct (recontextualisation).
>
> "*Obomi ngumzamo, Umzamo bubomi* (Zulu) [*"Life is about trying your best"*]. *This, to me, is very ethical*".
>
> "*Live beautiful*". ("*Do the beautiful things you so want to do. Approach life in a beautiful way*".)
>
> "What is your own motto? How do you want others to quote you in the future?"
>
> "*Know thyself*". ("*It talks to how important it is to know, be, and appreciate oneself. Do what 'is you.' This is the biggest achievement one can attain; a prerequisite to having happy and successful life*".)

(iv) What were the three most hurtful experiences in your youth (experiences that you don't want others to go through)?" (His "deep secrets"; the pain he suffered passively in early life that he needs to undo actively in the course of time.)

> "*I was in boarding school throughout my school years and had to learn to adapt to foreign cultures and circumstances repeatedly. That taught me to become resilient and turn failure into success*".
>
> "*I never got to spend much time with my parents or was at home as I wanted to*". "*I overly identified with sports and neglected my parents; did not show my appreciation, which I regretted*".
>
> "*My grandpa was shot in front of me while my caretaker and I sat helplessly in his care*". (There is a substantial correspondence between these responses and the themes uncovered by his three earliest recollections.)

(v) What are your three earliest recollections?" (These "stories" were interpreted and analysed by us together. They revealed his central career-life themes and his advice to himself about how to deal with his preoccupations. I analysed the three earliest recollections according to the guidelines proposed by Adler

(1958), Clark (2002), Hartung (2015), Maree (2013, 2018), Savickas (2019), and Sweeney (1975). I then asked Lionel to add a headline to each of the three "stories", and we together modified each heading until Lionel was satisfied that it conveyed the exact meaning of what he was trying to share with me.

Night Travel Murder Traumatises Vulnerable Boy

(a) "*At the age of four, my grandpa, my nurturer, and I were travelling home from his work in the evening*". "*Well, we were basically ambushed. A car drove up from behind, forced us to stop, and parked on my grandfather's side. My nurturer and I were sitting on the other side of the car. A man shot my grandpa several times. So my nurturer protected me. We escaped unharmed but grandpa passed away that night*". (TEARS WELL UP IN HIS EYES) (Emotions associated with this recollection: "*Mmm ... I am feeling severely traumatised, although consciously I don't want to think how traumatised I was*". [Dissociation])

Puzzling Cultural Blending Initiates Joyful Celebration

(b) "*Mmm ... I was about four or five years old when this incident happened. An uncle of mine got married to a white woman and my family and I attended the wedding. It was a huge thing in our village. The two different cultures getting together and people not understanding what was going on. They had never seen anything like this before. Macaroni and pasta were served as the woman was of Italian descent, and people were puzzled, wondering what kind of worms was served and why*". (Happiness, having fun; excitement about learning about other cultures.)

Mother's Selflessness Inspires Loving Son

(c) "*When I was three years old, my mother and I needed to travel to a remote village quite far away to visit an ailing family member. We walked all the way there; a very long distance that took us many hours. I did not want her to do so, but she often picked me up and carried me to protect me. I felt so sorry for her, carrying the heavy load, me and our stuff. I insisted on walking by myself but she refused*". (Resilience, sympathy, compassion.)

Lionel then insisted that he wanted to "share another recollection with [me]; one that is quite similar to the previous story". I, of course, I allowed him to do so.

Lionel's meta-reflections enabled him to establish a connection between his subconscious motivations (things he was not aware of on a conscious level) and his conscious thoughts or cognitions and feelings: a bridge between his past, present, and future. Reflexivity was thus promoted. Among other things, he said that "completing the intervention was especially interesting on a personal level because of the 'natural' way in which we integrated the generated information and the consistency of the themes that emerged. I was intrigued by how the researcher-counsellor-therapist sat down and jointly with me, the participant, interpreted, analysed and integrated the information without the former sounding like he was advising a 'real' client".

The first story reveals Lionel's subconscious anxiety about the daunting challenges associated with travelling back to his "roots". Currently, he is considering going back to his roots, but the question that needs to be answered is (in his own words): "Should I realistically go back home and do practice there? Is that really what I want to and should do now? Is it wise? It is so unsafe there these days … ". The second story reveals his advice to himself: there is no need to fear the "unknown". However, here there is a subtle shift in emphasis: subconsciously, he seems to be assuring himself that there is no need to fear culturally unfamiliar circumstances (symbol of travelling abroad). The third and fourth recollections are reminders from the subconscious that travelling in order to help others and/or access education and training opportunities is feasible provided one is willing to make sacrifices. Together, the three recollections reveal themes of compassion, selflessness, a desire to travel and help others, as well as the will to develop, grow as a human being, cross new boundaries, and be positive.

The first verb Lionel used was "travel". In fact, all the stories talk to a deep-seated desire to travel in an effort to grow, construct himself, help others, and learn more about other cultures. He struggled to make up his mind about what he wanted to do once he had qualified—his subconscious seemed to reassure him that, despite the sacrifices that would have to be made in his life journey, travelling was what he really wanted to do. He was also reassured that it was not important whether he first returned to his roots where he could serve others or went abroad to fulfil his need to meet people from different cultures. These two options were not mutually exclusive either. If he decided to travel abroad first to discover new shores, he would eventually return to his roots a "wiser" and more experienced person, able to serve his community even better. (However, it would also be all right to return to his roots first, devote time to his community, and gain experience while researching options for working abroad as a counselling psychologist before deciding where to go to.)

At the end of the session, I asked Lionel how he had experienced the intervention.

(a) "What did you enjoy?"

> "*I feel inspired; my spirit has been lifted. Participating in this intervention put me in touch with myself; allowed me to 'check in' with myself with the assistance of the counsellor/researcher. I enjoyed the opportunity and the safe space the counsellor provided that allowed me to open up and really reflect on my answers to the questions*". (Reflexivity.)

(b) "What did you not enjoy?"

"There was nothing that I did not enjoy. I enjoyed that the interview was about me, asking questions that prompted reflections that revealed the REAL me".

(c) "Is there anything else that I need to know about you?"

"Once more: I always wanted to be 'free' but I could just not get myself so far to act accordingly".

(d) "What has changed and how do you explain the change(s)?" (The purpose of this question was to establish qualitatively the usefulness of the intervention.)

"Before taking part, I was unsure which steps to take, how to structure goals and plans for my future and if my plans were realistic. The intervention confirmed my plans and provided direction: above all, I want to help my people but, above all right now, travel the world. I will strive to find a way to combine the two matters and practically plan my future".

We agreed that pursuing his master's degree studies in counselling psychology was well aligned with his key life themes and consistent with his interests, values, and traits. Second, we agreed that he should focus on finding a way to help underserved and disadvantaged populations (young sports people in particular). Because of his desire to travel, he could eventually also investigate the possibility of joining the management component of national sports teams, which would enable him to meet his need to travel, to work with disadvantaged youths, and to practise as a psychologist working with a national sports team. I obtained permission to give him the contact details of a young black woman from a neighbouring country (a counselling psychologist, also with a keen interest in sport) who had such a position—she could provide him with information on the requirements for such a job. After having completed his master's degree, he could consider writing a book on his life story—a book that could inspire people finding themselves in a position similar to his earlier in his career-life.

Third, we agreed that Lionel should participate actively in sport and listen to music that uplifted and inspired him.

Next, the two of us "co-authored" his mission and vision statement (Maree, 2013; Savickas, 2011, 2019). The aim was to impress upon him the importance of fully understanding the meaning and purpose of the intervention and to instil in him a sense of hope in his projected career trajectory:

"**I want to** *complete my studies in counselling psychology and later specialise in sports psychology and then find opportunities to travel abroad; maybe as the psychologist of a national or provincial sports team. And later return to South Africa so I can serve the needs of my community even better,*

so that I can (personal meaning; mission statement) *acquire the skills needed to work with other people, especially the poor and the disadvantaged youth; work as a member of a larger team with people, help them to reach their goals and solve their personal problems, actualise their potential* [construct themselves and their careers], *and learn more about themselves and about other people,*

and in the process (social meaning; vision statement) *travel the world and meet many interesting people from other cultures, touch personal lives and the lives of communities, and enjoy listening to music in new environments that will always remind me of those places".*

The intervention thus helped Lionel find answers to his most profound (almost existential) career-life-related questions: "Why do I live?", "Why am I here?", and "What is the purpose of my being here, of what I am doing, of life itself?". These questions can also be seen as his "real" reasons for seeking career counselling.

I ended the session by asking Lionel to reflect carefully on the intervention and to stay in touch with me for as long as he felt the need to do so.

12.4 Discussion

The goal of this chapter was to establish the influence of career construction counselling (based on co-constructive conversations) on the career choice indecision of a master's degree student in psychology. The following specific research questions guided the research.

(a) What was the influence of the intervention on the participant's career indecision?
(b) What changes did the intervention foster?
(c) What factors prompted these changes?

These three questions are considered in succession and the findings related to the literature on the influence of career construction intervention on people's ability to manage changing occupational contexts.

How Did the Participant Experience the Career Construction Counselling?

Lionel declared that he had found the experience "literally life-changing". He appreciated the fact that he was afforded the opportunity to narrate key aspects of his autobiography and that his deepest thoughts and ideals were validated (authorised) (Maree, 2013; Savickas, 2005, 2019). He believed that his reflections enhanced his self-insight and uncovered ways in which he could make personal meaning in his future career. He could now come to terms with his indecision and move forward resolutely: "The most challenging part was when we read the sentences out loud. I felt a lot of feelings moments before reading mine aloud but afterwards I felt liberated and good about myself. Talking to myself about myself is not something I have had an opportunity to do a lot lately, so it was a challenging but pleasant activity".

What Was the Influence of the Intervention on the Participant's Career Indecision?

Lionel initially stated that he needed "validation" that he was on the "right path" regarding his career. He needed confirmation of the alignment between his master's degree studies in counselling psychology and what he wanted to do in the future. He was unsure whether he should venture into the field of community-based empowerment work (with youth sports teams). He wanted to travel and work with youth sports teams abroad but was hesitant and perhaps even afraid to do so. After the intervention, he declared that participating in the intervention had shown him that previously—when deciding on his future career and employment—he had tended to look for work just as a means of making money and putting food on table (a metaphor for safety). However, he now realised that just making money should not be the basis of career decision-making. He also learned that a lot more was involved in career decision-making than career-related interests only and that his "personal story" was not something separate but was actually an integral part of his career story. What he was requesting therefore was guidance concerning a current career-life project (Cochran, 1992; Maree & Di Fabio, 2018; Savickas, 2019). In operational terms, he needed counselling on how to construct a life project comprising (a) a key life theme involving his need to travel, see the world, and help his disadvantaged fellow countrymen and countrywomen; (b) a concomitant life task involving reigniting his career decision and making meaning of his studies, especially as his busy schedule meant that he could no longer participate in sport satisfactorily; and (c) doing what he could to enjoy life again by, for instance, participating in sport more often, setting and achieving academic goals, socialising with friends, and clarifying his career-life identity (Cochran, 1992; Savickas, 2019).

What Changes Did the Intervention Bring About?

In Lionel's words: "The career construction intervention felt like a divine intervention from God, my ancestors, and other supernatural forces that had all combined to give me this critical knowledge about myself". Even though he still needed to do a thorough job analysis to fine-tune aspects of his decision, he displayed signs of improved career decision-making readiness (adaptive readiness) and of actually adapting to ever-changing circumstances in his career-life by being willing to explore career possibilities actively (Savickas & Porfeli, 2012). This finding on his progress is consistent with the findings of Maree (2018) and Wehrle et al. (2019) on the power of career construction counselling to effect change in the lives of young people.

What Factors Prompted These Changes?

Lionel said that he felt that the "genuine unconditional positive regard" demonstrated during the intervention, coupled with the subconscious advice obtained from "feeling and thinking about some earlier painful experiences and other childhood memories", had helped him understand why he had previously been unable to connect and identify with the idea of travelling—an activity that appealed to him and would enable him to construct his career and himself adequately, experience purpose in his career-life, and make social contributions.

The Influence of Career Construction Counselling (Based on Co-constructive Conversations) on the Career Choice Indecision of a Master's Degree Student in Psychology

Summarised, Lionel was helped to reflect critically on key aspects of his career-life story by considering and reconsidering past memories, encounters, and thoughts. This helped him connect his past and his present with his career future (Savickas, 2016). Recounting his career-life story, coupled with his reflections and meta-reflections, shed light on how he made meaning of his career-life and how this enhanced his career decision-making. Lionel's responses and reflections clarified his interests, attributes, and motivations in relation to some of his idiosyncratic and general life tasks and key career-life themes and jointly comprise a life project (Cochran, 1992; Savickas, 2019). It also revealed his sense of meaning and hope and how he endeavoured to design a purposeful life that would enable him to make meaningful social contributions. In mapping the outline of his career-life story, we identified the difference between his conscious view of a possible future career-life trajectory and his subconsciously regulated perspective on that trajectory. Together, we co-constructed his possible career-life trajectory by eliciting inner advice from himself in regard to why he had sought career counselling.

Lionel's experiences demonstrated how a person's conscious knowledge could be connected with his subconscious advice through career construction counselling to resolve career choice indecision. This finding is consistent with that of Peng and Chen (2014) on the power of career construction counselling to help people struggling with career indecision (re)discover a sense of purpose in their career-lives. It also confirms Clark's (2002) view that a closely intertwined sequence of early experiences elicited indirectly (by eliciting early memories) and semi-directly (by eliciting hurtful experiences) can shape how people respond to occurrences that continue to influence their actions subconsciously. The specific culture contents of this case study confirm McAdams' (2013) assertion that a person's culture accounts for "the basic forms, metaphors, motifs, and plots out of which narrative identity is made" (p. 287). The findings demonstrate also how Lionel, as an emerging adult (Arnett, 2000) and as an autobiographical author (McAdams, 2013), "work[ed] to formulate a meaningful narrative for life, integrating the reconstructed episodic past and the imagined episodic future in such a way as to explain, for the self and for others, why the actor does what it does, why the agent wants what it wants, and who the self was, is, and will be as a developing person in time" (p. 273).

The present research also supports McAdams and Olson's (2010) view that "[t]o construct a narrative identity, the person must envision his or her entire life—the past reconstructed and the future imagined—as a story that portrays a meaningful sequence of life events to explain how the person has developed into who he or she is now and may develop into who he or she may be in the future" (p. 528). The intervention promoted reflexivity in that Lionel was helped to consider and reconsider past memories, encounters, and thoughts and to connect the past and the present with his career future (Savickas, 2016).

Limitations

The cultural differences between Lionel and me may have caused some misunderstanding between us. Certain cultural nuances may have gone unnoticed during the intervention. In addition, my deep interest in the plight of disadvantaged, rural people may have resulted in my being somewhat less "objective" than perhaps I could have been. Third, the duration of the research together with the need for extensive training in the mode of intervention described here may deter others from conducting this kind of research (Essays, UK, 2018). Fourth, the fact that I fulfilled the dual roles of counsellor and researcher may have influenced the outcomes of the research (Hay-Smith et al., 2016) although I did all I could to prevent such a possibility by planning the research methodology (including the assessment and intervention) carefully up front and then implementing and monitoring it closely. Below, some of the limitations listed above are transformed into recommendations.

12.5 Conclusion

This case study confirmed the effectiveness of the career construction counselling style (using the *CCI* in conjunction with the *CIP*) in eliciting and integrating various career-life stories of a master's degree student struggling with career indecision. The approach reported on here holds much promise as a way of helping people struggling with career indecision to revisit their "stifled" dreams and plans, meet their decision-making challenges, and enable them to move forward with renewed vigour, motivation, and hope in the workplace (Savickas, 2019). The intervention promoted connecting the participant's conscious knowledge with his subconscious advice to enable him to resolve his career choice indecision. However, success depended on uncovering and examining the roots of the participant's career indecision.

Lastly: It seems important to (a) examine people's conscious knowledge about themselves and the world of work, (b) uncover their subconscious insights, and, subsequently, (c) connect their conscious knowledge about themselves with their subconscious insights to (d) enhance their "self-advising". Based on the findings reported on in this chapter, such a strategy may rekindle their sense of self-efficacy and of being in control of fluid work contexts.

Acknowledgements I thank the participant for taking part in the study. I also thank Tim Steward for his editing of the text.

References

Adler, A. (1958). *What life should mean to you*. Capricorn Books.

Arnett, J. J. (2000). Emerging adulthood: A theory of development from the late teens through the twenties. *American Psychologist, 55*(5), 469–480. https://doi.org/10.1037/0003-066X.55.5.469

Career Indecision. (n.d.). Available at https://psychology.iresearchnet.com/counseling-psychology/career-counseling/career-indecision/. Accessed 23 June 2019.

Clark, A. J. (2002). *Early recollections: Theory and practice in counseling and psychotherapy*. Brunner-Routledge.

Cochran, L. (1992). The career project. *Journal of Career Development, 18*(3), 187–197. https://doi.org/10.1177%2F089484539201800303.

Cochran, L. (1997). *Career counseling: A narrative approach*. Sage.

Di Fabio, A., Palazzeschi, L., Asulin-Peretz, L., & Gati, I. (2013). Career indecision versus indecisiveness: Associations with personality traits and emotional intelligence. *Journal of Career Assessment, 21*(1), 42–56. https://doi.org/10.1177%2F1069072712454698.

Erikson, E. H. (1968). *Identity: Youth and crisis*. W. W. Norton.

Essays, U. K. (2018). *Limitations of case study approach*. Available at https://www.ukessays.com/essays/psychology/limitations-of-case-study-approach-psychology-essay.php?vref=1. Accessed 23 June 2019.

Freud, S. (1963). *The standard edition of the complete psychological works of Sigmund Freud*. Hogarth Press.

Garcia, P. R. J. M., Restubog, S. L. D., Toledano, L. S., Tolentino, L. R., & Rafferty, A. E. (2012). Differential moderating effects of student-and parent-rated support in the relationship between learning goal orientation and career decision-making self-efficacy. *Journal of Career Assessment, 20*(1), 22–33. https://doi.org/10.1177%2F1069072711417162.

Gati, I., & Asher, I. (2001). The PIC model for career decision making: Prescreening, in-depth exploration, and choice. In F. T. L. Leong & A. Barak (Eds.), *Contemporary models in vocational psychology: A volume in honor of Samuel H. Osipow*. Lawrence Erlbaum Associates.

Gati, I., Gadassi, R., Saka, N., Hadadi, Y., Ansenberg, N., Friedmann, R., & Asulin-Perez, L. (2011). Emotional and personality-related aspects of career decision-making difficulties: Facets of career indecisiveness. *Journal of Career Assessment, 19*(1), 3–20. https://doi.org/10.1177%2F1069072710382525.

Gianakos, I. (1999). Patterns of career choice and career decision-making self-efficacy. *Journal of Vocational Behavior, 54*(2), 244–258. https://doi.org/10.1006/jvbe.1998.1668

Guay, F., Senécal, C., Gauthier, L., & Fernet, C. (2003). Predicting career indecision: A self-determination theory perspective. *Journal of Counseling Psychology, 50*(2), 165–177. https://doi.org/10.1037/0022-0167.50.2.165

Guichard, J. (2009). Self-constructing. *Journal of Vocational Behavior, 75*(3), 251–258. https://doi.org/10.1016/j.jvb.2009.03.004

Hartung, P. J. (2011). Barrier or benefit? Emotion in life-career design. *Journal of Career Assessment, 19*(3), 296–305. https://doi.org/10.1177%2F1069072710395536

Hartung, P. J. (2013). The life-span, life-space theory of careers. In S. D. Brown & R. W. Lent (Eds.), *Career development and counseling: Putting theory and research to work* (2nd ed.). John Wiley & Sons.

Hartung, P. J. (2015). Life design in childhood - Antecedents and advancement. In L. Nota & J. Rossier (Eds.), *Handbook of life design: From practice to theory and from theory to practice*. Hogrefe Publishing.

Hay-Smith, E. J. C., Brown, M., Anderson, L., & Treharne, G. J. (2016). Once a clinician, always a clinician: A systematic review to develop a typology of clinician-researcher dual-role experiences in health research with patient-participants. *BMC Medical Research Methodology, 16*, 95. https://doi.org/10.1186/s12874-016-0203-6

Holland, J. L. (1997). *Making vocational choices: A theory of vocational personalities and work environments* (3rd ed.). Psychological Assessment Resources.

Kelly, K. R., & Lee, W. C. (2002). Mapping the domain of career decision problems. *Journal of Vocational Behavior, 61*(2), 302–326. https://doi.org/10.1006/jvbe.2001.1858

Krumboltz, J. (2003). Creating and capitalizing on happenstance in educational and vocational guidance. In R. Kunz, J. Dauwalder, & J. Renz (Eds.), *Quality development in vocational counseling and training*. Zürich, Switzerland.

Maree, J. G. (2013). *Counselling for career construction: Connecting life themes to construct life portraits. Turning pain into hope*. Sense.

Maree, J. G. (2016a). Career construction as a way of resolving career indecision. *South African Journal of Higher Education, 30*(3), 170–192. https://doi.org/10.20853/30-3-665

Maree, J. G. (2016b). Career construction counseling with a mid-career black man [Special issue]. *The Career Development Quarterly, 64*(1), 20–34. https://doi.org/10.1002/cdq.12038

Maree, J. G. (2016c). How career construction counseling promotes reflection and reflexivity: Two case studies. *Journal of Vocational Behavior, 97*, 22–30. https://doi.org/10.1016/j.jvb.2016.07.009

Maree, J. G. (2017a). *Career interest profile* (Version 6). JvR Psychometrics.

Maree, J. G. (2017b). Life design counselling. In G. B. Stead & M. B. Watson (Eds.), *Career psychology in the South African context* (3rd ed.). Pretoria, South Africa.

Maree, J. G. (2018). Promoting career development in the early years of a person's life through self- and career construction counselling (using an integrated, qualitative + quantitative approach): A case study. *Early Child Development and Care, 188*(4), 437–451. https://doi.org/10.1080/03004430.2017.1365361

Maree, J. G. (2019). Postmodernising career counselling in a deep rural context. In J. G. Maree (Ed.), *Handbook of innovative career counselling*. Springer. https://doi.org/10.1007/978-3-030-22799-9

Maree, J. G. (2020). Connecting conscious knowledge with subconscious advice through career construction counselling to resolve career choice indecision [Special issue]. *South African Journal of Education, 40*(S1) Art. #1863, 13 pages. https://doi.org/10.15700/saje.v40ns1a1863

Maree, J. G., & Di Fabio, A. (2018). Integrating personal and career counseling to promote sustainable development and change. *Sustainability, 10*(11), 4176. https://doi.org/10.3390/su10114176

Maree, J. G., & Sommerville, J. (2008). Development of a post-modern career interest questionnaire for South African school settings. *Journal of Psychology in Africa, 18*(1), 125–132. https://doi.org/10.1080/14330237.2008.10820179

McAdams, D. P. (2013). The psychological self as actor, agent, and author. *Perspectives on Psychological Science, 8*(3), 272–295. https://doi.org/10.1177%2F1745691612464657.

McAdams, D. P., & Olson, B. D. (2010). Personality development: Continuity and change over the life course. *Annual Review of Psychology, 61*, 517–542. https://doi.org/10.1146/annurev.psych.093008.100507

Merrill, B., & West, L. (2009). *Using biographical methods in social research.* Sage.

Peng, H., & Chen, M. S. (2014). The study of spirituality, work value and career decision-making between Christian and non-Christian college students. *The Journal of Happiness & Well-Being, 2*(1), 63–74. Available at http://www.journalofhappiness.net/frontend/articles/pdf/v02i01/Huiling-1.pdf. Accessed 22 June 2019

Reid, H. (2018). Prompting reflection and learning in career construction counseling: Three research studies evaluating process and practice in the UK. In P. J. Hartung (Chair). *Reflexivity and change in life designing: Perspectives from process research. 29th International Congress of Applied Psychology (ICAP) conducted at the Palais des Congrès, Montréal, Canada, 26–30 June.*

Richardson, L. (2020). Writing: A method of inquiry. In N. K. Denzin & Y. S. Lincoln (Eds.), *Handbook of qualitative research* (2nd ed.). Sage.

Savickas, M. L. (2005). The theory and practice of career construction. In S. D. Brown & R. W. Lent (Eds.), *Career development and counseling: Putting theory and research to work.* John Wiley & Sons.

Savickas, M. L. (2011). Prologue: Reshaping the story of career counselling. In K. Maree (Ed.), *Shaping the story: A guide to facilitating narrative career counselling.* Rotterdam, The Netherlands.

Savickas, M. L. (2015a). Career counselling paradigms: Guiding, developing, and designing. In P. Hartung, M. Savickas, & W. Walsh (Eds.), *The APA handbook of career intervention* (Vol. 1). APA Books.

Savickas, M. L. (2015b). *Life-design counseling manual.* n.p.

Savickas, M. L. (2016). Reflection and reflexivity during life-design interventions: Comments on Career Construction Counseling. *Journal of Vocational Behavior, 97*, 84–89. https://doi.org/10.1016/j.jvb.2016.09.001

Savickas, M. L. (2019). *Career counselling* (2nd ed.). American Psychological Association.

Savickas, M. L., Nota, L., Rossier, J., Dauwalder, J. P., Duarte, M. E., Guichard, J., Soresi, S., Van Esbroeck, R., & Van Vianen, A. E. M. (2009). Life designing: A paradigm for career construction in the 21st century. *Journal of Vocational Behavior, 75*(3), 239–250. https://doi.org/10.1016/j.jvb.2009.04.004

Savickas, M. L., & Porfeli, E. J. (2012). Career Adapt-Abilities Scale: Construction, reliability, and measurement equivalence across 13 countries. *Journal of Vocational Behavior, 80*(3), 661–673. https://doi.org/10.1016/j.jvb.2012.01.011

Sieck, B. C. (2012). Obtaining clinical writing informed consent versus using client disguise and recommendations for practice. *Psychotherapy, 49*(1), 3–11. https://doi.org/10.1037/a0025059

Stewart, H., Gapp, R., & Harwood, I. (2017). Exploring the alchemy of qualitative management research: Seeking trustworthiness, credibility and rigor through crystallization. *The Qualitative Report, 22*(1), 1–19. Available at https://nsuworks.nova.edu/cgi/viewcontent.cgi?article=2604&context=tqr/. Accessed 22 June 2019

Super, D. E. (1957). *The psychology of careers.* Harper & Row.

Sweeney, T. (1975). *Adlerian counseling.* Houghton-Mifflin.

Urbanaviciute, I., Udayar, S., & Rossier, J. (2019). Career adaptability and employee well-being over a two-year period: Investigating cross-lagged effects and their boundary conditions. *Journal of Vocational Behavior, 111*, 74–90. https://doi.org/10.1016/j.jvb.2018.10.013

Wehrle, K., Kira, M., & Klehe, U. C. (2019). Putting career construction into context: Career adaptability among refugees. *Journal of Vocational Behavior, 111*, 107–124. https://doi.org/10.1016/j.jvb.2018.08.007

Welty, E. (1998). *One writer's beginnings.* Harvard University Press.

Chapter 13
Acceptance of Change Among Workers for Sustainability in Organizations: Trait Emotional Intelligence and Insight Orientation

Alessio Gori, Andrea Svicher, Letizia Palazzeschi, and Annamaria Di Fabio

13.1 Introduction

The psychology of sustainability and sustainable development (PSSD) (Di Fabio, 2017a, b; Di Fabio & Rosen, 2018) represents a recent research area settled within the transdisciplinary domain of sustainability science. Opening the black box of psychological processes in the field of sustainability represents a stimulating challenge (Di Fabio & Rosen, 2018) to tangible progress in the now and the future. The PSSD introduced a new perspective that underlines positive challenge and pays attention to new resources, new opportunities, and regenerating processes. From this perspective of the PSSD, different environments are considered—from the natural to the personal, social, organizational, community, cross-cultural continuing up to global environments. In this framework, the psychology of harmonization (Di Fabio & Tsuda, 2018) emerged as a vital pillar, introducing "harmonization in geographical and temporal perspectives, including meaningful construction processes from the past, to the present, and into the future" (p. 1). Furthermore, this perspective aims to identify and foster new strengths from a primary prevention perspective (Di Fabio & Kenny, 2015, 2016; Hage et al., 2007) and strength-based prevention perspectives (Di Fabio & Saklofske, 2021), promoting well-being in different environments, including the organizational environment (Di Fabio, 2017a; Di Fabio et al., 2020; Peiró, 2008; Tetrick & Peiró, 2012).

A. Gori
Department of Health Sciences, University of Florence, Florence, Italy
e-mail: annamaria.difabio@unifi.it

A. Di Fabio (✉) · A. Svicher · L. Palazzeschi
Department of Education, Languages, Intercultures, Literatures and Psychology (Psychology Section), University of Florence, Florence, Italy

A. Di Fabio (ed.), *Cross-cultural Perspectives on Well-Being and Sustainability in Organizations*, https://doi.org/10.1007/978-3-030-86709-6_13

The primary prevention perspective (Di Fabio & Kenny, 2015, 2016; Hage et al., 2007; Kenny & Hage, 2009) and strength-based prevention perspectives (Di Fabio & Saklofske, 2021) are centered on the promotion of psychological strengths that could be increased through training for effectively answering to the challenges of the current labor market (Blustein et al., 2019). In organizations, these perspectives highlight that fulfilling work can offer important contributions to the overall health and well-being of workers (Blustein, 2011; Peiró, 2008; Peiró et al., 2020; Tetrick & Peiró, 2012) and underline the importance of focusing on enhancing workers' strengths through early intervention to expose variables that could be amenable to specific training.

In the sphere of prevention perspectives (Di Fabio & Kenny, 2015, 2016; Di Fabio & Saklofske, 2014a, b, 2018, 2019a, b, 2021), the present study aims to investigate the possible contribution of trait emotional intelligence and insight as primary prevention resources for the acceptance of change referred to the personal environment. In particular, this study aims to analyze the mediation contribution of insight in the association between trait emotional intelligence and acceptance of change.

Acceptance of Change

Within the PSSD framework (Di Fabio, 2017a, b, Di Fabio & Rosen, 2018), acceptance of change is very promising for sustainability in organizations, since this framework changes the perspective toward sustainability from one of damage/threat to one of opportunity/change.

Acceptance of change is defined as the "tendency to embrace rather than shy away from change" (Di Fabio & Gori, 2016, p. 2). Individuals who accept change consider that change could have a positive effect on their lives; they also think they have the possibility to learn from change (Di Fabio & Gori, 2016). Acceptance of change is also seen as positive toward a person's well-being (Di Fabio & Gori, 2016) and a worker's job satisfaction (Gori et al., 2020). Acceptance of change has five dimensions: "predisposition to change, the perceived ability people have to learn from change and to employ change to enhance the quality of their lives; support for change, the perceived social support from other people when facing changes; change seeking, the propensity to search for change, and the perceived ability to obtain and retain information and to express a necessity for new stimuli; positive reaction to change, the perception of positive emotional reactions to change seen as positive and considering to have benefits from it; cognitive flexibility, mental ability to switch between different concepts or to adapt cognitive processing strategies" (Di Fabio & Gori, 2016, p. 3).

Emotional Intelligence

In the literature, two models of emotional intelligence are delineated (Stough et al., 2009): ability-based models (Mayer & Salovey, 1997) and a trait emotional intelligence model (Bar-On, 1997; Petrides & Furnham, 2001). Ability-based models focus on cognitive aspects of emotional intelligence as the ability to interpret emotional information in an accurate manner (Mayer et al., 2000). Contrarily, trait emotional intelligence models refer to the self-evaluated experience of emotional and social competences (Bar-On, 1997; Petrides & Furnham, 2001). According to Bar-On (1997), a self-reporting framework refers to self-reported emotional intelligence considering the self-perceptions of one's emotional and social skills that determine how individuals recognize their emotions, interact with others, and face the challenges posed by the environment. Trait emotional intelligence (Petrides & Furnham, 2001) integrates the dimensions included in the previous model by Bar-On (1997) and expands upon them by considering expression of emotions, regulation of emotions, and self-motivation. The model of Petrides and Furnham (2001) describes trait emotional self-efficacy in terms of a constellation of self-perceptions relative to emotions associated with the domain of personality.

In the literature, trait emotional intelligence is positively associated with a positive attitude toward organizational change (Vakola et al., 2004). Furthermore, trait emotional intelligence is inversely associated with resistance to change among Italian workers (Di Fabio et al., 2014).

Insight

From the primary prevention perspective (Di Fabio & Kenny, 2015, 2016: Hage et al., 2007; Kenny & Hage, 2009) and strength-based prevention perspectives (Di Fabio & Saklofske, 2021), insight could also represent a promising primary prevention resource due to its central role in promoting change (Gori et al., 2015). The concept of insight stems from psychotherapy (Gori et al., 2015; Connolly Gibbons et al., 2007). Change is created by promoting a safe and productive space where clients can learn to manage their emotions (including expressing negative emotion) and where they can provide insight (Greenberg & Safran, 1987). This process of insight is considered a fundamental action of change in psychotherapy (Brewin et al., 1996; Foa & Kozak, 1986; Greenberg, 2002; Samoilov et al., 2000; Stiles et al., 1990; Whelton, 2004). In different paradigms, insight is theorized as "a dynamic variable which could range from a superficial consciousness to a more complex comprehension leading to a change in point of view, and these conceptions also represent different levels of the elaboration of emotional material" (Gori et al., 2015, p. 298). Emotions are thus intertwined with insight-laden processes (Høglend & Hagtvet, 2019; McCarthy et al., 2017).

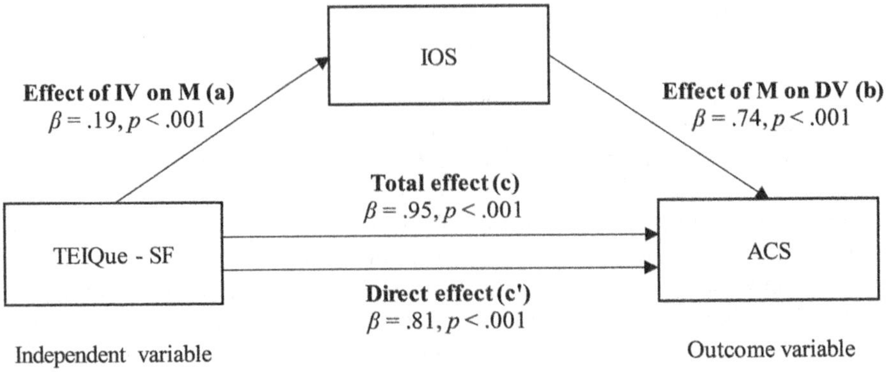

Fig. 13.1 Relationship between trait emotional intelligence (TEIQue-SF) and acceptance of change with IOS as mediator

13.2 Aim and Hypotheses

The present study examines the relationships between trait emotional intelligence, insight, and acceptance of change—specifically, the possibility of mediating the effect of insight in the relationship between trait emotional intelligence and acceptance of change.

The following hypotheses were formulated:

- H1. *Trait emotional intelligence* will be positively correlated with *acceptance of change* (estimating path *c* in Fig. 13.1) to establish whether there is the possibility of a mediated effect.
- H2. *Trait emotional intelligence* will be positively correlated with *insight*, the mediator variable (estimating and testing path *a* in Fig. 13.1).
- H3. *Insight* will be positively correlated with *acceptance of change* (estimating and testing path *b* in Fig. 13.1).
- H4. The association between *emotional intelligence* and *acceptance of change* will be mediated by *insight*.

These hypotheses will be verified by analyzing the effect of *trait emotional intelligence* on *acceptance of change*, controlling for *insight* (path *c'* in Fig. 13.1).

13.3 Method

Participants

The participants in this study were 182 workers in public Italian organizations (66.4% men, 33.6% women). The mean age was 49.07 years ($SD = 8.86$).

Measures

Trait Emotion Intelligence Questionnaire Short Form (TEIQue-SF)

To evaluate trait EI, the Italian-language version by Di Fabio and Palazzeschi (2011) of the TEIQue-SF (Petrides, 2009) was used. This questionnaire is composed of 30 items on a seven-point Likert scale from 1 = *completely disagree* to 7 = *completely agree*. The questionnaire detects four factors (well-being, self-control, emotionality, sociability) and a total score. The Cronbach alphas were 0.82 for well-being, 0.80 for self-control, 0.81 for emotionality, and 0.82 for sociability; the total score's Cronbach alpha was 0.81.

Insight Orientation Scale (IOS)

The IOS (Gori et al., 2015) included seven items on a five-point Likert scale from 1 = *not at all* to 5 = *a great deal*. The IOS investigated four specific aspects of insight as conscious change: "surprise (making connections), restructuring (change ability, conviction about a new meaning), level of consciousness (ability to experience feelings), problem solving (problem finding, problem shaping), complexity (abstraction, depth), self-reflectiveness (thoughtfulness), awareness (attention, ability to perceive)" (IOS; Gori et al., 2015, p. 300). The Cronbach alpha was 0.77.

Acceptance of Change Scale (ACS)

The ACS (Di Fabio & Gori, 2016) includes 20 items on a five-point Likert scale from 1 = *not at all* to 5 = *a great deal*. The scale has five dimensions: predisposition to change ($\alpha = 0.83$); support for change ($\alpha = 0.79$); change seeking ($\alpha = 0.80$); positive reaction to change ($\alpha = 0.72$); and cognitive flexibility ($\alpha = 0.72$). The Cronbach alpha coefficient for the total score was 0.88.

Procedure and Data Analysis

Trained psychologists administered the self-report measures and counterbalanced the sequence of the questionnaires to control for the potential effect of presentation. Norms relative to privacy and informed consent of Italian law were respected during the administration.

Descriptive statistics, Pearson's r correlations, and mediation analysis Hayes (2013) were calculated. A simple mediation model was hypothesized considering the relationship of an independent variable (x) with a dependent variable (y) mediated by a third variable (m).

13.4 Results

Table 13.1 presents the means, standard deviations, and correlations of the TEIQue-SF, IOS, and ACS.

Trait EI showed correlations > 0.30 with both *insight* and *acceptance of change*. A mediation analysis was carried out with *trait EI* as the independent variable and *insight* as the mediator.

Trait EI was associated in a direct and positive way with *acceptance of change* and was also inversely related to *acceptance of change* through *insight*. Participants who scored a higher *trait EI* showed more *insight* (path *a* in Fig. 13.1; $\beta = 0.19$) and in turn perceived themselves to have more *acceptance of change* (path *b* in Fig. 13.1; $\beta = 0.74$). A bias-corrected bootstrap confidence interval for the indirect effect (path *ab* in Fig. 13.1; $\beta = 0.14$) based on 5000 bootstrap samples was entirely above zero (0.08 to 0.21). The effect of *trait EI* on *acceptance of change* decreased after controlling for *insight*, but continued to be statistically significant (path *c'* in Fig. 13.1; $p < 0.001$). These findings showed a partial mediation model (R^2 mediator effect size = 0.17, $p < 0.001$).

13.5 Discussion

The present study aimed to analyze the contribution of insight in mediating the relationship between trait emotional intelligence and acceptance of change in Italian workers. The hypotheses were confirmed. Trait emotional intelligence was positively and directly associated with acceptance of change analogous to the results in the literature describing positive attitudes toward organizational change (Vakola et al., 2004), which highlights the possible contribution of emotions in change processes (H1). Trait emotional intelligence was also positively associated with insight, emphasizing that emotions could be involved in insight processes (H2) (Gori et al., 2015; Høglend & Hagtvet, 2019; McCarthy et al., 2017). There is also a route via insight toward acceptance of change (H3). Individuals with higher emotional intelligence have higher acceptance toward change. Furthermore, trait emotional intelligence appears to influence acceptance indirectly through insight. Participants with a higher level of emotional intelligence are more likely to have greater insight, and this is associated with an increase in acceptance of change. Hence, these results suggest the partial role of insight as a mediator between trait emotional intelligence and

Table 13.1 Correlations for the TEIQue-SF, IOS, and ACS

	1	2	3
1. TEIQue-SF	–		
2. IOS	.41**	–	
3. ACS	.50**	.57**	–

Note. $N = 182$. * $p < 0.05$, ** $p < 0.01$

acceptance of change (H4) in terms of tendency to embrace changes rather than avoid them (Di Fabio & Gori, 2016).

Despite these reliable obtained results, this study showed some limitations. The participants were a limited group of Italian workers who did not represent all Italian organizational contexts. The relationships among the variables investigated in the present study should be further examined, considering different types of organizations within the Italian context. It would be also interesting to reproduce the study involving different targets (e.g., adults in transition, etc.) and to replicate this study in other countries. Future studies should analyze associations between the different EI models: for example, the Bar-On model (Bar-On, 1997) and the ability-based emotional intelligence model (Mayer & Salovey, 1997).

13.6 Conclusion

If the results of this research are to be replicated in future studies, promising gateways toward beneficial interventions could also be opened. From a primary prevention perspective (Di Fabio & Kenny, 2015, 2016; Hage et al., 2007; Kenny & Hage, 2009) and strength-based prevention perspectives (Di Fabio & Saklofske, 2021), insight could represent a promising primary prevention resource with a central role in promoting change (Gori et al., 2015). Insight could be promoted using early specific intervention to facilitate workers in accepting changes that characterize the current unstable world of work (Blustein et al., 2019). From the psychology of sustainability and sustainable development perspective, insight could represent a primary prevention resource for acceptance of change referred to in the personal environment, which could underline the contribution of psychological variables in opening the black box of psychological processes of sustainability from wide-ranging perspectives (Di Fabio & Rosen, 2018).

References

Bar-On, R. (1997). *The emotional intelligence inventory (EQ-I): Technical manual.* Multi-Health Systems.

Blustein, D. L. (2011). A relational theory of working. *Journal of Vocational Behavior, 79*, 1–17.

Blustein, D. L., Kenny, M. E., Di Fabio, A., & Guichard, J. (2019). Expanding the impact of the psychology of working: Engaging psychology in the struggle for decent work and human rights. *Journal of Career Assessment, 27*, 3–28.

Brewin, C. R., Dalgleish, T., & Joseph, S. (1996). A dual representation theory of posttraumatic stress disorder. *Psychological Review, 103*(4), 670–686.

Connolly Gibbons, M. B. C., Crits-Christoph, P., Barber, J. P., & Schamberger, M. (2007). Insight in psychotherapy: A review of empirical literature. In L. G. Castonguay & C. E. Hill (Eds.), *Insight in psychotherapy* (pp. 143–165). American Psychological Association Press.

Di Fabio, A. (2017a). Positive Healthy Organizations: Promoting well-being, meaningfulness, and sustainability in organizations. *Frontiers in Psychology. Organizational Psychology, 8*, 1938. https://doi.org/10.3389/fpsyg.2017.01938

Di Fabio, A. (2017b). The psychology of sustainability and sustainable development for well-being in organizations. *Frontiers in Psychology. Organizational Psychology, 8*, 1534. https://doi.org/10.3389/fpsyg.2017.01534

Di Fabio, A., Bernaud, J.-L., & Loarer, E. (2014). Emotional intelligence or personality in resistance to change? Empirical results in an Italian health care context. *Journal of Employment Counseling, 51*, 146–157.

Di Fabio, A., Cheung, F., & Peiró, J.-M. (2020). Editorial Special Issue Personality and individual differences and healthy organizations. *Personality and Individual Differences, 166*. https://doi.org/10.1016/j.paid.2020.110196

Di Fabio, A., & Gori, A. (2016). Developing a new instrument for assessing Acceptance of Change. *Frontiers in Psychology. Section Organizational Psychology, 7*, 802. https://doi.org/10.3389/fpsyg.2016.00802

Di Fabio, A., & Kenny, M. E. (2015). The contributions of emotional intelligence and social support for adaptive career progress among Italian youth. *Journal of Career Development, 42*, 48–59.

Di Fabio, A., & Kenny, E. M. (2016). From decent work to decent lives: Positive Self and Relational Management (PS&RM) in the twenty-first century. In A. Di Fabio & D. L. Blustein (Eds.), From meaning of working to meaningful lives: The challenges of expanding decent work. Research Topic in *Frontiers in Psychology. Section Organizational Psychology, 7*, 361. https://doi.org/10.3389/fpsyg.2016.00361

Di Fabio, A., & Palazzeschi, L. (2011). Proprietà psicometriche del Trait Emotional Intelligence Questionnaire Short Form (TEIQue-SF) nel contesto italiano [Psychometric properties of the Trait Emotional Intelligence Questionnaire Short Form (TEIQue-SF) in the Italian context]. Counseling. *Giornale Italiano di Ricerca e Applicazioni, 4*, 327–336.

Di Fabio, A., & Rosen, M. A. (2018). Opening the Black Box of Psychological Processes in the Science of Sustainable Development: A New Frontier. *European Journal of Sustainable Development Research, 2*(4), 47. https://doi.org/10.20897/ejosdr/3933

Di Fabio, A., & Saklofske, D. H. (2014a). Comparing ability and self-report trait emotional intelligence, fluid intelligence, and personality traits in career decision. *Personality and Individual Differences, 64*, 174–178.

Di Fabio, A., & Saklofske, D. H. (2014b). Promoting individual resources: The challenge of trait emotional intelligence. *Personality and Individual Differences, 65*, 19–23.

Di Fabio, A., & Saklofske, D. H. (2018). The contributions of personality and emotional intelligence to resiliency. *Personality and Individual Differences, 123*, 140–144. https://doi.org/10.1016/j.paid.2017.11.012

Di Fabio, A., & Saklofske, D. H. (2019a). Positive relational management for sustainable development: Beyond personality traits- The contribution of emotional intelligence. In A. Di Fabio (Ed.), Psychology of sustainability and sustainable development. Special Issue in *Sustainability* MDPI, *11*(2), 330. https://doi.org/10.3390/su11020330

Di Fabio, A., & Saklofske, D. H. (2019b). The contributions of personality traits and emotional intelligence to intrapreneurial self-capital: Key resources for sustainability and sustainable development. Special Issue in *Sustainability* MDPI, *11*, 1240. https://doi.org/10.3390/su11051240

Di Fabio, A., & Saklofske, D. H. (2021). The relationship of compassion and self-compassion with personality and emotional intelligence. PAID 40th Anniversary Special Issue. *Personality and Individual Differences*, 169. https://doi.org/10.1016/j.paid.2020.110109

Di Fabio, A., & Tsuda, A. (2018). The psychology of harmony and harmonization: Advancing the perspectives for the psychology of sustainability and sustainable development. In A. Di Fabio (Ed.), Psychology of Sustainability and Sustainable Development. Special Issue in *Sustainability* MDPI, *10*(12), 4726. https://doi.org/10.3390/su10124726

Foa, E. B., & Kozak, M. J. (1986). Emotional processing of fear: exposure to corrective information. *Psychological Bulletin, 99*(1), 20–35.

Gori, A., Craparo, G., Giannini, M., Loscalzo, Y., Caretti, V., La Barbera, D., … Dewey, D. (2015). Development of a new measure for assessing insight: Psychometric properties of the insight orientation scale (IOS). *Schizophrenia Research, 169*(1-3), 298–302.

Gori, A., Topino, E., Palazzeschi, L., & Di Fabio, A. (2020). How can organizational justice contribute to job satisfaction? A chained mediation model. *Sustainability, 12*(19), 7902.

Greenberg, L. S. (2002). *Emotion-focused therapy: Coaching clients to work through their feelings*. American Psychological Association Press.

Greenberg, L. S., & Safran, J. D. (1987). *Emotion in psychotherapy: Affect cognition and the process of change*. Guildford Press.

Hage, S. M., Romano, J. L., Conyne, R. K., Kenny, M., Matthews, C., Schwartz, J. P., & Waldo, M. (2007). Best practice guidelines on prevention practice, research, training, and social advocacy for psychologists. *The Counseling Psychologist, 35*(4), 493–566.

Hayes, A. F. (2013). *Introduction to mediation, moderation, and conditional process analysis: A regression-based approach*. Guilford Press.

Høglend, P., & Hagtvet, K. (2019). Change mechanisms in psychotherapy: Both improved insight and improved affective awareness are necessary. *Journal of Consulting and Clinical Psychology, 87*(4), 332–344. https://doi.org/10.1037/ccp0000381

Kenny, M. E., & Hage, S. M. (2009). The next frontier: Prevention as an instrument of social justice. *The Journal of Primary Prevention, 30*(1), 1–10.

Mayer, J. D., & Salovey, P. (1997). What is emotional intelligence? In P. Salovey & D. Sluyter (Eds.), *Emotional development and emotional intelligence: Educational implications* (pp. 3–31). New York.

Mayer, J. D., Salovey, P., & Caruso, D. R. (2000). Selecting a measure of emotional intelligence: The case of ability scales. In R. Bar-On & J. D. Parker (Eds.), *The handbook of emotional intelligence* (pp. 320–342). Jossey Bass.

McCarthy, K. L., Caputi, P., & Grenyer, B. F. (2017). Significant change events in psychodynamic psychotherapy: Is cognition or emotion more important? *Psychology and Psychotherapy: Theory, Research and Practice, 90*(3), 377–388.

Peiró, J. M. (2008). Stress and coping at work: new research trends and their implications for practice. In K. Näswall, J. Hellgren, & M. Sverke (Eds.), *The individual in the changing working life* (pp. 284–310). Cambridge University Press.

Peiró, J. M., Bayonab, J. A., Caballer, A., & Di Fabio, A. (2020). Importance of work characteristics affects job performance: The mediating role of individual dispositions on the work design-performance relationships. *PAID 40th Anniversary Special Issue. Personality and Individual Differences, 157*. https://doi.org/10.1016/j.paid.2019.109808

Petrides, K. V. (2009). Psychometric properties of the Trait EI Questionnaire. In C. Stough, D. H. Saklofske, & J. D. Parker (Eds.), *Advances in the assessment of EI*. Springer. https://doi.org/10.1007/978-0-387-88370-0_5

Petrides, K. V., & Furnham, A. (2001). Trait EI: Psychometric investigation with reference to established trait taxonomies. *European Journal of Personality, 15*, 425–428.

Samoilov, A., Goldfried, M. R., & Shapiro, D. A. (2000). Coding system of therapeutic focus on action and insight. *Journal of Consulting and Clinical Psychology, 68*(3), 513–514.

Stiles, W. B., Elliott, R., Llewelyn, S., Firth-Cozens, J., Margison, F. R., Shapiro, D. A., & Hardy, G. (1990). Assimilation of problematic experiences by clients in psychotherapy. *Psychotherapy, 27*(1), 411–420.

Stough, C., Saklofske, D., & Parker, J. (2009). *Assessing EI: Theory, research, and applications*. Springer.

Tetrick, L. E., & Peiró, J. M. (2012). Occupational Safety and Health. In *The Oxford handbook of organizational psychology* (Vol. 2). Oxford University Press.

Vakola, M., Tsaousis, I., & Nikolaou, I. (2004). The role of emotional intelligence and personality variables on attitudes toward organisational change. *Journal of Managerial Psychology, 19*(2), 88–110.

Whelton, W. J. (2004). Emotional processes in psychotherapy: Evidence across therapeutic modalities. *Clinical Psychology & Psychotherapy: An International Journal of Theory & Practic, 11*(1), 58–71.

Chapter 14
Are They Competitive? Hungarian Business People and Competition Three Decades After the Regime Change

Márta Fülöp

14.1 Introduction

The general characteristics of today's economic environment, the fierce competition, the continuous technical development and the uninterrupted transformation of the economic and social environment necessarily have the consequence that companies must constantly develop and adapt their activities and technology to new needs. In order to remain competitive, companies must become more and more entrepreneurial: that is, they must detect and take advantage of, and create, the business opportunities that arise (Faragó, 2008; Wang et al., 2018).

Competition therefore is an important factor in the survival of organizations and central phenomenon in business life. As Brandts et al. (2005) state, there is a strong consensus among mainstream economists that "competitive spur" (p. 1) is a key ingredient for improving efficiency and welfare. Not only companies and organizations have to compete for their competitive advantage, but also those who have leadership roles in the business life, the business owners, business managers and chief executive directors. Market competition requires leaders to constantly pay attention to and be knowledgeable of competitors and their products (Kirkpatrick & Locke, 1991).

Within the company, competitive attitude, as a kind of workplace attitude, may also contribute to job performance, because those high in trait competitiveness devote more effort and energy to perform at their best (Wang et al., 2018). Business life in most of its segments provides a competitive environment, and it is a question

M. Fülöp (✉)
Institute of Cognitive Neuroscience and Psychology, Research Centre of Natural Sciences, Budapest, Hungary

Institute of Psychology, Károli Gáspár University of the Reformed Churches, Budapest, Hungary
e-mail: fulop.marta@ttk.hu

A. Di Fabio (ed.), *Cross-cultural Perspectives on Well-Being and Sustainability in Organizations*, https://doi.org/10.1007/978-3-030-86709-6_14

how much individuals self-select to work in that kind of environment, how much their personality characteristics match that environment or in contrast to this how much they try to avoid or shy away from competition (Bartling et al., 2009).

Di Fabio (2017a) postulates that healthy work climate contributes to a healthy organization and healthy business which as a consequence establishes the foundation of sustainability in organizations. There are no absolute attributes of a healthy work climate. Certain aspects of a climate which may be detrimental to the health of some can be enjoyable and conducive to psychological and somatic health for others. This largely depends on the person-environment fit (Kristof-Brown & Guay, 2011). If this is optimal, then it also raises the quality of life because instead of consuming considerable energy, it rather regenerates personal resources. If this is the case with those who are in leadership and managerial positions, then it may contribute to organizational effectiveness and successful business practice. Di Fabio (2017a, b) emphasizes that sustainable organizational profitability is dependent on worker's well-being. The research presented here is conducted with participants who work in an intensely competitive sector; therefore, their well-being – based on the literature – should be particularly at risk (Brandts et al., 2005). However, it was found that the great majority of them in fact enjoys competition and it rather gives them joy than overwhelms them (Urbig et al., 2020), potentially because they feel that their work culture is compatible with their personal strivings; thus, it rather may contribute to better life satisfaction. To those who are competitive-minded, a competitive environment facilitates bringing out their best, unleashing their intellectual potentials and achieving high performance which may contribute to higher subjective well-being. Positive healthy organizations are based on building resources and strengths with success as the criterion (Di Fabio, 2017b). However, for those who don't experience a comfortable overlap between their personality and the competitive sector, they may rather want to avoid competition, if they are still in that kind of work environment may consider themselves less successful, may have lower self-esteem and may experience burn-out. Compatibility of trait competitiveness and competitive climate among those in leadership positions in the business life from the perspective of the participants has never been studied.

Characteristics of Those Who Choose the Business Life

Entrepreneurs and enterprising behaviour are key factors in allowing organizations to compete successfully in a complex and changing environment. In order to become more competitive and increase national wealth, all countries need people with an interest in the business context. In high-growth, global, competitive economies, business-mindedness is regarded as a crucial determinant of economic growth and prosperity, and it is closely linked with efforts to create wealth (Fülöp & Berkics, 2002).

It is a commonly shared view that those with a propensity towards the business career have certain characteristics which distinguish them from those who are not attracted by that context.

The earliest and probably the most influential paradigm emphasizes that entrepreneurs have a higher than average achievement motivation and activity level (McClelland & Winter, 1969). They are characterized by a greater ambition and need for personal success (Kirkpatrick & and Locke, 1991). Another early paradigm connected entrepreneurship with creativity and innovation (Atkinson, 1957). Kirzner (1979) developed a theory of entrepreneurial alertness, which asserts that entrepreneurs are more alert to new opportunities and use information differently from others. It was also argued that entrepreneurs are more likely to believe in the internal rather than in external locus of control (Koh, 1996). According to Lumpkin and Dess (1996), entrepreneurial orientation consists of five characteristics: innovativeness, proactivity, risk-taking, autonomy and aggressive competitiveness. Camerer and Lovallo (1999) described a good deal of confidence or even overconfidence as a salient characteristic of the successful business people. Embracing risk proved to be one of the most important requirements (Caliendo et al., 2009; Wu & Knott, 2006; Rauch & Frese, 2007; Al Habib, 2012), but vigorously striving for and achieving profit and business growth; seeking out new opportunities; propensity to create business organizations (Cromie, 2000); initiative, flexibility and imagination (Gibb, 1987); and being pioneering, adventurous, venturesome, daring and go-ahead (Chell et al., 1991) were also emphasized. The vigorous strive and the daring proactive behaviour may sometimes devolve into aggressiveness or even ruthlessness (Miller, 2014). Sustaining goal-directed action, communicated vision, self-efficacy, perseverance, passion and energy even when faced with obstacles was found to have effects on venture growth as well (Baum & Locke, 2004). The meta-analysis of Rausch and Frese (2007) found that the following personality traits correlated with entrepreneurial spirit, starting a business and business success: performance motivation, general self-efficiency, innovation, stress tolerance, need for autonomy and proactivity.

Several of the characteristics listed in relation to those who are willing to become active in business life and have entrepreneurial intentions may have strong connection with competitiveness. For example, risk propensity was found to correlate significantly with self-developmental and hypercompetitiveness, while there was a significant negative correlation between these competitive attitudes and risk cautiousness (Fülöp et al., 2016). Trait competitiveness was considered by McClelland et al. (1953) a dimension of achievement motivation. The Achievement Scale of Personality Research Form (Jackson, 1974) which is frequently used to study entrepreneurs' personality characteristics measures besides aspiring to accomplish difficult tasks, maintaining high standards, willing to work towards distant goals and willing to put effort to attain excellence also the positive response to competition as an indicator of high need of achievement. Some associated the aggressive behaviour of entrepreneurs with their motivation to compete (Utsch et al., 1999; Hmielski & Lerner, 2016; Levine & Rubinstein, 2017). Performance motivation and performance goals which were found by Rausch and Frese (2007) typical among business people reflect competitive motives and the pursuit of outperforming others and display superior performance (Harackiewicz et al., 2000; Kerr et al., 2018).

Entrepreneurs and Competitiveness

While the studies mentioned in the following section may make the impression that among the thousands of studies on the necessary traits of entrepreneurs and business people their competitiveness has been also extensively studied, this is not the case. Most of the studies are like Baum & Locke, (2004) who studied the relationship of entrepreneurial traits, skill and motivation to subsequent venture growth, but they did not include the competitive attitudes of the participants.

Schumpeter (1934), one of the most influential economists of the first half of the past century already in 1934, stressed that those who like to work in a competitive environment and who like to fight and want to prove that they are better than others will become entrepreneurs, i.e. the love of competition is one of the major motivations to become an entrepreneur. Other researchers have also associated the love and joy of competition with the entrepreneur. In his book *Competition and Entrepreneurship* published in 1973, Kirzner states that competition is inseparable from the market process and entrepreneurship is essentially competitive: "the two notions of competition and entrepreneurship are... analytically inseparable. And regardless of which terms one chooses to use, these two notions must be recognized, and must be perceived as being at all times merely two sides of the same coin" (p. 16).

In a blog series about the top 10 characteristics of successful entrepreneurs, Mark Suster (2010a), an American businessman who was one of the biggest venture capital investors in Los Angeles at that time, underlines that, in addition to being persistent, resilient and capable of inspiring others, taking risks, making decisions and having an integral personality, the most successful entrepreneurs are essentially also *hypercompetitive*: they love to fight for victory, never give up and hate to lose. This is the quality that venture capitalists look for in the first place in entrepreneurs. They are looking for persons who want to win, who have a strong fighting spirit and who want and are able to recover immediately from a loss and learn from the mistakes that had caused it (Suster, 2010b). According to Suster, the ideal businessman partly takes competing very seriously and partly enjoys it very much. Urbig et al. (2020) confirmed this in an experimental study. They found that the enjoyment of competition can explain why those being successful in the business world agree to adopt business decisions where the chances of success are very low (apart from being overconfident and risk-taking). They investigated whether potential and revealed entrepreneurs were more likely to seek competition than non-entrepreneurs. Three facets of individual competitiveness were examined: desire to win, striving for personal development and enjoyment of competition. Shoppers in a shopping mall in a large German city were asked to participate in the investigation. Participants were divided into three groups: those who currently have or had a business in the past, those who wanted to start a business and those who have never been interested in working in the business life. It was found that persons who actually worked in business in the past or do at present and are also potential businesspeople chose competition much more frequently than those who had never got close to business life and do not desire to do so. Of the three competition factors under study, potential "entrepreneurs" were typically characterized by the enjoyment of competition in the first place; they were motivated by the excitement concurrent with competition rather than victory or the possibility of personal development. Based on this result, the authors drew the conclusion that, in addition to the traits of excessive self-confidence and risk-taking described previously, the enjoyment of competition could be the third most important personality component determining the market decisions of entrepreneurs and businessmen. Their indirect conclusion was that entrepreneurship was a definitely attractive career for the competitive individual. The Gallup Management Consulting Group conducted a study in which more than half a million salespeople were interviewed, and it was concluded that one of the personality characteristics of great salespeople is that they are competitors, people who "want to beat colleagues or the competition" (Brewer, 1994, p. 82).

According to Suster (2010a), competition is in the DNA of the successful entrepreneur. In fact, there is a line of research which investigated the role of biological factors, the early life determinants of the individual's selection into entrepreneurship. Bönte et al. (2016) examined the role of prenatal testosterone exposure (PTE) in entrepreneurship as a career choice later in life. They used the second-to-fourth digit ratio (2D:4D or DR) as a marker for PTE, and they argue that they have found empirical evidence for the association with risk-taking and entrepreneurship. According to several studies, entrepreneurial success or as a contrast giving up entrepreneurial activities is in connection with this early hormonal effect (Medland et al., 2010; Guiso & Rustichini, 2011). Prenatal testosterone has been related to competitive behaviour as well (Salvador et al., 1999; Salvador, 2005). Branas Garza et al. (2019) aimed to reveal how preferences for competition are shaped in utero. They found that those with a low DR ratio (high prenatal testosterone exposure) are more aggressively competitive. It was also found that low DR people are more successful in competitive scenarios like financial trading (Coates et al., 2009). Bönte et al. (2017) found that low DR correlated with self-reported competitiveness, trading styles and profits among financial professionals. High DR however correlated negatively with risk-taking in relation to financial incentives (Coates et al., 2009).

As claimed by self-selection theory (Bartling et al., 2009), the more competitive persons, who are at the same time risk-takers and overconfident, are more likely to choose a more competitive career such as entrepreneurship (Nicolaou et al., 2018), financial trading (Sapienza et al., 2009) or high-prestige academic careers (Buser et al., 2014). In contrast to these, individuals who have more egalitarian views self-select less into competitive environments. According to Holm et al. (2013), Chinese company managers prefer to be paid in the competitive structure rather than the non-competitive one. Berge et al. (2015) found that competitive behaviour under laboratory conditions was positively related to the actual competitive choices of the entrepreneurs in business life. Entrepreneurs keener on competition are more likely to compete in the field of investment and are more successful in terms of profits and sales. This is particularly interesting because in this study, risk-taking, self-confidence or cognitive skills proved less predictive of actual business success, whereas the competitive attitude showed close correlation with it. Bönte and Piegler (2013) based on the 2009 Eurobarometer "Entreprise" survey covering 32 European countries and China, Japan, South Korea and the USA, and representative for those above the age of 15 in all countries, found that individuals who liked the situations where they could compete with others were more likely to want to start a business.

It is a question what comes first: those who start a business may anticipate competition, and thus people who like to/enjoy competing are more likely to choose such careers, i.e. they express their preference for competition by self-selecting into a competitive setting instead of a non-competitive setting. Alternatively, business people, entrepreneurs and those working in leadership positions in companies being in the competitive market are exposed more to competition so they are more used to it and they are socialized to compete, i.e. in order to be successful, they inevitably become competitive even if they had not been before entering the field.

The person-environment, person-organization and person-job fit theories (e.g. Kristof-Brown & Guay, 2011; Boon et al., 2011) emphasize less the sole role of individual characteristics of business people in their success but rather focus on the interaction, namely, how much the personality of those who choose the business world as their professional area matches the competitive environment they work in, how much their values are compatible with the values of the organization and how much the environment is able to shape their attitudes (Chatman, 1991).

Even if the professional sector is characterized by intensive competition, there are segments and also particular positions which do not necessarily require competition-friendly and especially trait competitive employees. Some employees are in strategic positions and have to make decisions in relation to the market processes, while others are not. Some compete energetically to get promoted, while there may be others who are not interested in competition and getting ahead in spite of companies creating a dynamic in which employees compete with each other for positions, pay rise, bonus and recognition. For example, Utsch et al. (1999) found that entrepreneurs were more competitive than managers. There are also those who are so-called forced entrepreneurs, who are persons who are not driven to such careers by their competitive character, but act under some labour market constraint. Wang et al. (2018) propose a dynamic model with two separate components of employee competitiveness based on a study with salespeople in a large insurance company: trait competitiveness and state competitiveness, the latter demonstrated as a manifest competitive behaviour. According to this theory, if somebody is not competitive as a trait still can behave competitively if the environment requires that, therefore the background motivation and preference and the actual, manifest behaviour are separated.

The attraction-selection-attrition theory (Schneider, 1987) states that individuals are attracted to and seek to work for organizations where they perceive high levels of person-organization fit. It is generally assumed that person-environment fit leads to positive outcomes, such as improved satisfaction, adjustment, performance and overall well-being (Ostroff & Schulte, 2007), to higher organizational commitment and reduced turnover and increased organizational citizenship behaviours (Andrews et al., 2010). On the contrary, as a consequence of poor person-environment fit, it has been demonstrated that the stress level increased (Kristof-Brown & Guay, 2011). This may be a crucial question in relation to the fit between personal competitiveness and competitive professional climate.

While there is a relatively extensive literature on the different attitudes towards competition, this is hardly present in the organizational context. This means that in the literature which discusses that working in this sector most of the time goes together with a preference towards competition, the nature of this preference is not explained in detail. Most of the recent studies in economic psychology have focused on gender differences in choosing competitive or non-competitive reward structures with the aim to understand why women are underrepresented in the business world. The method they applied were laboratory experiments (e.g. Niederle & Vesterlund, 2007, 2011). Research on the competitive attitudes of entrepreneurs and non-entrepreneurs has been carried out also in an experimental context focusing on behavioural choices (e.g. Urbig et al., 2020). Competitive attitude was defined by the number and structure of choices of competitive or non-competitive reward allocations, and the underlying motives have been rarely investigated, and different patterns of competitiveness were not identified.

However, in the personality and individual differences literature, there have been a number of classifications which aimed at distinguishing qualitatively different competitive attitudes and patterns. Therefore, it can be stated that the existing research on competitiveness in the business life and the research on competitiveness as a personality trait or orientation have not really met each other. It has been hardly studied what kind of different competitive attitudes can be identified among those working in the business world and what kind of potential significance these differences may have, for example, in an organizational context.

The most extensive studies on competitiveness in relation to organizational managerial positions had been mainly within the framework of Type A behaviour (Morrison, 1997; Pedler et al., 2006). Type A individuals were described as competitive, ambitious, hostile, impatient and hard driving, while Type B individuals as relaxed, laid back and not preoccupied with achievement. Type A was generally associated with greater achievement and success (Thornton et al., 2011). It was generally agreed that the sort of work in leadership roles in the business world requires Type A behaviours because leading and managing requires competitive skills and would be impossible to effectively manage with a Type B personality (Pedler et al., 2006). Initial research also demonstrated that this behavioural pattern, especially the competitive component, had positive correlates with cardiovascular diseases; therefore, it was commonly called the "manager disease". Several decades of research however proved that the concept of "Type A behaviour" has to be deconstructed and must be divided into different components, which may behave differently in relation to psychological and somatic health (Gomez, 1998; Keltikangas-Jarvinen & Raikkönen, 1991), and it is not the competitiveness component of Type A behaviour that is responsible for its relationship with cardiovascular diseases but the impatience/hostility factor (see Keresztes et al., 2015).

In the literature which focused on competitiveness as a personality trait, the simplest definition was "the enjoyment of interpersonal competition and the desire to win and be better than others" (Spence and Helmreich 1983:41). In contrast to this, most of the researchers have identified two main types of competitive strivings among which one was considered a more adaptive, positive, healthier attitude and the other a less beneficial in terms of mental health and interpersonal relationships. Griffin-Pierson (1990) distinguished competitiveness emphasizing doing or being better than others ("interpersonal competitiveness") and competitiveness to obtain a goal and achieve excellence ("goal competitiveness"), the difference between the two being in their emphasis and focus. Both goals embody the idea of pursuing a goal and the idea of being better than others, but in the first case, the focus is on being better than others; in the second case, the focus is on the goal. Franken and Brown (1995) stated that if the focus or goal of the competitive process is winning, it is not adaptive. When the focus or goal is to master an ability, the competitive process is a healthy one. Others also considered it to be more favourable if the competitive desire is to excel and obtain the goal ("goal competitiveness"), bringing out the best one can do, mastering the task ("mastery competitiveness") and developing oneself. One less favourable type of competitiveness was characterized by a strong desire to win ("competing to win"), to dominate or to be superior to others ("superiority competitiveness") and to strive for success compared to others and exceed them ("competitive excellence") (Heggestad & Kanfer, 2000; Kayhan, 2003; Hibbard & Buhrmester, 2010; see also in Orosz et al., 2018). Tassi and Schneider (1997) distinguished "other-referenced competition" (i.e. to be proven superior to others) and "task-oriented competition" (i.e. competing in order to do well at something), and they found that peer rejection was only linked to competitive behaviour if the behaviour was fuelled by a desire to assert one's superiority over others and not if the goal was improvement of one's own performance.

Ryckman and his colleagues (Ryckman et al., 1990, 1996, 2009) in a series of studies differentiated two competitive orientations, "hypercompetitiveness" and "personal development competitiveness", and studied their correlates with psychological and somatic health. Hypercompetitiveness was described as a hostile, aggressive, manipulative and exploitative behaviour towards others (Ryckman et al., 1990), while personal development competitiveness as a healthy and positive competitive orientation, in which the main emphasis is on one's own personal development, on discovery of one's potentials via competition (Ryckman et al., 1996). For personal development competitors, competition is a motivation which makes them put more effort into an activity and unleashes their creativity. Personal development competitiveness was shown to be associated with fewer health problems, whereas hypercompetitiveness was shown to be associated with greater self-reported health problems (Ryckman et al., 2009; Thornton et al. 2011). Newby and Klein (2014) identified four dimensions of competitiveness: general competitiveness (the degree to which an individual considers him- or herself to be a competitive person), dominance (desire to win in interpersonal situations), competitive affectivity (enjoyment of competition, but also feelings of superiority and powerfulness from competition) and personal enhancement (demonstrating self-competence, mastery, achievement and self-improvement).

Non-competiveness or competition avoidance has been less studied. For a long time, simply the lack of the above-mentioned desires was implicitly considered non-competitiveness. Among the rare examples were Franken and Prpich (1996) who studied why people dislike competition and differentiated between "self-image concerns" and "performance concerns". Competition avoidance was described in a more in-depth way by Ryckman et al. (2009) and was characterized by lack of confidence in achievement situations and engaging in self-handicapping behaviour. Kanfer and Heggestad (2000) as opposed to competitive excellence and personal mastery as achievement traits identified achievement anxiety as a condition that also leads to the avoidance of competition. It was also found that competition avoidance had correlates with psychological ill-health (Ryckman et al., 2009).

Orosz et al. (2018) in their study of multiple competitive orientations identified four different attitudes:

The self-developmental which describes an orientation which focuses on the positive effect of competition on the self in the long-run: "Competitive situations allow me to bring the best out of myself".

The hypercompetitive which places instead of the process the result, i.e. victory in the focus of the competitive process, and because of the excessive importance attributed to victory allows any kind of means to reach it: "I will do anything to win, even nasty things".

The indifference towards competition when somebody does not enjoy competing and is not motivated by competitions: "I don't care about competitions".

The fourth orientation is competition avoidance, when somebody wants to avoid any kind of competitive situation because it causes too much stress and anxiety: "I feel distressed in a competitive environment, so I avoid them whenever I can".

Menesini et al. (2018) found that in some cases personal developmental competitiveness and hypercompetitiveness can overlap and correlate positively with each other. It has been more and more evident that these orientations do not characterize a person solely but they can be mixed and constitute qualitatively different competitive patterns. In a recent study of Fülöp and Varga (2020 manuscript), cluster analysis was carried out with the competitive orientation scales of Orosz et al.'s (2018), and five different personal patterns of competitiveness were identified: the self-developmental, the hypercompetitive, the competition avoidant, the indifferent and a fifth attitude which was called complex competitive attitude because persons characterized by this attitude are high in both self-developmental competitiveness and hypercompetitiveness representing a competitive striving which places winning or being the best into the focus but not by any means possible, but mainly relying on own hard work and effort. They place both the result and the process into the focus. Because not only winning is important but also self-improvement, the aggressive, manipulative, exploitative aspects of hypercompetitiveness are less pronounced (Fülöp & Varga, 2020, manuscript in preparation). These five patterns have been confirmed on a Hungarian representative sample, a sample of adolescents, a sample of college students and a sample of old people above 65 years of age (Fülöp & Varga, 2020, manuscript in preparation).

The study of trait competitiveness in the personality psychology literature has been almost exclusively based on surveys and inquired about a general attitude of mainly university students. The economic psychology literature applied mainly experimental methods and focused on situational choices of those who plan to be involved or already are involved in the business life. Only Urbig et al. (2020) studied three different motivations of competition among past, present and future entrepreneurs: the desire to win, striving for personal development and enjoyment of competition. There has not been any other systematic study so far to reveal what kind of competitive attitude characterizes the entrepreneurs, business managers, CEOs, project managers and business owners and how they perceive their own fit to their competitive professional environment. While in the personality psychological research it was found that an excessive desire to win, "goal competitiveness" or hypercompetitiveness is not as favourable in terms of psychological and somatic well-being and causes more interpersonal conflicts as the desire to improve and grow by competition. i.e. self-developmental competitiveness (e.g. Thornton et al., 2011), in the competitive business world, a strong but not excessive desire to win may have a special significance and can be more adaptive, especially if it is interpreted as a fit between the personality and the context.

Sommer (1995) differentiated between social or informal or intentional competition (personal or attitudinal) which derives from the inner drive to apply social comparison and to strive for being better than others or being the best and structural or formal competition (i.e. which is inherent in the situation, e.g. an institutionalized contest or the competition-driven economy characterized by intra and inter-organizational competition. Brown et al. (1998) suggested that a key determinant of motivation and performance is the interaction between the employees' personalities and perceptions of organizational environments and found that salespeople set higher goals and perform better if these two coincide. The competitive climate may drive many to be competitive, no matter whether they personally like to compete or not, because in order to be able to be successful in their job, they have to be able to function at least at the behavioural level pragmatically and effectively as a competitive person. It has been not studied (1) if goal competitiveness isn't healthier in a context which requires exactly this kind of goal setting and (2) how are those able to cope with an intense competitive work environment who themselves do not consider themselves competitive and also not able to manifest a pragmatically competitive behaviour either.

The Hungarian Business Environment

All studies related to the competitive attitudes of business people have been carried out in countries which have had a centuries-old market economy with competition being the central concept of market functioning. The competitive attitudes of those who were brought up and socialized in countries where competition has been explicitly encouraged and expected in the business life may differ from the attitudes of those who chose the business sector in a post-socialist societal context like the Hungarian. Therefore, it may be particularly relevant to study the competitive attitudes of business people in the Hungarian business life which for more than 40 years banned competition, and people having competitive traits could not self-select into a professional sector where competition was openly and explicitly encouraged. If the sociocultural context changes, it is a question if there is a kind of acculturation and adaptation process to meet the new and altered expectations or if there is an evident struggle to fit in into this new environment, if there is already a new generation of business people brought up who are fully compatible with the market economy, with competitive tenders, competitive reward structures and an intensively competitive working climate.

Hungary was a socialist country from 1948 to 1989 for more than 40 years, and during the communist-socialist era, competition was not structurally in-built in the economic and the political life. The economy had been mostly state-controlled and largely excluded market processes, while the political life was characterized by a one-party system which lacked democratic competitive elections. In addition to this, before 1989, there has been full employment in Hungary; therefore, there was no institutionalized competition on the job market; in fact, no such thing as job market existed (Fülöp, 2002a). Obviously, this did not mean that competition totally disappeared from the society. There was competition for positions in the Hungarian Communist Party, in the higher positions providing more power and access to resources and also in workplaces for positions and prestige. As the proportion of those who could be admitted into higher education was also low and limited, there was a competition for higher education places as well. There have been also school, district, city and nation-wide contests of different school subjects, and also the young communist, so-called "pioneer" movement, organized numerous competitions. Therefore, a kind of competitive spirit existed and survived in spite of the ideological condemnation of competition as a characteristic of the "declining imperialism" (Fülöp, 2005). Hungary however in 1989 changed its political and economic system to democracy and market economy, and together with these changes, competition as a guiding principle appeared. As a high rate of unemployment characterized the first 10 years of the introduction of market economy, potential employees had to start to compete for jobs, and the workforce in the organizations was expected to be competitive, to strive for success, to understand the mechanisms of a competitive market economy and to be able to function successfully among the newly established business circumstances (Fülöp & Berkics, 2002). As Gumpel (1993) stated, it is not enough simply to change the social structure; there has to be a kind of "mentality" that maintains and develops it. In other words, the way of thinking that characterizes a socialist planned economy had to change to a market economy way of thinking. This required a kind of cultural change and the development of new competencies including being comfortable with competition, and also a new group of people had to emerge who are entrepreneurial and business-minded and are able to lead companies in the newly established markets. In other words, the new competitive market environment needed a competition-friendly workforce.

Derecskei and Hurta (2007) carried out in-depth interviews with 71 entrepreneurs almost 20 years after the regime change. Only 14% of them considered Hungarians competitive, another 14% believed that they are "forced" competitors who compete because they have to, and another 25% thought that Hungarians are slowly learning to be competitive. Faragó (2008) studied 52 Hungarian senior managers of different companies with surveys and found that those who worked for a successful company were risk-oriented and were also characterized by both pro-active and aggressive competitiveness. The adjective "aggressive" rather meant intensity, an extraordinary drive to respond to competitive actions taken by the rival. Those senior managers however who worked for unsuccessful companies reported less competitive striving.

Research on the perception of the role of competition in the Hungarian society among students and teachers 10 years after the political changes showed that the Hungarian respondents viewed the role of competition in Hungary in a more negative way than their US American or Japanese or English counterparts in their own country (Fülöp, 1999; Fülöp, 2002a, b; Fülöp & Berkics, 2002). It was also found that Hungarian business people had a very dark view of the role of competition in the Hungarian business life (Fülöp & Orosz, 2006).

Garcia et al. (2015) studied if competitive attitudes change parallel to sociocultural change. They repeated experiments carried out in the 1970s 40 years later in the 2010s and found that Mexican children became much more competitive as a result of urbanization and industrialization. Similarly, by 2020, the Hungarian business life has had 30 years to become accustomed to the competitive economic structure and learn how to be effective in it. Although it is widely shared in the societal discourse that those who work in the business life need to have a strong competitive drive and a business person is often compared to a sports champion (Kenyeres, 2017), Hungarians still have a kind of negative attitude towards the free market economy (Biró-Nagy et al., 2016) and don't believe in a meritocratic competitive society and market processes (Hunyady, 2018).

The Competitive Attitudes of Business People in Hungary

Unfortunately, as it was mentioned before, the existing research on the competitiveness of those who are interested in working or already work in the business life has been mainly studied experimentally, and competitiveness was mainly defined by the willingness (indicated by the number of choices) to choose competition when there is also an alternative to choose a reward allocation which is not dependent on how the person's achievement compares to others. On the other hand, the qualitatively different competitive attitudes which were identified in the personality psychology literature have not been examined among business people and not among those who are in leadership positions in the business world. Also in the literature of competitive attitudes, the studies focused on decontextualized, general attitudes without paying attention, for example, to professional environments which may have certain expectations and constraints in terms of those who work in them, i.e. what kind of competitive attitudes are needed and efficient and lead towards the goals. As it was discussed previously, different competitive attitudes have been related to different psychological and somatic health outcomes; however, this relationship has never been discussed in a person-environment or person-job fit context either. There have been no previous research on how entrepreneurs and business people perceive and reflect on their own competitiveness and how they perceive the significance or need to be competitive in the business life and the business organization. It is especially meaningful to study this in a society which for 40 years banned and denied competitiveness and then with a shift suddenly needed people in the business sector with competitive spirit and ability to cope successfully with its mostly cut-throat climate (Fülöp, 1999; Fülöp & Orosz, 2006).

Research: An Interview Study with Hungarian Entrepreneurs and Managers in the Business Life

The interest in business leaders' self-perception and perception of competitive attitudes was part of a bigger research focusing on different aspects of competition and cooperation in the Hungarian business life (Fülöp & Szarvas, 2012; Fülöp & Berkics, 2015). In order to have a more personal and in-depth understanding of these aspects, a semi-structured interview was carried out, recorded and verbatim transcribed, and content analysis was carried out along the various research questions. The categorization of the competitive attitudes was carried out by two researchers independent of each other based on pre-existing categories of previous classification of attitudes in the psychological literature; however, based on the self-characterization of the business people, the pre-existing category system was modified by consensus to accommodate all the qualitatively different orientations.

Altogether 201 in-depth interviews were carried out with 137 men (68%) and 64 women (32%). The gender distribution according to the statistics of Bisnode Hungary (the digital data and market research company) corresponds to the Hungarian statistics on women in leadership positions which was 34%. Among all business owners (from small to large), the proportion of women was higher, namely, 42% (https://www.bisnode.hu/tudastar/gondolatok/lassulo-novekedes-a-noi-vezetok-es-tulajdonosok-aranyaban/). The participants' age range was 23 to 65 (average age 42 years, STD: 11.31). Regarding the profession of respondents, they were entrepreneurs/business owners (42%), company presidents/CEOs (22%), senior executives (17%), marketing and sales experts (10%), bankers (5%) and accountants and economists working at big national or international companies (5%). All of the participants work in a complex, risky and competitive environment with a high degree of autonomy and responsibility to attain objectives.

Among the participants, there were 77 (38%) from Budapest, the capital of Hungary, and 126 (62%) from other cities in the country. The participants were randomly selected from a pool of business persons whose contact details were provided by students of economics of a large university in south Hungary. All participants signed an informed consent, and their interviews were anonymized.

14.2 Results

The Perception of Competition in the Business World: The Climate

The subjective perception of the competitiveness of the business sector and the professional environment is an indicator how our participants perceive their professional climate and how much they perceive their company's and their personal rewards to be contingent on comparisons of their performance against that of other companies, businesses and other professionals. This has importance because those are the perceptions and interpretations of an environment that mediate people's behavioural responses to it (Brown et al., 1998).

All of the participants in our interview study emphasized that competition is a key element of the economic life. There is no economy without competition, and the business world is highly competitive, and those who participate in it have to aim for winning, because profit depends on the relative position in the competition. Those in leadership position have to make decisions which lead to an increase in the company's value. Business leaders have to find the competitive advantage of their company and products or services and have to be able to think competitively and understand the logic of the competition in the market:

> *This is not like a sports contest or an intellectual contest where achieving second or third place is a most honourable result. Let's assume, a third place at a national recitation contest, it is quite an achievement. But being second instead of third at a tender means absolutely nothing, because only the winner, the one who is the first, will be the winner of the project.* (Man, age 54, CEO)

The Competitive Attitudes of the Business People

Based on their personal accounts on how they relate to competition, the participants were divided into three main groups: the trait competitors, the pragmatic competitors and the trait non-competitors. Trait competitiveness was further divided into five sub-categories, three of them with a clear focus on winning; one with the focus on the process, i.e. self-development; and one with both result and process orientation. Those attitudes which place winning in the focus represent different degrees in the importance to win, different energy levels to be devoted to make one reach the goal and different means to be applied in reaching the goal, i.e. to win over the business rivals. These three attitudes are the *balanced*, the *extraordinary* and the *hypercompetitive*. It is not winning but improvement by competition which is in the focus of the *self-developmental competitors*. *Complex competitors* however integrate two orientations, the process, i.e. developing themselves, and the result, i.e. to win over the rivals. All these patterns represent a personal competitive motive, in other words trait competitiveness which guides the participating business people. However, it was possible to identify a group of participants, named *pragmatic competitors*, who denied having a personal motive to compete or any joy that would derive from competing or winning over a rival; they consider themselves "forced competitors" or with Wang et al.'s (2018) expression "state" or "behavioural competitors", who because of their profession are in an intensively competitive environment and cannot avoid competing, i.e. they adapt to it at the behavioural level as a pragmatic choice. The non-competitive business people belonged to three different groups: The first are those who characterize themselves as having a *high need of achievement* without the interest to compare themselves to others and wanting to be better or win over them. They focus on the goal and would like to do their work, and they can and consider this to be enough to survive in a competitive arena. The other group was the *indifferent* towards competition, who generally are not motivated by anything; they explain that they have no competitive drive either. Competing does not give them joy or excitement at all. The third group of non-competitive business people are those of the *competition avoidant* who explicitly try to avoid competition partly due to a fear of failure, partly because they are disillusioned by the non-meritocratic nature of the Hungarian business life and partly due to their age.

In terms of the proportions of these attitudes, the great majority of the respondents characterized themselves as competitive by character (81%). The most frequent attitude was the *balanced* one (31%), followed by the *complex competitive* (23%) and with almost the same proportion by the *extraordinary competitive* (20.5%). *Hypercompetitiveness* (3.5%) and *self-developmental* competitiveness (3%) were not characteristic to this group of actors in the business life. An additional 8% of the respondents reported to be a *pragmatic* or forced competitor meaning that they are competitors not by "nature" but by their recognized interest. Only 11% of the participating business people portrayed themselves as non-competitive both by personality and by behaviour. The biggest group among them was the *high achievement motivated* (6%), secondly the *competition avoidant* (3%) and lastly the *indifferent* (2%). (See Table 14.1.)

Gender Differences

In order to analyse whether the attitudes are gender specific, we used a chi-squared test which allows for determining whether or not the categorical variables are associated at a 5% level of significance. If necessary, we collapsed categories so we could use 2 × 2 contingency tables. If those who are not trait competitive but pragmatically competitive were handled together with the trait competitive, no statistically significant association was found, even if the proportion of *high achievement motivated* non-competitive was almost three times higher among women than among men. However, if the *pragmatic*

Table 14.1 Attitudes towards competition

Attitudes towards competition	Female	Male	Altogether
Competitive by trait	**45/70.5%**	**117/85.5/%**	**162/81%**
Balanced competitive	18/28%	44/32%	62/31%
Extraordinary competitive	15/23%	26/19%	41/20.5%
Hypercompetitive	2/3%	5/4%	7/3.5%
Self-developmental competitive	2/3%	4/3%	6/3%
Complex competitive	8/12.5%	38/28%	46/23%
Competitive by behaviour-non-competitive by trait			
Pragmatically competitive	8/12.5%	9/6.5%	17/8%
Non-competitive	**11/17%**	**11/8%**	**22/11%**
High achievement motivated	7/11%	6/4%	13/6%
Competition avoidant	2/3%	3/1.5%	5/3%
Indifferent-amotivated	2/3%	2/1.5%	4/2%
Altogether	**64/100%**	**137/100%**	**201/100%**

competitive (not competitive by "nature", but competing as a constraint of their profession) were handled together with the non-competitive, a statistically significant association was found between gender and competitive and non-competitive attitudes ($X^2(1, N = 201) = 7.120, p = .008$), meaning that there were significantly more men who reported themselves to be competitive than women. This is due to the fact that the proportion of *pragmatic competitors* was higher among women, but due to the small sample size, this difference in itself did not reach statistical significance, but when grouped among the trait non-competitive, it contributed to the significant result. Among the different competitive attitudes, only the *complex* competitive attitude differed significantly ($X^2(1, N = 201) = 5.284, p = .022$), being more typical to men than to women. In the case of the *balanced*, the *extraordinary*, the *hypercompetitive* and the *self-developmental* competitive attitudes, businessmen and businesswomen did not differ.

Business People Self-Portrayal of Their Competitive Attitudes

If the business people's own voice is presented, it is possible to gain a kind of insider view in what way these attitudes differ from each other. The category system of the different personal relationships with competition provides a useful framework to understand the different degrees of person-environment fit, which consequently may influence effective performance and subjective well-being related to work.

The Balanced Competitors

Almost one-third (31%) of the participating business people considered themselves balanced competitors. Balanced competitors emphasize that they are competitive and enjoy competing, but this does not mean that they compete at everything or compete excessively. In their profession, it is important for them to get ahead and not fall behind, and they want to be competent, they want to reach their goal, and they want to win. But they do not necessarily compete in their private life. They are also able to withdraw from competition in case they believe it to be a better strategy. In other words, they are able to exert control over their competitive drive, and the importance of victory is not enough justification for them to apply aggressive or unfair means in order to win:

> *One question is where do you want to be in the race, you want to be the first or lag behind; another is if you want to defeat the others at all price; to put it more forcefully, if you want a situation where there is only a first and all the others are last. Now, in the previous sense, I consider myself a competitor, but not in the second sense.* (Male, age 58, head of financial department)

> *Well, I started playing sports when I was 11 years old. Now, I will be 40, which means that for almost 3 decades now, competitive sports have been an integral part of my life. Of course, it also puts a stamp on my work. If you ask me if there is one area where I don't like to compete, that is, I think it's more about family life or private life....Of course for me it is a priority, and it is an eternal criterion for competition to be fair play. I can't stress this enough and enough that honesty is first and foremost. I believe that in the long run, honesty always pays off whether in the business life, in private life, or in sports, when I compete precisely this honesty and fair play is what characterizes it.* (Male, age 40, entrepreneur)

> *I can no longer imagine my life without daily competition, constant competitive situations. It brings dynamism and impetus to the life of the company.* (Male, age 29, CEO)

The Extraordinary Competitive

The adjective "extraordinary" refers to the extreme strong competitive drive which characterizes one-fifth (20.5%) of the participants. In the business world, these extraordinary competition-driven people are called "hypercompetitive" because of the "hyper"-intensity of their motivation to win (e.g. Jeff Bezos and Elon Musk are labelled hypercompetitive by Brandenburger & Nalebuff, 2021); however, the psychological literature uses this label to identify a rather controversial personality constellation (Ryckman et al., 1996). In the present categorization, the *extraordinary* competitive consider themselves "born" competitors. This elevated drive is present in whatever they do or with whoever they are in relationship. The results, i.e. victory, business success and profit, are highly important, and they strive for them relentlessly and energetically. They want themselves and their company to a leading player among the rival players. Their strong competitiveness leads to innovation, to successful sales and financially to successful outcomes:

> *Yes, I have been very competitive in my whole life, and I think this is also fortunate. If I had not been competitive, I could not have achieved the things I did. If I did not want to win, I do not know who would inspire my business team instead of me. If I were not competitive, and I would not compete, we would be swept off the market.* (Woman, age 50, top manager, insurance company)

> *I compete in every field, in work, in every respect. Also in sports, I am competitive to the maximum, there is no such thing that this is only for the sake of playing, e.g. in tennis, it is*

important to hit every ball and be better than the other. In swimming, I want to make one or two more laps than on the previous occasion and certainly another man cannot beat me. In your workplace it is the same and it's not only up to you, it is also expected. (Man, age 45, managing director)

I love competition. It has been in my blood since I was born. I always learn something new from it. Sure, I hate being second, but it only further increases the spirit of competition in me. However, sometimes I feel like I can't take it anymore. There's a point when I'm tired of stressful situations, the perpetual compulsion to comply, and that's when I have to let go of the steam a little bit. But it usually doesn't take long and I address the next challenge. (Male, age 28, financial manager)

Sport has taught me that you have to stand up even if you failed, you have to think over what you did not do well, what you still need to improve. Actually, I compete in every field of life. In the most diverse fields, sometimes even where I shouldn't, where I have no chance or for example it is not necessary. Like all people I have a desire of some sort to win, only this is stronger in some areas and less strong in others. Many times, I unnecessarily enter into a "battle" with someone, like at the red traffic light. I do not get any benefit from winning, no prize or anything, it just feels good to win. It feels good to know you are better. And there is the excitement of competing that, in my opinion, is also present in everyone. (Man, age 22, entrepreneur)

The Hypercompetitive

Only a small proportion of the respondents (3.5%) were characterized as hypercompetitive having characteristics that correspond to the descriptions of Ryckman et al.'s (1996) or Stalk and Lachenauer's (2004) "hard ball players". This small group of business people in our sample are overly or extremely competitive with a strong quest for self-enhancement. They constantly compete and compete at everything, i.e. basically anything can become the object of competition, and they must always be winners whatever it takes. They are compelled to win because losing is a serious narcissistic injury that must be avoided at all costs. Their intense need for power when taken to the extreme can result in aggressiveness, ruthlessness and manipulative, deceptive and unethical behaviour. Their philosophy is "the goal justifies the means". As Stalk and Lachenauer (2004) suggest, they are the rough players, who focus relentlessly on competitive advantage, devastate their rivals and are not afraid to challenge legal borders. While extraordinary competitive rely mainly on their capacities to win, hypercompetitive may use any kind of means against the rival in order to win:

I hate to lie, I hate lying. Except for business. Then I can lie fluently, without batting an eye, I have never been caught at it … I always play to win, I cannot stand losing. Whether in a board game, even when my child was 6, I would never let him win. Why would I? Just because he is such a nice little boy? He should win not because I let him win, but because he is better than me. The same goes for the tennis court: I would rather die, but I must win. (Man, age 56, entrepreneur)

In the market sector where I work, competition is rather fierce and I think this can only be achieved by pursuing your activity with very great intensity. Well, I say I also use any opportunity to be able to beat my rivals, sometimes less fair things also occur in that context. (Man, age 37, executive of financial consulting firm)

You must think of the future, and of course business life is like that, once up, then down. I may also land in a situation where I am vulnerable. But the rival should never be in a position to be able to hurt me. The essential thing is that that the other party must be ruthlessly crushed, because this is what life in Hungary is about, nothing else. There is no fair play in anything. (Man, age 35, business owner)

Self-Developmental Competitors

While in the personality psychology literature, self-development competition is considered to be the most adaptive and constructive way of competing, and in Fülöp and Varga's representative study of the Hungarian population (2020 manuscript in preparation) study, it correlated the most with positive health outcomes, in our sample, there were hardly any business people who presented this type of competitive attitude. Self-improvement competition focuses on long-term development of the individual by competition. Although constant learning, improvement of products and innovation are crucially important in the business life, the effect has to be "immediate". Learning and developing by failure, which may be acceptable in general and at the individual level, is not an option in the Hungarian business context where the choice is rather "swim or sink". For self-development competitors, the rival is considered to be not so much a person to win over but rather as an inspiration to develop themselves. The final aim of competing this way is to achieve change inside the person; therefore, it may have rather indirect than direct effect on business success. Only 3% of the business people described themselves this way:

Construction is my fundamental law. To create something, to always do something, never stop. Construction is like putting one more brick in place. That is, symbolically, reaching a target is a new brick in the wall that you are constructing ...

I: And what is it that you are constructing?

Myself. I do not measure success in terms of money or the number of property acquired, or cars, vehicles, any material thing. Self-realisation is success. If you do what you would like to do. (Man, age 29, managing director)

I wouldn't single out an area where I would compete so much. I'd rather try to do my job as best I can. So there is no area for which I would say that this is okay, I don't have to learn further, I don't have to improve myself here. I have to constantly evolve in all areas to be on parity with my partners. (Male, age 25, research development director)

Complex Competitive

Complex competitive is the second biggest group of respondents in our study (23%). These business people are characterized by both self-developmental and extraordinary (hyper) competitive features. The "classical" hypercompetitive person wants to win at any cost and is not afraid to be aggressive or manipulative or to be a cheater to be victorious; however, the strong desire to win in the case of the complex competitive does not lead to such questionable acts; they rather work hard and do whatever they can in their competencies to win mostly by developing themselves, bringing out their best potentials and finding new and creative solutions to problems. Complex means that improvement of personal capacities, finding the competitive advantages and win, i.e. the process and the result are equally important and motivating. In other words, it is of prime importance for complex competitors to develop personally, but competition is used not only to make them better than themselves, but they also have a very strong desire to be better than the rivals and to win and obtain their goals effectively in a competitive environment.

Those who are complex competitive make it clear that although they have a strong competitive drive, they are not willing to apply certain means, but they want to achieve business and financial success by focusing on their own strategic thinking. They want to win opportunities in order to improve themselves, and by improving themselves, they increase their chances to win new opportunities and expand their company and business:

> *Direct, cut-throat competition, fighting, pushing, open humiliation are not typical of me. But if we call it competition that I have always wanted to keep pace with my acquaintances and I even always wanted to be better and better, for example if there was an opportunity to win a scholarship abroad or to work abroad, I always applied and accepted that, so from this point of view I am competitive.... I have always tried to do somewhat different things from the others, to show something special, exotic, and of course be good at it. This is why I went to Malaysia on scholarship, this is why I wanted to work also abroad by all means, this is how I got to Germany for two years. I knew this was more, and also better than what a graduate in economics usually does.* (Man, age 30, senior manager)

> *I also love challenges, I like to measure myself compared to others, to compare what I know in both business and private life. I love to win, I love to be successful too, I like to make a lot of money, but I don't think victory is so important at all costs.* (Male, age 29, entrepreneur)

The Pragmatic Competitor: Non-competitive by Trait and Competitive by Behaviour

People who are not competitive as a trait can also exhibit competitive behaviour if the circumstances require, particularly in such an intensively competitive climate like business life. A group of respondents expressed that they are not interested in competition, competition is not a motivating agent in their life but they still demonstrate competitive behaviour because it is inevitable and crucial to keep the business successful. They are an example of those who do not self-select to the competitive environment because that climate attracts them, but they rather adapt to it and change their behaviour to achieve the person-job fit. There was a relatively small group of respondents (8%) who belonged to this category. The accounts reflect a tension that is between the comfort zone of the person and the professional constraints. Because they do not quit, they do not shy away from competition, they speak about a high stress level, and their working life is a constant battle with themselves:

> *...I do not like competing. That is, regardless of the fact that for 15 years I have had to live in an environment where you must compete, and it is necessary, otherwise, you cannot earn a living, you cannot get a salary, take care of your family, children, parents, anyone, that is, a relative.* (Man, age 52, marketing manager)

> *I realise it, with my rational mind, that I must take that step in my own interest, even if I don't feel like to compete. That is, I have to advertise and always find out something new. I must do this by all means. But truly, if I was not forced to do it, I think I would not deal with it, not half as intensively. Some would do it anyway even then, because it is in their blood, I am not this type. I do it because I must, because I have to.* (Man, age 53, business owner)

> *I am not the racer type. I am not competing with other entrepreneurs because I want to, but because to support my family. Competitors force you to compete....I have to compete with competitors for my mere existence. I run on a moving treadmill that is "driven" faster and faster under me, and I always have to run faster so I don't fall down. I run, but, I don't move forward, because the better riders get ahead of me.* (Male, age 44, owner of a small trading company)

> *I consider competing a necessary evil, but I take part in it. If you ask whether I enjoy it, whether ever in my life I wanted to compete for a higher position for example to become from construction manager a chief construction manager, from director a chief executive, I have never wanted to compete for these. This was my potential not my motivation and I got it out to the maximum, but not because I wanted to compete. Even today I do not like to compete. If it is about tenders, then of course I do it, but I hate it.* (Male, age 60, managing director)

Non-competitive Business People

This relatively small group of the participants (11%) insisted that they are not only non-competitive but do not even behave competitively. There were three different types of non-competitive respondents: The first are those who are *highly motivated* but not by competition which requires a comparison between the individual and the rival, but by their high expectations of achievement, which they conceptualize between themselves and their goals without any other party, i.e. a rival or comparative other involved. Another group, called the *indifferent*, are not getting excited by competition, it doesn't mean joy for them, and they claim that they stay non-interested or they do not speak about being particularly motivated to do their job well either. The third group is the *avoidant*, and those who belong to this category hate competing and try to actively stay away from competition.

Non-competitive with a High Need of Achievement

Among the non-competitive, this was the biggest group (6%). These business people are highly motivated to achieve and reach their goals, but they explicitly denied in the interview any competitive striving. They considered professionalism, competence and performing high very important; however, they are non-interested in social comparison, and they are autonomous and believe that to be successful in business life, it is perfectly enough to do what you do well; there is no need to compete. These are business people who have some competence which is highly demanded in their field, and they get what they want by doing exceptionally well in what they do:

Because you know they said about me this is a good professional and we recommend her for example. No propaganda, nothing was needed on my side then. So it is not sure you must be competitive, I have never been competitive. (Woman, age 59, auditor, managing director)

Ever since I was little. I've always wanted more, but it's not that I'm competing with someone. I was always doing more and better to achieve more. (Woman, age 59, director of a bank)

My only strategy is to meet the expectations. Whose expectations? Primarily mine not what my bosses think about my job .I'm a maximalist, I like to be satisfied with what I've done myself, and I know that if I'm happy, my boss will definitely be happy! (Man, age 27, top manager)

Indifferent Towards Competition

The indifferent (3%) describe themselves as somebody who does the job and what is necessary to do, but does not feel any special urge to be pro-active or achieve more than that. They do not have high achievement motivation and no desire to be better than others or to win over others. They however recognize that this may be a disadvantage in the particular profession they have:

And the disadvantage is that I'm not a competitive type, and it's much harder for me to achieve anything in life.....I may have been in a competitive position, and I'm sure I was, but as I mentioned, I wasn't the competitive type, and I've never competed in these situations. At least not so far. (Man, age 50, company co-owner)

Well, truth be told, I don't consider myself a competitive type because... (thinking for a long time) *This is mostly... I try to do my job the way I need to, as I need to, but I don't usually do much more than that.* (Man, age 29, owner of a small enterprise)

I don't consider myself a competitor. There have always been others who are better than me and that has never bothered me, I'm used to it, and I don't feel like I have to be the best, to compete.... So if they leave me alone to do what I love, I won't compete for anything. (Woman, age 48, economic expert)

Some participants who belong to the older generation explained the lack of a competitive spirit with socio-historical reasons. They believe that the years they spent working during the socialist time strongly influenced their attitude towards competition. Due to the lack of a market economy at that time, they could not develop the ability to compete and practice it:

I do not consider myself competitive due to my age, I don't have a competitive spirit, because I was raised in the socialist regime, when efforts were made to radically eliminate this trait and, unfortunately, this has been achieved almost completely, as a result of which I am at a great disadvantage compared to entrepreneurs who start out with the currently approved approaches. (Male, age 60, co-owner of a company)

Competition Avoidant

There were very few among the respondents (2%) who clearly try to avoid competition mainly because by competing, they risk failure and losing. Some spoke about a series of business failures which reduced their self-esteem:

> *I'm a born pessimist. After a victory, my first thought is what kind of failure will follow, what disappointment will come... So life more and more points out that when something small good happens to you, it is followed by several smaller or bigger evils. That's why you're afraid to rejoice in the good, because you know how bad comes next.* (Woman, age 40, small entrepreneur)

> *I am not competitive because I'm afraid of the risk of being among the losers. Sure, everyone is afraid of that, but I might be more afraid than others. So in my work I try to avoid such situations if possible.* (Woman, 31, chief accountant)

A couple of the participants became disillusioned by the immoral nature of the Hungarian business life in which they felt it is not worth to compete and the result of the competition is outside of their efforts or achievements. Therefore, they try to avoid competition and seriously consider to quit from the business sector. Their avoidance is less a personality trait, but rather a learned and reactive "state" of mind:

> *Well, because it's not fair competition. So it has evolved from our history that competition is not fair. There are strong but not really professionally good competitors in a monopoly position. The winner in the competition is the one in a dominant position, and unfortunately there are many of them here, as large state-owned companies have been privatized.* (Man, age 59, co-owner of an enterprise)

14.3 The Necessity of a Competitive Character in the Business Life

It is a widely shared perception that those who are trait competitive who self-select into the highly competitive business sector and those who are effective and successful in this area possess a competitive character. The interviews aimed to reveal the participants' own subjective perception and personal experience whether a competitive personality is truly necessary to achieve business success or not.

The overwhelming majority (99%) of the participants, even those who portray themselves as non-competitive, believes that the ability to compete is fundamentally important for those who work in the business sector and those who cannot cope with this environment will fail, will "drown" or "will be killed":

> *In general I could say that today there is a fierce fight or competition going on not only in Hungary but perhaps in the whole world in business life. I think that those who cannot keep up with this pace will drown in the market. Of course, everyone has to finesse with their own means.* (Man, age 50, company manager)

> *In my opinion if someone in the competitive sector is not the competitive type, he will be killed, and he may feel awful. This is like a musician getting sick because he realizes that he*

will not be the best musician in the world. If somebody is not strongly competitive it is better not to start this profession. (Man, age 50, owner of IT company)

Most of the respondents believe in the person-environment fit, that the business world is attractive for only those who have competitive personalities:

If one gets into the competitive sector and is not competitive he would drop out. Either he would say that this is the end, it is not worth doing, not good for him, and exit himself, or the business sector as a whole would not let him stay on. (Man, age 27, banker)

The people who somehow get there, especially those who choose an economics career, are those whose competitive spirit is stronger, whose will to win is stronger than that of others. Compared to doctors, teachers, the personality of those who work in the business life, especially that of businessmen, is quite different in my opinion from that of a doctor. (Man, age 24, owner of auditor company)

Business life is considered to be a constant battlefield with strict rules, and only those who have a healthy competitive spirit can meet the expectations and stay healthy:

Without a healthy competitive spirit, without a positive attitude to take challenges, this will certainly not work. So this is not tepid water in which we sit and the boat takes us along, here you have to work hard. (Man, age 50, company manager)

The business environment itself reinforces the characteristics that fit into the environment; therefore, those who are competitive by working in this sector will become even more competitive, and those, who are not, will develop endogenously this trait after becoming involved in the business life:

I think that those who have a business enterprise in the competitive sector and can keep it up, obviously and by all means have a competitive spirit, because if they do not, the business will go bankrupt. Or the business fails because they could not shoulder this kind of competition. I think that after a while competitiveness becomes an integral part of your personality. (Woman, age 42, business lawyer)

Whoever isn't preparing to win usually has a harder time winning. If someone works in the competitive sphere and spends more time there, will involuntarily develop the "racehorse syndrome" here. After all, this world is about winning. (Man, age 59, regional director)

Business owners state that a competitive personality and mind-set are not only needed but also that is the most preferred when thinking about future staff members, because by employing such people, a higher performance can be achieved even if it may have some psychological price for the individual:

It is much easier to work together with those who are the competitive type. They would get results even from under the ground, by digging with their ten fingers. That is, they may become their own enemies because constant competition may cost them their health, their peace of mind, their family life. But, no doubt, if you are surrounded by people like this, your business or anything you started in any field, if you are with such people, your business will be successful. (Woman, age 56, business director)

According to some, not necessarily all the employees have to be competitive; there are different requirements for those who occupy different positions in the hierarchy; those in the leadership positions must be competitive, but those who they employ, not necessarily:

There is no winner company in the competitive sector without a competitive leader. That is, if the company manager has no motivation to enhance competition, if there is no relative performance measurement by which the leader suggests a value system, if there is no rank order among the staff and no measures of where the company stands by benchmarking, no external comparison continuously in the competition, to see what needs to be improved -- this leads, for sure leads to failure in the competitive sector. That is, without a competitive leader, you can only achieve success in the competitive sector in exceptional cases. (Man, age 39, director)

If somebody is an employee in a business, then competitiveness is not essential. Some workplaces are naturally more competitive than others – the respondents say. For example, sales and marketing are more competitive, but in a number of positions within the business world, competitiveness is not part of the job description:

Not everyone can be a leader, not everyone can be a 'jaguar', but that's why there are employee types and non-employee types. And I could basically distinguish here also the predator from the non-predator. (Man, age 29, MD)

There is a need for competitive leaders in the private sector. And there is a need for those who can work very well under a creative personality. This type of person can be a useful member of a team, an effective piece of a group, but he's not a competitive spirit himself. (Male, age 43, director of company)

There is a conviction that even if somebody is not competitive by personality, he/she still cannot avoid to participate in or lead a competitive process:

Working in the competitive sector forces people to be able to be compete. You must get used to it, prepare for the competition at the highest level possible and adopt an attitude that is as free of emotions as possible. You must represent your company, or whatever you want to sell in a pragmatic way. (Man, age 60, MD)

Although the great majority of the interviewees believes that an important component of the business success is that those who work in the business life be comfortable with competition, they do not consider each form of competing desirable. According to many, the aggressive hypercompetitive, hard player character may easily become a competitive disadvantage instead of a competitive advantage of the company:

It is not absolutely necessary for someone to push violently and knock over the rival; on the contrary, the higher your position, the bigger the company you work for, the more it is a disadvantage if you are known as an unpleasant, rude person with whom no one can work. For the sake of higher goals, you must at least seem to be tolerant and cooperative. Aggression should not be shown: when something like this is seen, it immediately spreads that one cannot control oneself, which is in no way advantageous for a company. (Man, age 30, CEO)

14.4 Discussion

Our study took place more than two decades after the political system change in Hungary. Competition was a key phenomenon of the regime changes, being structurally introduced to the political life by the democratic elections and to the economic life by market economy. Competition as a previously banned phenomenon became highly emphasized and expected primarily from those who started to run businesses or continued leading companies which had to survive and produce profit among the profoundly changed circumstances. There was suddenly a high demand for competitive, entrepreneurial-minded business professionals. A great number of NGOs started to prepare young people for the competitive market economy, for example, the American Chamber of Commerce ran Junior Achievement Hungary (Fülöp & Berkics, 2002). Our study shows that even if there had been 40 years of anti-competitive ideology (1948–1989), when the political and economic and cultural context changed in 1989, the needed competitive mentality gradually also emerged. Szabó and Kocsis (2003) interviewed Hungarian business people who worked for multinational companies established after the political changes in Hungary about their adaptation to the culturally new organizational climate. The title of the book was *Learning and Unlearning*, meaning that they had to learn fast new ways of functioning while unlearning those old automatisms and practices which were no longer adequate in the new context. The study presented here demonstrates that a great majority of the interviewed business leaders and entrepreneurs managed to acquire the needed competitive culture and felt highly comfortable with living in a non-stop and fiercely competitive work climate. Others could learn to function accordingly, recognizing their pragmatic interest to do so even if their personality did not match the sector's competitive climate. However, a minority of our participants experienced a gap between their personal attitude towards competition and the requirements of the sector, and this led to feeling unsuccessful, anxiety and avoidance.

Unfortunately, there are not many studies on business people's competitive attitudes in other post-socialist countries. Utsch et al. (1999) compared small entrepreneurs and managers from the ex-socialist East and from West Germany in terms of different personality orientations among them competitive aggressiveness, which they interpreted as a Machiavellian attitude. They did not find striking differences in autonomy, innovativeness, competitive aggressiveness and achievement orientation between the two groups, and they explained it with a kind of self-selection in the case of the East German participants. Given the fact that these business people had had no role models in terms of entrepreneurship or systematic socialization to market economy, it had to be their personality striving that led them to choose competitive business as a work environment.

The existing studies in economic psychology have been exclusively carried out in countries with uninterrupted, centuries-old market economies. Our study clearly demonstrates that in the Hungarian business life, in spite of a several decades-long systematic ideological indoctrination against explicit competitiveness, competitive-minded business people emerged relatively fast. This may also be a rebound effect, meaning that the entrepreneurial characteristics had to be suppressed but got "green light" after the political changes. Obviously, business people are only a small percentage of the Hungarian population who – like in the former East Germany – might have self-selected into the competitive environment because they are the few who happened to have an elevated competitive motivation. Previous studies showed that right after the political changes, even students who participated in systematic courses preparing them for the market economy had reservations, and teachers also had rather negative views about the role of competition in Hungary (Fülöp, 2002a; Fülöp & Berkics, 2002).

When trait competitiveness is studied in the personality and individual differences literature, the context is an intrapsychic one, namely, it is studied how competitive attitudes relate to other psychological constructs of the individual. The situational and cultural context has not been taken into consideration. In this research tradition, it was found that those competitive attitudes which place into the focus winning over the rival instead of striving for personal excellence and improvement via competition are less healthy and adaptive. However, what is adaptive is also defined by the particular situation and context. Evidently, in the business life, competition is strongly result-oriented, just like in sports. The tender must be won; otherwise, it is possible that employees will not be paid. A company has to be present in the market with its products and gain more market share compared to competitors. This requires specific, often short-term goals and outcomes. The research in the field of competitive attitudes and the psychological and health correlates lacked the situational, personality-environment fit framework. If there is a compatibility between the person and the organization and if fundamental characteristics are shared that may contribute to better job satisfaction, life satisfaction and better psychological and somatic health and also better performance and sustainability (Brown et al., 1998; Di Fabio, 2017a, b), this is beneficial for the enterprise and for the organization as well.

In our study, participants articulate their relationship to competition and how much they consider themselves competitive in the context of business. This framework makes it easily interpretable why the most common attitudes that were identified were those (the balanced, the extraordinary, the hypercompetitive and the complex competitive) which place, at least partly, in the focus the result which is victory in the different aspects of their business life. William et al. (1982) describe functional congruence in the person-job interaction which means that the competitive climate is able to make people to compete even if they are low in trait competitiveness. This is what was found among a sizable minority of the business people of this study as well, represented by the pragmatic competitors. Kerr et al. (2018) mention two broad types of entrepreneurs: growth-driven entrepreneurs who seek opportunity and innovation and necessity-driven entrepreneurs who open new businesses when options are meagre. The pragmatic competitors in our study are not necessity-driven entrepreneurs as it is understood be Kerr et al. (2018) but necessity-driven because they lack the personal motivation to do so, but try to meet the requirements in a conscious and determined way. Self-improvement competitiveness not in combination with extraordinary or balanced or hypercompetitiveness was hardly identifiable among business people. In business life, the measurable external performance is such an unavoidable demand that it does not allow to focus solely on long-term goals like self-perfection as a result of competing with the rivals. On the other hand, when self-developmental competitiveness which aims at becoming better and better is combined with a sharp orientation to outperform the rivals in the marketplace, i.e. complex competitiveness, it may be a combination which integrates the demands of the market and the personal subjective well-being in relation to competition.. Fülöp and Varga (2020 in preparation) found in a representative study that those who are characterized with the complex competitive attitude (measured as a general and not contextualized attitude) are as healthy mentally and somatically as the self-developmental competitors as opposed to the hypercompetitive, the indifferent and the avoidant. Just very few of the business people portrayed themselves in a way which could be categorized as hypercompetitive or "hard ball" player representing an aggressive and Machiavellian attitude. The so-called dark side of entrepreneurial personality (Miller, 2014) just rarely was identifiable. Hypercompetitiveness consists of characteristics which people not necessarily acknowledge about themselves: cruelty, manipulation, exploitation, unfairness, lying and corruption in order to win. Therefore, the participants in the interviews may have had the conscious or unconscious motivation to conceal the less salutary aspects of their personalities and conduct. In a study by Fülöp and Orosz (2006), Hungarian business people described the competition in the Hungarian business life largely as immoral due to the immoral competitive behaviour of the participants; however, they typically characterized the "generalized other" business people's practices rather than their own. However, they also emphasized that this kind of behaviour is almost inevitable if a business wants to survive in the market, the behaviour being more of a pragmatic choice or a constraint than a personality trait (Fülöp & Orosz, 2006).

Among the non-competitors, the biggest group were those with the high need of achievement. High achievement motivation is one of the most frequently identified entrepreneurial traits. In our study, both those who considered themselves competitive and this particular non-competitive group aimed for the best possible achievement; the difference was the presence or lack of and disinterest in a constant reference to the competitors. Those with a non-competitive personality but high achievement needs felt comfortable and safe in their position in spite of being non-competitive. However, those who described themselves as indifferent or avoidant felt that they are in a disadvantageous position and demonstrated a kind of burn-out, demotivated and in case of the avoidant highly stressed psychological state. They explicitly felt that they do not fit in.

In our study, gender differences were not striking. Economic psychologists (e.g. Niederle & Vesterlund, 2007, 2011) regularly found that women are less likely to choose competitive reward structures. However, Fulton et al. (2006) in their study with potential congresswomen found that women carefully consider chances before they enter a competition, but when they enter, they participate fully. This means that the gender difference which is frequently found in competitiveness is partly related to avoidance of competitions with high risk of losing. Once the risks are estimated to be manageable, the gender difference fades away. In this study, businesswomen in leadership position participated, meaning that they already had to go through a series of competitions and risky decisions successfully. Their competitive strategies might have been successful to compete with men as well (Fülöp & Bekics, 2015). In spite of this, there was a clear tendency among women to present themselves as not trait competitive; there were proportionately more among them who considered themselves as pragmatic competitors or high achievement-oriented non-competitors, and when the non-competitive and the pragmatic competitive categories were collapsed, the difference also became significant. There were also significantly more complex competitive among the men, who enjoy each possible aspect of competition. The results may indicate a slightly less successful person-job fit among businesswomen, than among businessmen.

The great majority of business people in our interviews took for granted that competition is prevalent and a competitive mind-set and spirit are necessary to survive and succeed. Given that this is considered so self-evident by the participants, it is rather stunning that a complex understanding of the internal dynamics of competitive attitudes of those who run the business life is a largely unexplored area in the field. As it was presented previously, several traits can be connected to the qualitatively different kinds of competitive attitudes. These traits like high need of achievement, risk-taking, self-efficacy and overconfidence are extensively studied, but they have not been studied as potential components of qualitatively and quantitatively different competitive personality and behavioural patterns. As so far there have not been systematic studies to reconcile the economic psychological and personality psychological literature in relation to competitiveness, it has not been possible to study among business people what kind of subjective well-being and psychological and somatic health are related to these attitudes, how they effect work-life balance, professional interpersonal relationships and relationships in the private and family life and how success and failure and victory and losing in business are handled by those with different attitudes towards competition.

In this study, business people in leadership position participated. Different competitive attitudes may go together with different leadership styles which may result in different degrees of healthy sustainability of organizations. Di Fabio and Peiró (2018) identified four different leadership styles that converge into the higher-order construct of Human Capital Sustainability Leadership which may contribute to sustainability and to create a healthy organization. Ethical leadership means that a leader strives for ethical goals. Apparently, this is not the case when a business leader has a hypercompetitive attitude towards competition as there is the danger that the person may apply unethical means to obtain the goals. Sustainable and mindful leadership refer to the shared responsibility not to exhaust the organization's human and financial resources and to restrict social and environmental damage as far as possible and to pay attention to the effect of the leader on others. This is also not characteristic of our respondents. One of them explicitly mentions that the best is if an employee works hard in spite of damaging his/her health. Servant leadership, putting others' growth and interests above the aims of the organization or of the leaders, is also not emerging. In her 2017 paper, Du Fabio (2017b) described decent leadership as a combination of, e.g. the sustainable, ethical and servant or benevolent leadership styles. Di Fabio and Peiro (2018) labelled the combination of the ethical, sustainable, mindful and servant leadership, Human Capital Sustainability Leadership. Based on the interviews, the participant business leaders cannot be characterized by decent or sustainability leadership. In order to be successful in the fiercely competitive Hungarian market, to reach a bigger market share is an imperative. In the European Value Survey/World Value Survey, Hungarians have been characterized by survival/materialistic values as opposed to, e.g. Western Europeans and Americans who are characterized by post-materialistic/self-expression values. Although, since 1990 to 2020, there has been a gradual move to the direction of self-expression/post-materialistic values in Hungary, still the difference is very significant (EVS/WVS, 2020). This may explain that the interviewed Hungarian business people did not really demonstrate any concern of sustainability and sustainable leadership. Their main focus is to win, create profit and reach financial success in the market, by good strategic decisions and innovations and sometimes by immoral means. They also rather conceptualize their attitude towards their competitive work in the motivational (largely extrinsic rewards, but also inner satisfaction) paradigm and not in the meaning paradigm (Di Fabio & Blustein, 2016). In terms of personal well-being, they have positive narratives; they put emphasis on experiences of success and focus on positive aspects of their performance and on winning resulting in positive energizing psychological effects in terms of self-esteem and self-efficacy. They can also face new challenges, and from this aspect, they have a healthy work-life attitude.

Our study has a number of limitations. First of all, it is not a representative study, but is based on random sampling among business people in Hungary. Compared to the fact that semi-structured interviews were applied, the number of participants however is high, more than 200. Secondly, the categorization of the competitive attitudes were based on a qualitative analysis of the answers which may have been distorted by social desirability and self-presentation in a positive light, resulting in a potentially lower number of those who can be categorized as hypercompetitive or Machiavellian and aggressive competitor. Nevertheless, we believe that the study is able to provide an insight into potential qualitatively different competitive attitudes of business people in a certain sociocultural context, in Hungary decades after the regime change.

14.5 Conclusions

Based on this qualitative interview study, if the socio-economic context changes, it requires a different workforce especially in the positions which are crucial in running the economy. If the business life is fiercely competitive, there is a high probability that those who self-select themselves and fit into this environment, who mostly enjoy this climate and can cope with its challenges and stress and who do not break down in case of failure but able to continue to strive for winning, i.e. who are competitive, may emerge into leadership positions. If there is such a fit between the person and the climate, psychological health does not seem to be compromised. In a society of materialistic and survival values, the necessity to make a company profitable and to maximize financial success may not promote a decent type of sustainable leadership but rather a leadership which places winning and self and company interest and growth into the focus (sometimes even at the expense of fairness), and this may compromise the care for the personal growth and needs of those with whom these goals are to be achieved, namely, the employees.

Acknowledgements The author, Márta Fülöp, was supported by grant NKFIH-OTKA-K (135963) while writing this chapter.

References

Akinson, J. W. (1957). Motivational determinants of risk taking behaviour. *Psychological Review, 64*, 359–372. https://doi.org/10.1037/h0043445

Al-Habib, M. (2012). Identifying The Traits of Entrepreneurs In A University Setting: An Empirical Examination Of Saudi Arabian University. *International Business & Economics Research Journal, 11*(9), 1019–1028. https://doi.org/10.19030/iber.v11i9.7185

Andrews, M. C., Baker, T., & Hunt, T. G. (2010). Values and Person –Organization Fit: Does Moral Intensity Strengthen Outcomes? *Journal of Leadership and Organization Development, 32*(1), 5–19. https://doi.org/10.12691/jbms-1-4-2

Bartling, B., Fehr, E., Maréchal, M. A., & Schunk, D. (2009). Egalitarianism and competitiveness. *American Economic Review: Papers and Proceedings, 99*(2), 93–98. https://doi.org/10.1257/aer.99.2.93

Baum, J., & Locke, E. (2004). The relationship of entrepreneurial traits, skill, and motivation to subsequent venture growth. *Journal of Applied Psychology, 89*(4), 588. https://doi.org/10.1257/aer.99.2.93

Berge, L. I. O., Bjorvatn, K., Pires, A. J. G., & Tungodden, B. (2015). Competitive in the lab, successful in the field? *Journal of Economic Behavior & Organization, 118*, 303–317. https://doi.org/10.1016/j.jebo.2014.11.014

Biró-Nagy, A., Dolmai, D., Kadlót, T., & König, A. (2016). *Rendszerváltás és demokrácia és a magyar társadalom* (System Change and Democracy and Hungarian Society). Friedrich Ebert Stiftung. https://library.fes.de/pdf-files/bueros/budapest/13268.pdf

Boon, C., Den Hartog, D. N., Boselie, P., & Paauwe, J. (2011). The relationship between perceptions of HR practices and employee outcomes: Examining the role of person–organisation and person–job fit. *The International Journal of Human Resource Management, 22*(01), 138–162. https://doi.org/10.1080/09585192.2011.538978

Bönte, W., & Piegeler, M. (2013). Gender gap in latent and nascent entrepreneurship: Driven by competitiveness. *Small Business Economics, 41*(4), 961–987. https://doi.org/10.1007/s11187-012-9459-3

Bönte, W., Procher, V., & Urbig, D. (2016). Biology and selection into entrepreneurship: The relevance of prenatal testosterone exposure. *Entrepreneurship, Theory & Practice, 40*(5), 1121–1148. https://doi.org/10.1111/etap.1216(pdf)

Bönte, W., Lombardo, S., & Urbig, D. (2017). Economics meets psychology: Experimental and self-reported measures of individual competitiveness. *Personality and Individual Differences, 116*, 179–185. https://doi.org/10.1016/j.paid.2017.04.036(pdf)

Bönte, W., Procher, V., & Urbig, D. (2018). Gender differences in selection into self-competition. *Applied Economics Letters, 25*(8), 539–543. https://doi.org/10.1080/13504851.2017.1343441(pdf)

Bönte, W., Procher, V., Urbig, D., & Voracek, M. (2017). Digit ratio (2D:4D) predicts self-reported measures of general competitiveness, but not behavior in economic experiments. *Frontiers in Behavioral Neuroscience, 11*, 238. https://doi.org/10.3389/fnbeh.2017.00238 (open access)

Brañas-Garza, P., Chowdhury, S.M., Espín, A., & Nieboer, J. (2019). *Born this Way'?* Prenatal Exposure to Testosterone May Determine Behavior in Competition and Conflict. SSRN: https://doi.org/10.2139/ssrn.3351310

Brandenburger, A., & Nalebuff, B. (2021). The rules of co-opetition. *Harvard Business Review, January-February* retrieved January 12, 2021. https://hbr.org/2021/01/the-rules-of-co-opetition

Brandts, J., Riedl, A., & van Winden, F. (2005). *Competition and well-being* (IZA Discussion Papers, No. 1769). Institute for the Study of Labor (IZA). http://hdl.handle.net/10419/33629

Brewer, G. (1994). Mind reading: What drives top salespeople to greatness. *Sales and Marketing Management*, (May), 82–92. https://www.elibrary.ru/item.asp?id=2085198

Brown, S. P., Cron, W. L., & Slocum, J. (1998). Effects of trait competitiveness and perceived intra-organizational competition on salesperson goal setting and performance. *Journal of Marketing, 62*(October), 88–98. https://doi.org/10.2307/1252289

Buser, T. (2016). The impact of losing in a competition on the willingness to seek further challenges. *Management Science, 62*, 3439–3449. https://doi.org/10.1287/mnsc.2015.2321

Buser, T., Niederle, M., Oosterbeek, H. (2014). Gender, Competitiveness, and Career Choices. *The Quarterly Journal of Economics, 129*(3), 1409–1447, https://doi.org/10.1093/qje/qju009

Caliendo, M., Fossen, F. M., & Kritikos, A. S. (2009). Risk attitudes of nascent entrepreneurs–new evidence from an experimentally validated survey. *Small Business Economics, 32*(2), 153–167. https://doi.org/10.1007/s11187-007-9078-6

Camerer, C., & Lovallo, D. (1999). Overconfidence and excess entry: An experimental approach. *The American Economic Review, 89*(1), 306–318. https://doi.org/10.1257/aer.89.1.306

Chatman, J. A. (1989). Improving interactional organizational research: A model of person-organization fit. *Academy of Management Review, 14*, 333–349. https://doi.org/10.2307/258171

Chatman, J. A. (1991). Matching People and Organizations: Selection and Socialization in Public Accounting Firms. *Administrative Science Quarterly, 36*, 459–484.

Chell, E., Haworth, J. M., & Brearley, S. A. (1991). *The entrepreneurial personality: Concepts, cases and categories*. Routledge. https://doi.org/10.1177/026624269100900406

Cromie, S. (2000). Assessing entrepreneurial inclinations: Some approaches and empirical evidence. *European Journal of Work and Organizational Psychology, 9*(1), 7–30. https://doi.org/10.1080/135943200398030

Coates, J. M., Gurnell, M., & Rustichini, A. (2009). Second-to-fourth digit ratio predicts success among high-frequency financial traders. *Proceedings of the National Academy of Sciences, 106*(2), 623–628. https://doi.org/10.1073/pnas.0810907106

Derecskei, A., Hurta, H. (2007). A versengés és együttműködés megjelenése a magyar gazdaságban (Competition and Cooperation in the Hungarian Economic Life) 5th International Conference on Management, Enterprise and Benchmarking June 1-2, 2007, Budapest,Hungary. https://kgk.uni-obuda.hu/sites/default/files/33_Derecskei.pdf

Di Fabio, A. (2017b). The psychology of sustainability and sustainable development for well-being in organizations. *Frontiers in Psychology. Organizational Psychology, 8*, 1534. https://doi.org/10.3389/fpsyg.2017.01534

Di Fabio, A., & Blustein, D. L. (Eds.). (2016). *E-book research topic from meaning of working to meaningful lives: The challenges of expanding decent work*. Frontiers Media. https://doi.org/10.3389/978-2-88919-970-9

Di Fabio, A., & Peiró, J. M. (2018). Human Capital Sustainability Leadership to promote sustainable development and healthy organizations: A new scale. In A. Di Fabio (Ed.) *Psychology of Sustainability and Sustainable Development. Special Issue in Sustainability* MDPI, *10*(7), 2413. https://doi.org/10.3390/su10072413

Edwards, J. (2019). *Mastering strategic management*. Creative Commons. https://opentextbc.ca/strategicmanagement/chapter/entrepreneurial-orientation/#fig2.9

EVS/WVS (2020). *European Values Study and World Values Survey*: Joint EVS/WVS 2017-2021 Dataset (Joint EVS/WVS). JD Systems Institute & WVSA. Dataset Version 1.0.0, https://doi.org/10.14281/18241.2

Faragó, K. (2008). Siker, kockázatvállalás és versengés a szervezetekben. *Alkalmazott Pszichológia*, (Success, Risk Taking and Competition in Organizations). *Applied Psychology, 10*(1-2), 7–29.

Fulton, S. A., Maestas, C. D., Maisel, L. S., & Stone, W. J. (2006). The sense of a woman: Gender, ambition, and the decision to run for congress. *Political Research Quarterly, 59*(2), 235–248. https://doi.org/10.1177/106591290605900206

Fülöp, M. (1999). Students' perception of the role of competition in their respective countries: Hungary, Japan, USA. In A. Ross (Ed.), *Young Citizens in Europe* (pp. 195–219). University of North London.

Fülöp, M. (2002a). Intergenerational differences and social transition: Teachers' and students perception of competition in Hungary. In E. Nasman & A. Ross (Eds.), *Children's understanding in the new Europe* (pp. 63–89). Trentham Books.

Fülöp, M. (2002b). Competition in Hungary and Britain perceived by adolescents. *Applied Psychology in Hungary, 2001/2002*(3-4), 33–53.

Fülöp, M., & Berkics, M. (2002). Economic education and attitudes towards enterprise, business and competition among adolescents in Hungary. In M. Hutchings, M. Fülöp, & A. Van Den Dries (Eds.), *Young People's understanding of economic issues in Europe* (pp. 129–153). Trentham Books.

Fülöp, M. (2005). The development of social, economical, political identity among adolescents in the post-socialist countries of Europe. In M. Fülöp & A. Ross (Eds.), *Growing up in Europe today: Developing identities among adolescents* (pp. 11–39). Trentham Books.

Fülöp, M., Berkics, M. (2015). Perception of gender differences in competition in post-socialist Hungary. S. Safdar, N. Koszakowksa-Berezecka (Ed.) The Psychology of Gender through the Lens of Culture, (pp. 193–218). Switzerland: Springer International Publishing.

Fülöp, M., & Orosz, G. (2006). The perception of economic competition and the competitive strategies of Hungarian business people. In *IAREP-SABE Paris, July 5-9, Conference Proceedings* (pp. 1–12).

Fülöp, M., & Szarvas, H. (2012). The ability of cooperation in a competitive context as a civic competence. In S. Striker (Ed.), *Adult education. Historical and theoretical aspects* (pp. 21–35). Eötvös University Press.

Fülöp, M., Molnár, M., Orosz, G., & Ivaskevics, K. (2016). Coping with competition, winning and losing and risk taking in young and old age. *International Journal of Psychology, 51*, 1. http://onlinelibrary.wiley.com/doi/10.1002/ijop.12295/full

Fülöp, M., & Varga, B. A. (2020). Competitive Attitudes Revisited: Personalized Competitive Patterns. (manuscript in preparation).

Ganster, D. C., Schaubroeck, J., Sime, W. E., & Mayes, B. T. (1991). The nomological validity of the Type A personality among employed adults. *Journal of Applied Psychology, 76*, 143–168. https://doi.org/10.1037/0021-9010.76.1.143

Garcia, C., Rivera, N., & Greenfield, P. M. (2015). The decline of cooperation, the rise of competition: Developmental effects of long-term social change in Mexico. *International Journal of Psychology, 50*, 6–11. https://doi.org/10.1002/ijop.12120

Gibb, A. A. (1987). Enterprise culture - its meaning and implications for education and training. *Journal of European Industrial Training, 11*, 1–38. https://doi.org/10.1177/017084068800900315

Gomez, R. (1998). Impatience-agression, competitiveness and avoidant coping: direct and moderating effects on maladjustment among adolescence, Personality and Individual Differences, 25, 649–661

Griffin-Pierson. (1990). The competitiveness questionnaire: A measure of two components of competitiveness. *Measurement and Evaluation in Counselling and Development, 23*, 108–115. https://doi.org/10.1037/t06562-000

Franken, R. E., & Brown, J. D. (1995). Why Do People Like Competition? The motivation for winning, putting forth effort, improving one's performance, performing well, being instrumental, and expressing forceful/aggressive behavior. *Personality and Individual Differences, 19*, 175–184. https://doi.org/10.1016/0191-8869(95)00035-5

Franken, R. E., & Prpich, W. (1996). Dislike of competition and the need to win: Self-image concerns, performance concerns, and the distraction of attention. *Journal of Social Behaviour and Personality, 11*, 695–713.

Guiso, L., & Rustichini, A. (2011). *Understanding the size and profitability of firms: The role of a biological factor* (European University Institute Working Paper ECO 2011/01. 02). European University Institute.
Gumpel, W. (1993). The mentality problem in the transition process from centrally planned economy to market economy, in. R. Schonfeld (ed.) Transforming Economic Systems in East Central Europe, Munich: Sudosteuropa-Gesellschaft.
Harackiewicz, J. M., Barron, K. E., Tauer, J. M., Carter, S. M., & Elliot, A. J. (2000). Short-term and long-term consequences of achievement goals: Predicting interest and performance over time. *Journal of Educational Psychology, 92*, 316–330. https://doi.org/10.1037//0022-0663.92.2.316
Heggestad, E. D., & Kanfer, R. (2000). Individual differences in trait motivation: Development of the Motivational Trait Questionnaire. *International Journal of Educational Research, 33*, 751–776. https://doi.org/10.1016/S0883-0355(00)00049-5
Hibbard, D. R., & Buhrmester, D. (2010). Competitiveness. *Gender, and Adjustment Among Adolescents, Sex Roles, 63*, 412–424. https://doi.org/10.1007/s11199-010-9809-z
Hmieleski, K. M., & Lerner, D. A. (2016). The dark triad and nascent entrepreneurship: An examination of unproductive versus productive entrepreneurial motives. *Journal of Small Business Management, 54*(S1), 7–32. https://doi.org/10.1111/jsbm.12296
Holm, H. J., Opper, S., & Nee, V. (2013). Entrepreneurs under uncertainty: An economic experiment in China. *Management Science, 59*(7), 1671–1687. https://doi.org/10.1287/mnsc.1120.1670
Hunyady, G. (2018). Rendszerigazolás és rendszerkritika (System justification and system criticism). *Magyar Pszichológiai Szemle, (Hungarian Journal of Psychology), 73*(2), 269–298. https://doi.org/10.1556/0016.2018.004
Jackson, D. N. (1974 & 1984). *Personality research form manual*. Research Psychologists Press.
Kanfer, R., & Heggestad, E. (2000). Individual differences in trait motivation: Development of the Motivational Trait Questionnaire (MTQ). *International Journal of Educational Research, 33*, 751–776. https://doi.org/10.1016/S0883-0355(00)00049-5
Kayhan, E. (2003). *Two facets of competitiveness and their influence on psychological adjustment. Honors projects. Paper 4.* Available at: http://digitalcommons.iwu.edu/psych_honproj/4
Keltikangas-Järvinen, L., & Räikkönen, K. (1991). Type A behaviour and types of competitor in young adults. *European Journal of Personality, 5*, 61–69. https://doi.org/10.1002/per.2410050105
Kenyeres, A. (2017). *A sikeres üzletember olyan, mint az élsportoló* (The successful business person is like a sports champion) P/AC@PROF/T April 2, 2017. https://piacesprofit.hu/kkv_cegblog/a-sikeres-uzletember-olyan-mint-az-elsportolo/.
Keresztes, M., Pikó, B., & Fülöp, M. (2015). Does competitiveness count? The role of competitive attitudes in health risk and preventive health behaviors. *European Journal of Mental Health, 10*(1), 44–61. https://doi.org/10.5708/EJMH.10.2015.1.3
Kerr, S. P., Kerr, W. R., & Xu, T. (2018). Personality traits of entrepreneurs: A review of recent literature. *Foundations and Trends in Entrepreneurship, 14*(3), 279–356. https://doi.org/10.1561/0300000080
Kirkpatrick, S. A., & Locke, E. A. (1991). Leadership: Do traits matter? *Academy of Management Executive, 5*, 48–60. https://doi.org/10.5465/ame.1991.4274679
Kirzner, I. M. (1973). *Competition and entrepreneurship*. The University of Chicago Press. https://doi.org/10.2307/1056112
Kirzner, I. M. (1979). Perception, Opportunity and Profit: Studies in the Theory of Entrepreneurship. Chicago: University of Chicago Press.
Koh, H. C. (1996). Testing hypotheses of entrepreneurial characteristics. *Journal of Managerial Psychology, 11*, 12–25. https://doi.org/10.1108/02683949610113566
Kristof-Brown, A., & Guay, R. P. (2011). Person–environment fit. *APA handbook of industrial and organizational psychology, 3*, 3–50. https://doi.org/10.1037/12171-001
Levine, R., & Rubinstein, Y. (2017). Smart and illicit: Who becomes an entrepreneur and do they earn more? *The Quarterly Journal of Economics, 132*(2), 963–1018. https://doi.org/10.3386/w19276

Lumpkin, G. T., & Dess, G. G. (1996). Clarifying the entrepreneurial orientation construct and linking it to performance. *Academy of Management Review, 21*, 135–172. https://doi.org/10.2307/258632

McClelland, D. C., Atkinson, J. W., Clark, R. A., & Lowell, E. L. (1953). *The achievement motive*. Appleton-Century-Crofts. https://doi.org/10.1007/BF02260672

McClelland, D. C., & Winter, D. G. (1969). *Motivating economic achievement*. Free Press-Collier-Macmillan. https://doi.org/10.1002/bs.3830180110

Medland, S. E., Zayats, T., Glaser, B., Nyholt, D. R., Gordon, S. D., Wright, M. J., et al. (2010). A variant in LIN28B is associated with 2D:4D finger-length ratio, a putative retrospective biomarker of prenatal testosterone exposure. *The American Journal of Human Genetics, 86*(4), 519–525. https://doi.org/10.1016/j.ajhg.2010.02.017

Menesini, E., Tassi, F., & Nocentini, A. (2018). The competitive attitude scale (CAS): A multidimensional measure of competitiveness in adolescence. *Journal of Psychology and Clinical Psychiatry, 9*(3), 240–244. https://doi.org/10.15406/jpcpy.2018.09.00528

Miller, D. (2014). A downside of the entrepreneurial personality? *Entrepreneurship: Theory and Practice, 39*, 1–8. https://doi.org/10.1111/etap.12130

Morrison, K. A. (1997). Personality correlates of the five-factor model for a sample of business owners/managers. Associations with scores on self-monitoring, Type A behavior, locus of control, and subjective well-being. *Psychological Reports, 80*, 255–272. https://doi.org/10.2466/PR0.80.1.255-272

Nicolaou, N., Patel, P. C., & Wolfe, M. T. (2018). Testosterone and tendency to engage in self-employment. *Management Science, 64*(4), 1825–1841. https://doi.org/10.1287/mnsc.2016.2664

Niederle, M., & Vesterlund, L. (2007). Do women shy away from competition? Do men compete too much? *The Quarterly Journal of Economics, 122*(3), 1067–1101. https://doi.org/10.1162/qjec.122.3.1067

Niederle, M., & Vesterlund, L. (2011). Gender and competition. *Annual Review in Economics, 3*, 601–630. https://doi.org/10.1146/annurev-economics-111809-125122

Newby, J. L., & Klein, R. G. (2014). Competitiveness reconceptualized: Psychometric development of the competitiveness orientation measure as a unified measure of trait competitiveness. *Psychological Record, 64*, 879–895. https://doi.org/10.1007/s40732-014-0083-2

Orosz, G., Tóth-Király, I., Büki, N., Ivaskevics, K., Bőthe, B., & Fülöp, M. (2018). The four faces of competition: The development of the multidimensional competitive orientation inventory. *Frontiers in Psychology: Personality and Social Psychology, 9*, 779. https://doi.org/10.3389/fpsyg.2018.00779

Ostroff, C., & Schulte, M. (2007). Multiple perspectives of fit in organizations across levels of analysis. In C. Ostroff & T. A. Judge (Eds.), Perspectives on organizational fit (pp. 3–69). Lawrence Erlbaum Associates Publishers.

Pedler, M., Burgoyne, J., & Boydell, T. (2004). *A manager's guide to leadership*. McGraw-Hill. https://doi.org/10.1080/14767333.2011.581028

Pedler, M., Burgoyne, J., & Boydell, T. (2006). A Manager's Guide to Self-development. Maidenhead: McGraw-Hill.

Rauch, A., & Frese, M. (2007). Born to Be an Entrepreneur? Revisiting the Personality Approach to Entrepreneurship. In J. R. Baum, M. Frese, & R. A. Baron (Eds.), The psychology ofentrepreneurship (pp. 41–65). Lawrence Erlbaum Associates Publishers.

Ryckman, R. M., Hammer, M., Kaczor, L. M., & Gold, J. A. (1990). Construction of a hypercompetitive attitude scale. *Journal of Personality. Assessment, 55*, 630–639. https://doi.org/10.1080/00223891.1990.9674097

Ryckman, R. M., Hammer, M., Kaczor, L. M., & Gold, J. A. (1996). Construction of a personal development competitive attitude scale. *Journal of Personality. Assessment., 66*, 374–385. https://doi.org/10.1207/s15327752jpa6602_15

Ryckman, R. M., Thornton, B., & Gold, J. A. (2009). Assessing competition avoidance as a basic personality dimension. *The Journal of Psychology: Interdisciplinary and Applied, 143*(2), 175–192. https://doi.org/10.3200/JRLP.143.2.175-192

Salvador, A., Suay, F., Martinez-Sanchis, S., Simon, V. M., & Brain, P. F. (1999). Correlating testosterone and fighting in male participants in judo contests. *Physiology & Behavior, 68*(1-2), 205–209. https://doi.org/10.1016/S0031-9384(99)00168-7

Salvador, A. (2005). Coping with competitive situations in humans. *Neuroscience & Biobehavioral Reviews, 29*(1), 195–205. https://doi.org/10.1016/j.neubiorev.2004.07.004

Sapienza, P., Zingales, L., & Maestripieri, D. (2009). Gender differences in financial risk aversion and career choices are affected by testosterone. *Proceedings of the National Academy of Sciences, 106*(36), 15268–15273. https://doi.org/10.1073/pnas.0907352106

Schumpeter, J. A. (1934). *The theory of economic development*. Harvard University Press.

Schneider, B. (1987). The people make the place. *Personnel Psychology, 40*, 437–453. https://doi.org/10.1111/j.1744-6570.1987.tb00609.x

Sommer, S. M. (1995). Social competition: Identifying new perspectives and strategies for task motivation. *The International Journal of Conflict Management, 6*(3), 239–256. https://doi.org/10.1108/eb022764

Spence, J. T., & Helmreich, R.-L. (1983). Achievement-related motives and behavior. In J. T. Spence (Ed.), *Achievement and achievement motives: Psychological and sociological approaches* (pp. 10–74). Freeman. https://doi.org/10.2307/1422172

Stalk, G., Lachenauer, R. (2004). Hard ball. Five killer strategies for trouncing the competition. *Harvard Business Review*, April, 64-71. https://hbr.org/2004/04/hardball-five-killer-strategies-for-trouncing-the-competition

Suster, M. (2010a). *The best entrepreneurs are hyper competitive & hate losing*. https://bothsidesofthetable.com/the-best-entrepreneurs-are-hyper-competitive-hate-losing-7201972b2fbe

Suster, M. (2010b). *Embrace your losses – They will make you stronger*. https://bothsidesofthetable.com/embrace-losing-it-will-make-you-stronger-ba94b03ac7ad

Szabó, K., & Kocsi, É. (2003). *Tanulás és felejtés vegyes vállalatokban* (Learning and unlearning in multinational companies). Oktatási Minisztérium.

Tassi, F., & Schneider, B. H. (1997). Task-oriented versus other-referenced competition: Differential implications for children's peer relations1. *Journal of Applied Social Psychology, 27*, 1557–1580. https://doi.org/10.1111/j.1559-1816.1997.tb01613.x

Thornton, B., Ryckman, R. M., & Gold, J. A. (2011). Competitive orientations and the type a behavior pattern. *Psychology, 5*(2), 411–415. https://doi.org/10.4236/psych.2011.25064

Urbig, D., Bönte, W., Lombardo, S., & Procher, V. D. (2020). Entrepreneurs embrace competition: Evidence from a lab-in-the-field study. *Small Business Economics, 55*, 193–214. https://doi.org/10.1007/s11187-019-00141-0. (open access).

Utsch, A., Rauch, A., Rothfufs, R., & Frese, M. (1999). Who becomes a small scale entrepreneur in a post-socialist environment: On the differences between entrepreneurs and managers in East Germany. *Journal of Small Business Management, 37*(3), 31–42.

Utsch, A., & Rauch, A. (2000). Innovativeness and initiative as mediators between achievement orientation and venture performance. *European Journal of Work and Organizational Psychology, 9*(1), 45–62. https://doi.org/10.1080/135943200398058

Wang, H., Wang, L., & Liu, C. (2018). Employee competitive attitude and competitive behavior promote job-crafting and performance: A two-component dynamic model. *Frontiers in Psychology, 9*, 2223. https://doi.org/10.3389/fpsyg.2018.02223

William, J., Slocum, J. W., & von Glinow, M. A. (1982). Person-situation interaction: Competing models of fit. *Journal of Occupational Behavior, 3*(July), 265–280. https://doi.org/10.1002/job.4030030402

Wu, B., & Knott, A. M. (2006). Entrepreneurial risk and market entry. *Management Science, 52*(9), 1315–1330. https://doi.org/10.1287/mnsc.1050.0543

Chapter 15
Sustainability and Changing Care Programs in the South Asian Countries

Anwarul Hasan Sufi

15.1 Introduction

South Asian countries are thickly populated countries of the world. Very high density of population compared to land sizes of these countries is mainly because of the fertile flat lands, tropical climate, and relatively easygoing life attitude of the people. Though there are some specific differences in the living style, attitude, values, beliefs, personality traits, emotions, etc. of the people of different countries and within the same countries, it may be difficult for the western people to perceive the differences very easily unless they stay with the people of this region for some days.

The indigenous care programs of the people of South Asian countries are relatively different from the countries those developed significantly after the Second World War.

The concept of sustainability is a relatively new terminology among the people of South Asian countries as they did not face any manmade disaster until the decade of the 1970s of the last century. They faced only the natural disasters, epidemics, and famine. The warriors and emperors ruled these lands for thousands of years, and the large majorities of the people adapted to the rules without much resistance.

However, the issue of sustainability was there among the people of this region from the very beginning of the human settlements. Landslides, cyclones, famines, river erosions, superstitions, invaders, etc. forced them to migrate from place to place until the seventh century. Then the changes came with the introduction of the new religions, economic disparity, and promotion of social classes and cast systems. But the people of this region accepted everything and continued their lives in their own indigenous ways.

A. H. Sufi (✉)
University of Rajshahi, Rajshahi, Bangladesh
e-mail: sufi@ru.ac.bd

A. Di Fabio (ed.), *Cross-cultural Perspectives on Well-Being and Sustainability in Organizations*, https://doi.org/10.1007/978-3-030-86709-6_15

The remains of the urbanization of civilization of Indus Valley; the remains of the submerged road bridge between India and Sri Lanka; the oldest education center of Paharpur in Bangladesh; the temples of Nepal, Bhutan, and India; and palaces and magnificent buildings in Indian subcontinent all these depict the ancient culture of this region. Pictures and statues found in the caves of this region depict that there existed both the normal and in some cases the abnormal behavior including the caregiving process among the humans and animals in this region.

With the new inventions in science, industrialization, urbanization, modernization, etc. gradually changed the need and attitude of the people in caregiving. Though the large majorities of the people are yet involved in agriculture-oriented jobs, a huge number of the population are engaging themselves in industries, business, and service-oriented jobs. With the change of the jobs, the indigenous system of caregiving is gradually being changed. These changes significantly affected the caregiving of the elderly and the disabled people. The nature, number, and seriousness of sustainability issues of the people became observable and are going beyond the traditional perspectives.

The psychology of sustainability and sustainable development goes beyond the traditional perspective and introduces a framework focused not only on using increasingly smaller quantities of resources but also on regenerating resources and on a positive approach. That approach is based on new keywords such as promotion, enrichment, growth, and flexible change (Di Fabio, 2017a, b).

15.2 Sustainability and Changing Care Programs in the South Asian Countries

The current population of Southern Asia is 1,953,461,900 as of Thursday, February 4, 2021, based on the latest United Nations estimates. Southern Asian population is equivalent to 24.89% of the total world population. Southern Asia also ranks number 1 in Asia among sub-regions ranked by population. The average population density in Southern Asia is 303 per Km^2 (785 people per mi^2). The total land area is 6,400,127 square kilometers. In 2019, it was estimated that only 36.6 % of the population live in the urban areas. The median age in Southern Asia is 27.6 years (www.worldometers.info/world-population/southern-asia-population).

The seven countries of South Asia are changing more or less in the same ways. There are central governments, local governments, and the informal indigenous community control bodies for grassroots-level population in all the countries. The informal indigenous community control bodies are named as *Panchayat* or *Saamaj* in different regions. These are very powerful institutions (Johnson, 2003). The nature, role, and authorities of these informal indigenous community control bodies are briefly described below. To have an in-depth perception of the sustainability and changing care programs in the South Asian countries, we need to know a little about the *Panchayat* or *Saamaj*.

Panchayats can be called as an organ of village self-government, village council, and a group of influential older men acknowledged by the community as its governing bodies. Panchayat system has been the basic and critical informal judicial system for the villages for justice provision as the decentralized power and local governance, and it has never been completely adopted by national or local governments (Farooqi et al., 2019).

Panchayat is also in practice in Nepal. Kumar (1964) in his article entitled "The Panchayat Constitution of Nepal and Its Operation" mentioned that traditionally, it means village self-government through a committee of five representatives of the village assembly.

Udgaonkar (1986) of India mentioned that these popular courts are first mentioned by Yajnavalkya and then by Narada, Brihaspati, Somadeva, and Shukra. These writers covered a period of about a thousand years, c. 100 to 1950 A.D., and they could not have mechanically referred to the popular courts if they were not actually functioning. However, there are evaluations from different perceptual viewpoints; and an article published in *The Times of India* on September 9, 2007, by Mullick Rohit and Raaj Neelam mentioned that Panchayats turn into kangaroo courts.

The *Trinidad and Tobago Newsday*, on May 30, 2016, published an article entitled "The Panchayat system as an early form of conflict resolution in Trinidad" mentioning that *the Panchayat raj (panchayat "village council," raj "rule") is a political system, originating from the Indian subcontinent, found mainly in India, Pakistan, Bangladesh, Sri Lanka, and Nepal. It is the oldest system of local government in the Indian subcontinent, and historical mentions date to the 250 CE period. The contents of this news were earlier mentioned by* Udgaonkar in 1986, where it was mentioned that *the word raj means "rule" and panchayat means "assembly" (ayat) of five (panch) . Traditionally panchayats consisted of wise and respected elders chosen and accepted by the local community. However, there were varying forms of such assemblies. Traditionally, these assemblies settled disputes between individuals and between villages. And* Rohit and Raaj Neelam in 2007 mentioned in *The Times of India* that *the leader of the panchayat was often called the mukhiya, sarpanch, or pradhan, an elected or generally acknowledged position. The modern panchayati raj of India and its gram panchayats are neither to be confused with the traditional system nor with the extra-constitutional khap panchayats (or caste panchayats) found in parts of northern India.*

More or less, the same system exists in Bangladesh in the name of "Saamaj." The decision-makers of Saamaj are not elected persons, but may include a person officially elected in the local government (Sukumar et al., 2019). The Saamaj decides who will get the government pension, who will feed his neighbor in poverty, who will take care of the elderly people, what should be the treatment of a disabled persons, which legal issue must not go to the government court, how much can be told to a journalist, etc. The decision of the Saamaj is final, and nobody under the umbrella of that Saamaj can go against the decision. But there are processes for review. If the embankment of the river is damaged and the government delays to repair, the Saamaj initiates the emergency repair ordering all the able persons to join the works. Therefore, it can be said that Saamaj is the supreme authority in the grassroots level (Sufi et al., 1996).

To analyze the sustainability and changing care programs in the South Asian countries, it is pertinent to know the situation of these seven countries as shown in the following tables. The figures shown in the tables were extracted from the websites of the World Bank, WHO, UNICEF, UNDP, and other related organizations. The countries are presented in alphabetic order (Tables 15.1, 15.2 and 15.3).

In Table 15.4, the GDP per capita shows only the mean values, not the standard deviations. The economic conditions of the rich and the poor were merged to be averaged. If the standard deviation was available together with the mean values, it could be assumed whether the countries are developing with the mass population like the countries having strong social security systems.

The sustainability and changing care programs in the South Asian countries are significantly observed in the care programs of the disabled persons and the elderly persons.

Disability

The percentages of disability as shown in Table 15.5 were extracted from the reports published by national and international agencies and different NGOs working in the related fields in the South Asian countries. As the criteria of assessment of disabilities were not same, it seems that the figures do not project the reality of the countries. Every year thousands of people become permanently handicapped after different types of accidents, illnesses, and many other conditions. Large majorities of such persons need the help of others in most of their daily living activities. It is assumed that if all such persons were included, the percentage of disabled population will be around 10% in all the countries of South Asia.

Sufi and Tushar (2017) in their study entitled "Exploring the status of developmentally disabled children of Sukhandighi village" depicted the prevalence, status, and prevention of childhood disabilities in a typical Bangladeshi village named Sukhandighi. The village is under Joynagar Union of Upazila Durgapur in Rajshahi

Table 15.1

	Country	Area sq km	Population	Population density per sq km	% of agricultural land	Remarks
1	Bangladesh	148,460	164,689,383	1104	70.63	
2	Bhutan	38,394	771,608	20	13.61	
3	India	3,287,000	1,380,004,385	420	60.45	
4	Maldives	300	540,544	1800	26.33	
5	Nepal	147,516	29,136,808	203	28.75	
6	Pakistan	796,095	220,892,340	278	47.79	
7	Sri Lanka	65,610	21,413,249	326	43.69	

Table 15.2

	Country	Life expectancy in years	Population growth rate	Annual death rate per thousand	Children per woman	Remarks
1	Bangladesh	73	1.00	5.5	2.028	
2	Bhutan	72	1.10	6.3	1.973	
3	India	69	1.00	7.0	3.300	
4	Maldives	67	2.90	3.0	1.859	
5	Nepal	71	1.80	6.0	1.880	
6	Pakistan	79	2.00	7.0	3.425	
7	Sri Lanka	77	0.60	6.5	2.177	

Table 15.3

	Country	Forex reserves Billion US$	% of unemployment	% of population undernourished/acute hunger	Remarks
1	Bangladesh	42.97	4.19	11	
2	Bhutan	1.24	2.34	12	
3	India	590.19	6.60	14	
4	Maldives	0.98	6.40	10	
5	Nepal	11.50	1.51	23	
6	Pakistan	13.03	4.45	20	45% children stunted
7	Sri Lanka	5.55	5.40	17	

District. The distance is about 20 kilometers by road in the northern side from Rajshahi University campus. The researchers studied the village from July 2015 to May 2017. Door-to-door visit was done from time to time. The main objective was to study the nature, number, and seriousness of childhood disability. However, all demographic changes were observed and recorded. In July 2015, the population of the village was 2781 in 639 households. In May 2017, the population of the village is 2792 in 639 households. The total number of disabled children is 16 persons (0.57%). Out of these 16 cases, 2 children are Down's syndrome children, 2 are very severely handicapped, 4 are mild intellectually disabled, 1 has severe hearing impairment, and the other 7 possess neuromotor disorders. It was also found that not a single disabled child was born in this village since 2009. The achievement is mainly because of the awareness development and vigilance of the field-level health workers. The researchers observed the changing pattern of the problems of the caregivers, mainly the family members. Treatment patterns and special education were studied, too. It was found that family members attempted treatments when the cases were children. Then they gradually reduced treatment initiatives. Practically nothing was done toward special education, even home-based special education. However, from the psychological viewpoints, it was found that the lives of these disabled persons are better in the village compared to the lives of the disabled persons living in urban areas. All the mild and some of the moderately disabled persons

Table 15.4

	Country	Literacy rate %	Poverty (2018) in %	GDP per capita US$ (2019)	Remarks
1	Bangladesh	73.09	21.80	1855	
2	Bhutan	66.56	8.20	3316	
3	India	77.70	30.00	2099	
4	Maldives	97.33	46.50	10,626	
5	Nepal	67.09	25.20	1071	
6	Pakistan	58.00	24.30	1284	
7	Sri Lanka	91.71	4.10	3853	

Table 15.5

	Country	% of elderly (60+) population	% of disability	Remarks
1	Bangladesh	8.00	5.60	All disabilities: mental, visual, auditory, physical, etc.
2	Bhutan	7.80	2.80	
3	India	8.60	2.21	
4	Maldives	5.00	6.80	
5	Nepal	6.50	2.90	In hill districts, disability is more than 4%
6	Pakistan	5.60	2.38	
7	Sri Lanka	6.30	8.70	WHO revealed that approximately 9% of the population in Sri Lanka has loss of hearing in 2019

ultimately get the opportunities of entering the mainstream agricultural jobs. As factors of disabilities, birth hazards in the absence of qualified midwives, indiscriminate use of medicines by the pregnant mothers, and infantile illnesses were found as serious factors. The researchers studied the health, education, economic, housing, and social security status of the disabled children.

The criteria and concepts of disability are different in different countries and communities within the countries. In Bangladesh, in the rural areas, not only the mild but also the moderate mentally handicapped persons, with abilities to do physical works, both male and female, join the mainstream workforce, mainly in the agricultural works, in their adolescence period. They are not considered as disabled persons. Most of them get married, give birth to children, and lead a happy life. Similar is the condition in other countries of South Asia except Maldives. In Maldives, such persons if are not able to manage the fishing boats cannot join the mainstream workforce.

In all the South Asian countries, the persons with disabilities, living in the urban areas, possess miserable life compared to their rural counterparts. These disabled children cannot enjoy the games with playmates, cannot spend time with the opposite door neighbors in apartment houses, and are segregated. It was also found that approximately 3% of the total disabled children of the entire South Asia are affiliated with some day centers or special schools. The remaining 97% are beyond any scientific stimulation or services.

Now if we consider the issue of sustainable care programs for the disabled persons, we are to emphasize the importance of recognizing, respecting, and using the meaning of suitable/appropriate works for the disabled persons, as a key for growth and success, as mentioned by Di Fabio Annamaria (2017a) in her paper "Positive Healthy Organizations: Promoting Well-Being, Meaningfulness, and Sustainability in Organizations." However, the promotion of healthy organizations, promoting well-being, meaningfulness, and sustainability in organization to uplift the condition of the disabled, did not develop well in South Asian countries as the meaning and criterion of disability differ in different countries.

However, India is the first country toward the identification and establishment of the criteria of disabilities through an act in 2016. The act is known as the Rights of Persons with Disabilities (RPWD) Act, 2016. In this Act, 21 conditions were mentioned out of innumerable types of disabilities that can affect a human being. Some of these conditions are more common than others.

1. Blindness

   Blindness is defined as the state of being sightless. A blind individual is unable to see. In a strict sense, the word blindness denotes the condition of total blackness of vision with the inability of a person to distinguish darkness from bright light in either eye.
2. Low Vision

   Low vision means a condition where a person has any of the following conditions, namely:

   (1) Visual acuity not exceeding 6/18 or less than 20/60 up to 3/60 or up to 10/200 (Snellen) in the better eye with best possible corrections
   (2) Limitation of the field of vision subtending an angle of less than 40 degree up to 10 degree

3. Leprosy-Cured Persons

   Leprosy, also known as Hansen's disease (HD), is a chronic infectious disease caused by a bacterium called *Mycobacterium leprae*. The disease mainly affects the skin, the peripheral nerves, the mucosal surfaces of the upper respiratory tract, and the eyes. Leprosy is known to occur at all ages ranging from early infancy to very old age. About 95% of people who contact *M. leprae* do not develop the disease.
4. Hearing Impairment

   Hearing impairment is a partial or total inability to hear. It is a disability which is sub-divided into two categories of deaf and hard of hearing.

   - "Deaf" means persons having 70 dB hearing loss in speech frequencies in both ears.
   - "Hard of hearing" means person having 60 dB to 70 dB hearing loss in speech frequencies in both ears.

5. Locomotor Disability

   Strictly speaking, locomotor disability means problem in moving from one place to another—i.e., disability in legs. But, in general, it is taken as a disability related with bones, joints, and muscles. It causes problems in person's movements (like walking, picking, or holding things in hand, etc.)
6. Dwarfism

   Dwarfism is a growth disorder characterized by shorter than average body height.
7. Intellectual Disability

   Intellectual disability, also known as general learning disability and mental retardation (MR), is a condition characterized by significant limitation both in intellectual functioning (reasoning, learning, problem-solving) and in adaptive behavior which covers a range of every day, social, and practical skills.
8. Mental Illness

   Mental illness or mental disorder refers to a substantial disorder of thinking, mood, perception, orientation, or memory that grossly impairs judgment, behavior, capacity to recognize reality, or ability to meet the ordinary demands of life. But it *does not include* retardation which is a condition of arrested or incomplete development of mind of a person, specially characterized by subnormality of intelligence.
9. Autism Spectrum Disorder

   Autism spectrum disorder (ASD) is a neurological and developmental disorder which affects communication and behavior. Autism can be diagnosed at any age. But still it is called a "developmental disorder" because symptoms generally appear in the first 2 years of life. Autism affects the overall cognitive, emotional, social, and physical health of the affected individual.
10. Cerebral Palsy

    Cerebral palsy (CP) is a disabling physical condition in which muscle coordination is impaired due to damage to the brain. It occurs at or before child birth. Cerebral palsy is not a progressive condition, meaning it does not get worse with time. However, muscle disuse could increase the extent of disability over the period of time. At present, there is no cure available for this condition. Thus, cerebral palsy is an incurable and lifelong condition, at present.
11. Muscular Dystrophy

    Muscular dystrophy (MD) is a group of neuromuscular genetic disorders that cause muscle weakness and overall loss of muscle mass. MD is a progressive condition, meaning that it gets worse with the passage of time.
12. Chronic Neurological Conditions

    Examples of chronic neurological conditions are as follows:

    (1) Alzheimer's disease and dementia
    (2) Parkinson's disease
    (3) Dystonia
    (4) ALS (Lou Gehrig's disease)
    (5) Huntington's disease

(6) Neuromuscular disease
(7) Multiple sclerosis
(8) Epilepsy
(9) Stroke

13. Specific Learning Disabilities

    Specific learning disabilities are a group of disabling conditions that hampers a person's ability to listen, think, speak, write, spell, or do mathematical calculations. One or more of these abilities may be hampered.

14. Multiple Sclerosis

    In multiple sclerosis (MS), the immune system of the body attacks the central nervous system, which includes the brain and spinal cord. As a result of MS, the myelin sheath covering on neurons gets damaged. This exposes the nerve fiber and causes problems in the information flow through nerves. With time, MS can lead to the permanent damage to nerves.

15. Speech and Language Disability

    A permanent disability arising out of conditions such as laryngectomy or aphasia affecting one or more components of speech and language due to organic or neurological causes.

16. Thalassemia

    Thalassemia is a genetically inherited blood disorder which is characterized by the production of less or abnormal hemoglobin. As we know, hemoglobin is a protein found in red blood cells. Hemoglobin is responsible for carrying oxygen around in the body. Thalassemia results in large numbers of red blood cells being destroyed, which leads to anemia. As a result of anemia, person affected with thalassemia will have pale skin, fatigue, and dark coloration of urine.

17. Hemophilia

    Hemophilia is a blood disorder characterized by the lack of blood clotting proteins. In the absence of these proteins, bleeding goes on for a longer time than normal. Hemophilia almost always occurs in males, and they get it from their mothers. Females are rarely affected with hemophilia.

18. Sickle Cell Disease

    Sickle cell disease is a group of blood disorders that causes red blood cells (RBCs) to become sickle-shaped, misshapen, and break down. The oxygen-carrying capacity of such misshapen RBCs reduces significantly. It is a genetically transferred disease. Red blood cells contain a protein called hemoglobin. This is the protein that binds oxygen and carries it to all the parts of the body.

19. Multiple Disabilities Including Deaf Blindness

    Multiple disabilities are the simultaneous occurrence of two or more disabling conditions that affect learning or other important life functions. These disabilities could be a combination of both motor and sensory nature.

20. Acid Attack Victims

    An acid attack victim means a person disfigured due to violent assaults by throwing of acid or similar corrosive substance.

21. Parkinson's Disease

Parkinson's disease (PD) is central nervous system disorder which affects movement. Parkinson's disease is characterized by tremors and stiffness. It is a progressive disease, which means that it worsens with time. There is no cure available at present.

It is observed that after the implementation of this act in India, the neighboring countries also accepted the above criterion and undertaking programs more or less in the same strategies to sustain and uplift the life of the disabled persons. All the South Asian countries introduced monthly financial support systems for the disabled persons. Though the amount of money is not enough for the persons, this financial support has increased their status in their families. Previously they were considered as burden and liabilities of the families. Now, the family members are taking more care of them that they remain alive and the families receive the monthly grants from the governments.

In the South Asian countries, some awareness related to the physically handicapped people were observed during the last 100 years. But about the mentally disabled, the awareness started developing since the decade of the 1970s of the last century. Earlier, they were considered as mentally ill persons, and they faced many wrong treatments. Though the scientific attention was given to them 50 years ago, they were subjects of trials and errors until recently. Introduction of (DSM 5), the Diagnostic and Statistical Manual of Mental Disorders, Fifth Edition is the 2013 update to the Diagnostic and Statistical Manual of Mental Disorders, the taxonomic and diagnostic tool published by the American Psychiatric Association started changing the practice of both the professionals and family members, at least in the urban areas of the South Asian countries. Dependency on psychotropic drugs is being reduced nowadays in all these countries.

Elderly

In all the countries of South Asia, people aged over 60 years are considered as senior citizens or elderly people. Elderly or the term aging refers to the process of growing older. It is the aggregate result of the decrement processes at cellular, subcellular, or organ level that are associated with the passage of time. It is the end of life cycle. It is of course a biological reality which (despite medical intervention) has its own dynamic largely beyond human control (Gorman et al., 2019). Aging causes a functional deterioration and vulnerability that also creates physical changes as well. With the increasing of his age, older people become incapable to carry out his normal functioning as he did before. Thus, aging is a social process that involves individual and population aging, biological and psychological aging, individual and social change, individual and cultural differences within and between age cohorts, and cross-cultural and sub-cultural differences in values, beliefs, and norms (Mcpherson, 1991). Social thinkers are mostly concerned about social factors of aging that is termed as social aging, because they can contribute to the aged in this context. Social aging is a multidimensional and dynamic force. It includes the transition into and out of roles, expectations about behavior, societal allocation of resources and opportunities, negotiation about the meaning and implications of chronological age, and the experience of individuals traveling the life course and negotiating life stages (Morgan & Kunkel, 2001).

Common health problems that occur in elderly in Bangladesh are cardiovascular disease including hypertension, stroke, angina, myocardial infarction, or heart attack, diabetes mellitus and its complications, cataract, arthritis, osteoporosis, skeletal deformity, obesity, enlarged prostate in men, sleep disturbance, changes in behavior, and dementia. Cancer is another important disease. Elderly people suffer from cancer like prostate, lung, stomach, colon, breast, and cervix of the uterus. Arthritis affects 50% of people over 65 years and can lead to pain and lower quality of life for some of them. Heart disease remains the leading killer of adults over 65, cancer is the second, and chronic obstructive pulmonary disease (COPD) is the third most common cause of death among people over 65 years. A cross-sectional study carried out by house-to-house survey of all people aged over 60 years in an urban slum and a village in India found that use of tobacco is very high 58.9%. Visual impairment was the most common handicap in that study 83.29%. Uncorrected hearing impairment was also common. A large portion had dental problem. Almost half of the population had depression. Prevalence of cognitive impairment among elderly people which is about 22.4% was found in a Malaysian study. The most common mental and neurological disorders in this age group are dementia and depression which affect approximately 5% and 7% of the world's older people, respectively. In a study relating health problems of elderly patients attending the Combined Military Hospital, Dhaka, it is found that majority of elderly patients had diabetes mellitus (21.1%) followed by rheumatoid arthritis (17.6%), asthma (12.5%), cataract (11.2%), ENT problem (6.6%), malignancy (5.9%), and benign enlargement of the prostate (5.3%). Elderly people in lower and lower middle socioeconomic class suffers from financial crisis when they are in need of money for treatment. In a study in Dhaka district, only 14% of elderly had health insurance coverage (Islam & Rahman, 2017).

In Bangladesh, it has traditionally been the responsibility of the family to provide food and shelter for its elderly members. The poor families are unable to meet the basic needs of the older people including food, health care, clothing adequate shelter, and access to income or employment opportunities (Hossain et al., 2006).

India has nearly 120 million elderly people with various physical, psychosocial, economic, and spiritual problems. While the functionally and cognitively fit can access usual health-care facilities provided by the government, these people need active aging program to keep them independent. Health ministry has created geriatric centers and geriatric clinics in most of the states; however, these centers may not serve the functionally and cognitively impaired elderly. There is great need for mobile units, day-care centers and hospices, and training of personnel in home nursing. Routine care clinics cannot handle the burden of geriatric population to address their multimorbidity and several other age-related problems. There is a need for a rapid training of health-care professionals of various disciplines in geriatric care. Government must support nongovernmental organizations and other agencies which provide day care, home care, and palliative care so that these services become affordable to all the elderly (Adhikari, 2017).

The life of the elderly people in other countries of South Asia is more or less like India and Bangladesh. These countries could not develop old home facilities like Japan, Scandinavian countries, and EU countries. The elderly people are yet dependent on their family members. Newspapers of these countries frequently publish reports of the miseries of the elderly people of both rich and poor families. Nowadays, some reports were published that the children abandoned their sick aged parents in railway stations, public bus terminals, and other places without notice. Analyzing the reports, it was found that most of such cases of are in the urban areas, not the rural areas. It was also found that in rural areas, the Panchayat or Saamaj leaders remain vigilant that children take care of their aged parents. The governments of all the seven countries have introduced old age pension schemes, especially for the poor families. Though the amount is not significant for the living costs, this financial support system has increased the importance of the elderly persons in their families.

During interviews with the elderly people living in the rural areas of Bangladesh, India, Bhutan, and Nepal, the author came to know that many elderly persons possess sufficient money to pay the medical costs, but none accompanied them to travel to such centers. In Bangladesh, the average distance from home to medical centers is maximum 5 kilometers; but in India and Pakistan, in the rural areas, such centers are sometimes so far away that elderly people do not take the risks of traveling. In Nepal and Bhutan, in the mountain regions, large majorities of the elderly people avoid traveling to the medical centers unless there are roads and transports. Sustainability of the elderly people is a challenge in some regions.

SDG Outlook

The sustainability and changing care programs in the South Asian countries can also be considered from the 17 SDG goals adopted by the United Nations for the period 2015 to 2030. It was mentioned that the 17 SDGs are integrated—that is, they recognize that action in one area will affect outcomes in others and that development must balance social, economic, and environmental sustainability. Through the pledge to Leave No One Behind, countries have committed to fast-track progress for those furthest behind first. That is why the SDGs are designed to bring the world to several life-changing "zeros," including zero poverty, hunger, AIDS, and discrimination against women and girls. Everyone is needed to reach these ambitious targets. The creativity, know-how, technology, and financial resources from all of society are necessary to achieve the SDGs in every context. The following section of this chapter briefly reviews the changes taking place in the 17 areas in South Asian countries.

No Poverty

The criterion of poverty differs in different regions of the world. The World Bank defines poverty in absolute terms. The bank defines extreme poverty as living on less than US$1.90 per day (PPP) and moderate poverty as less than $3.10 a day. Percentages of people under poverty level in 2018 according to the World Bank criteria were shown in Table 15.4 at the beginning of this paper. It was found that the highest poverty level (46.50%) was found in Maldives and the lowest (4.10%) in Sri Lanka. It was also found that in Bangladesh (21.80%), Bhutan (8.20%), India (30.00%), Nepal (25.20%) and in Pakistan (24.30%) people live under the poverty line.

It is understood that the day laborers in Bangladesh earn about Tk 500 (about 4.5 US$) per day nowadays which was US$ 3 in 2018. And during harvesting period, it is about US $ 6.5 per day plus breakfast and two meals. Women are also working in the labor force in the country. During the last 2 years, scarcity of day laborers was observed. Therefore, the harvesting machines are becoming popular in Bangladesh. Similar is the condition in India, but the work opportunities are relatively lesser compared to Bangladesh.

Asadullah and Chaudhury (2012) in their paper entitled "Subjective Well-Being and Relative Poverty in Rural Bangladesh" reviewed the debate over the importance of absolute vs. relative income as a correlate of subjective well-being using data from Bangladesh, one of the poorest countries in the world with high levels of corruption and poor governance. Combining household data with population census and village survey records, they found that conditional on own household income, respondents report higher satisfaction levels when they experience an increase in their income over the past years. More importantly, individuals who reported their income to be lower than their neighbors in the village also report less satisfaction with life. They also found that relative wealth effect is stronger for the rich. Similarly, in villages with higher inequality, individuals reported less satisfaction with life. However, when compared to the effect of absolute income, these effects (i.e., relative income and local inequality) were found as modest.

Though the abovementioned study is relatively old in relation to the speedy advancements more or less in all sectors in Bangladesh, the perceived subjective well-being remained the same for the laborers in all sectors.

Zero Hunger

On a global scale, the simplest definition of hunger is a scarcity of food in a country. This occurs when the population of a country quite literally does not have enough to eat.

When we look to the old photographs of the South Asian people of the 1950s and 1970s of the last century, we see that the people were thin and wearing poor clothes. But nowadays, except some parts of India, in all the countries, the people are found relatively thicker and wear better clothes.

Bangladesh achieved food sufficiency more or less from 1998. In the decade of the 1970s of the last century, the lands growing 6 tons of rice now provide the output of 30 tons on the average. This has been achieved because of the excellent research of the agricultural scientists and vigilant supervisions of the field-level officers. Now all the farmers have mobile apps to consult directly with the concerned experts in relation to seeds, pesticides, and fertilizers and in all other problems. During the last 2 years, more varieties of hybrid seeds were invented, and it is assumed that the output may increase to 40 tons in the same-sized lands. Similar is the situation in Bangladesh in the production of fish, chicken, eggs, goats, vegetables, etc. Rarely stunted children are found in Bangladesh. The other countries are also progressing satisfactorily in this region, but not like Bangladesh, because they have mountains and difficulties in irrigation.

Besides, there are religious and cultural values that work in reducing hunger. Bangladesh is a liberal Muslim country, and the attitude and values are little different from other Muslim countries. Under the religious umbrella, it is a practice that no one will sleep unless the hungry neighbor is getting food. It does not matter what is the religion or social status of the neighbor. During COVID-19 pandemic, many new charities came to the limelights who provided food to the poor and yet are continuing their programs. Local governments and Panchayats are also vigilant to eradicate hunger. But in India, there is a cast system in the religion, and it is a little difficult to uplift such free food charities in some regions.

Good Health and Well-Being

Health means not only the physical illness or morbidity; it includes complete social and mental well-being. It is well known that the South Asian countries do not have medical facilities like the western countries, but people are surviving. In these countries, beside the modern allopathic treatments, homeopathy and Ayurvedic (herbal) treatments are in existence. And interestingly there are influential faith healers. Let us forget about the faith healers. The homeopathic and Ayurvedic treatments are cheaper in these countries. Indian Ayurvedic medicines are indigenous and significantly old. The oldest Ayurvedic books, written thousands of years ago, narrated the symptoms and treatments of different disabilities. Though rooted in ancient India, in Bangladesh, there is one Ayurvedic university, recognized by the government.

If we look to the COVID-19 pandemic situation, except India, all other countries handled their citizen relatively with efficiency. However, Bangladesh started vaccinating its people with the vaccines imported from India. Bangladesh also invented “BongoVax” vaccine and waiting for human trial. However, yet it is not a time to conclude, but to start research why the epidemic did not hit in large scale in this region except India.

If we consider the social and mental well-being, the people of these countries were found to have scored happy in the survey scale. But the question is how long they will score happy? Now these regions are agriculture oriented and mostly rural areas. When there will be an industrial revolution, what will they score in the well-being?

Practically, there are few diseases existing in these countries. The smallpox, polio (except Pakistan), diphtheria, diarrhea, tuberculosis, and other serious diseases were eradicated. The primary treatments are managed by the village quacks, the homeopaths, and Ayurvedic medicine men. Anyone can buy antibiotics and psychotropic drugs over the counters without any prescriptions. And the people sustained.

Quality Education

According to UNESCO, the quality education specifically entails issues such as appropriate skills development, gender parity, provision of relevant school infrastructure, equipment, educational materials and resources, scholarships, or teaching force. The meaning of a quality education is one that is pedagogically and developmentally sound and educates the student in becoming an active and productive member of the society. A quality education provides the outcomes needed for individuals, communities, and societies to prosper.

Though we find many famous scientists who obtained their basic education in South Asian countries and are now working in the significantly successful international research centers, we are to agree that quality education in higher levels of science were not achieved in South Asian countries. Lack of laboratory equipments, research funds, and supervisors exists in these countries. But in the other fields of philosophy, humanities, and social sciences, quality education exists in South Asian countries.

But above all, primary or basic education is the most important stage of education. Yet in many villages of all the South Asian countries, there is no primary school. In the urban areas, competition is very high to get admission in the reputed schools. In the absence of primary schools in their own neighborhoods, many children go to the religious schools in these countries. The curriculums of these religion-centered schools are not well developed and different from the mainstream government-controlled curriculum. Secondly, large majorities of the teachers are not well educated. Basic orientation of small children in these religion-based so-called schools mat develop in a conservative manner. If the child is not intelligent enough, there are chances of blockage of faculty development to perceive what is true and false.

If we consider the normal distribution curve of the statistical sciences, 3% children will become genius wherever they study. Therefore, the SDG objective should be focused toward the 68% children between the +2 and -2 standard deviations. Not the infrastructure, rather the curriculum, teachers, and teaching methods should be given more emphasis that all the children of these countries can obtain such education that will open their eyes to the universe, other countries, civilizations, ethical values, and fundamental science. If UNESCO and UNICEF give more emphasis toward the generalization of effective curriculum of the primary education in these countries, quality education will automatically be achieved. It does not matter whether the children seat under the trees or on a floating boat as their schools.

Gender Equality

The women of the Indian subcontinent, now the South Asian countries, had significant access to politics, war, and work from the beginning of Indian civilization. "Razia Sultana" became the sultan of Delhi in 1211 CE; "Rani Chennamma" was crowned ruler until 1671; "Ahilyabai Holkar" came to power in 1767 and ruled until 1795; "Begum Samru" was the feminist icon of the eighteenth century; "Rani Mangammal": The Builder of Highways (died in 1705) ruled the Madurai Nayak kingdom. In addition, "Rani Lakshmibai," "Rani Padmini," "Meerabai," "Akkadevi," "Maharani Jijabai," and many other famous rulers of different regions of the present-day South Asian countries not only ruled their regions but also contributed many issues to uplift the status of women in the subcontinent. In Bangladesh, Rani Bhabani significantly uplifted the educational institutions during the British Colonial Period.

Indira Gandhi, Former Prime Minister of India; Chandrika Kumaratunga, Former President of Sri Lanka; Sirimavo Bandaranaike, Former Prime Minister of Sri Lanka; J. Jayalalithaa, Former Chief Minister of Tamil Nadu; and Benazir Bhutto, Former Prime Minister of Pakistan, played significant roles in global diplomacy that changed many burning issues of the South Asian countries. Bidya Devi Bhandari is the current President of Nepal.

Bangladesh is being ruled by women since 1991. At present, not only the Prime Minister but also the Parliament Speaker, the Leader of the Opposition Party, many diplomats, bureaucrats, and other important decision-makers are women in Bangladesh.

When the women of the United Kingdom were fighting for women's right to vote in the Victorian era under the banner "women's suffrage" (1918–1928), the Indian women had the voting rights and decision giving powers in their local governments.

If we look to the women in the mainstream workforce, we find that In India, a large number of women do not work and by implication spend much of their time at home. In India, in 1991, only 22 percent of the women were in the workforce as compared to 52 percent of the men (Gopalan & Shiva 2000). These proportions are lower than in 1961 when 28 percent of the women were in the workforce (ibid.). Many of the women in 1991, however, worked in the unorganized sector and did not have regular employment.

According to ILO reports in 2020, in Bangladesh, 36% women work in the formal sectors, while it is 59% in Bhutan, 20% in India, 42% in Maldives, 83% in Nepal, 22% in Pakistan, and 35% in Sri Lanka.

Besides the formal sectors, the women work in their homes. They cook; take care of the children and animals; look after the elderly family members; and nowadays also manufacture handicrafts and earn some small cash money which she can spend according to her wish. There are some microcredit programs in all the countries for women, and many of them became entrepreneurs who are employing other women to run the business. Changes are being observable, but not significantly in all the countries, except Bangladesh.

Clean Water and Sanitation

Availability of safe drinking water was considered as one of the most important issues in the entire South Asia until very recently. The Himalayan countries like Bhutan and Nepal had water sources coming down from the water falls, but access to these water bodies was difficult, and bringing that water to home was hazardous or risky. The flat lands of India and Bangladesh, abandoned with rivers and canals, were the source of drinking water from the beginning of human settlements. In the dry lands of Pakistan and western part of India, collection of drinking water from the manmade wells was a difficult task. The excessive salinity in the sea and rivers near the sea in southern countries was considered scarcity of safe drinking water.

Thousands of people, mainly children, died every year in diarrhea and other water-borne diseases until the 1970s of the last century. Lack of knowledge was the main reason of sufferings and deaths. But people survived through boiling of the water and reserving rain water for drinking. But collection and preserving of the drinking water was not easy for most of the families.

The small kings and the emperors of Greater India financed digging of large ponds throughout the subcontinent. And those were the sustainable survival techniques until very recently.

Now with the scientific developments, introduction of deep tube wells, and invention of oral saline and other medicines, the quality of life is improving.

On the other hand, sanitation system was a serious and important problem until very recently. The large expensive palaces and building found in South Asian countries show well-decorated bedrooms but not the toilet facilities.

In Bangladesh, even in the Islamic girls' schools, there are vending machines to provide sanitary pads to their students at the cost of 0.059 US$ per packet. And the products are made in Bangladesh by the non-profit companies.

Affordable and Clean Energy

Until the decade of the 1960s of the last century, burning the trees of jungles, dry cow dunks, and coals from the mines for fire were the only three energy-producing means in all the countries of South Asia. Then came the crude oil, liquid petroleum gas, and then solar energy. Hydroelectric projects were initiated in the decade of the 1940s of the last century in these countries, but except Bhutan, all the countries are yet largely dependent on crude oil. If we consider the total population of South Asia, nearly 60% are using indigenous coal, firewood, etc. to produce energy and spoiling the environment.

Decent Work and Economic Growth

In the South Asian countries, in 2015, in the developing countries, the middle class population could engage themselves in approximately 34% of total employment. The number tripled between 1991 and 2015. Over the past 25 years, the number of workers living in extreme poverty has declined dramatically. However, engaging children in labor force has significantly declined in these countries. But due to corrupt international syndicates, human trafficking is taking places.

Industry, Innovation, and Infrastructure

Though there are conflicts among the countries, the South Asian countries are found exchanging their innovations. There are examples of trans-border infrastructure sharing and other issues mentioned by UN goals.

Reducing Inequality

Although not significantly, the governments are trying to reduce inequality and poverty. Wage inequality of male and female workers was significantly reduced. Homelessness is almost solved in Bangladesh, and other countries are in progress. Occupational sex segregation yet exists because of social, religious, and security issues. Racial gaps in education do not exist nowadays in all the countries, except some regions of India and Pakistan. Racial discrimination has significantly reduced in the labor market. Residential segregation does not exist in these countries, except some regions of Pakistan and India. Health insurance process was initiated by all the countries, but the achievement is not very encouraging. Bad jobs yet prevail in these countries, but in the process of elimination nowadays. Wealth inequality exists, and none of the international financial organizations have published the figures of standard deviations. Other issues like intergenerational income mobility, deregulation of the labor market, immigrants and inequality and incarceration, etc. are not considered important in these countries related to sustainable care programs.

Sustainable Cities and Communities

These courtiers are yet agriculture-oriented countries. Nearly 70% of the population lives in rural areas. All the governments are trying their best to provide electricity, road, Internet, better health care, social security benefits, quality education up to secondary level, etc. If these are achieved, urbanization process will significantly reduce. Secondly, the indigenous family care programs prevailing for the elderly and disabled people will continue. In Bangladesh, Bhutan, and Nepal, the achievement rate is significantly speedy.

Responsible Consumption and Production

Nowadays, especially after the COVID-19 pandemic, all the governments are seriously considering the responsible consumption and production issue. In all the South Asian countries, the goal was more or less achieved. People of this region were given information using mass media. The farmers were encouraged toward proportionate cultivation of different crops. Especially in Bangladesh and Bhutan, the concerned government officials were found to have remained vigilant. However, there are corrupt international business syndicates also in the South Asian countries. They tried to manipulate the markets, but could not succeed.

Climate Action

Through mass media, press, and social networks, now more or less, all the people of this region are aware of the climate action programs. It seems that people are becoming conscious day by day.

Shushant et al. (2019) stressed the need for environmental administration through the local government institutions in the protection of environment in India and Bhutan. They mentioned that among different levels of environmental administration in India and Bhutan, the most effective is the presence of local government institutions for the efficient utilization and management of natural resources. They evaluated the relevant policies and practices promoted by these institutions for preserving and protecting environment at the local government level. It was mentioned several mechanisms through which information regarding public welfare and environment conservation can be communicated to the villagers.

Life Below Water

South Asian governments especially Bangladesh, India, Sri Lanka, and Maldives are serious in the issue toward conservation and sustainable use of the oceans, seas, and marine resources for sustainable development. People are being informed that healthy oceans and seas are essential to our existences which cover 70 percent of our planet and how we should rely on these water bodies for food, energy, and safe water.

It is not well known what the other governments are doing, but the Bangladesh government is seriously vigilant in this issue. The actions include prohibiting the fishermen for some specific periods from fishing in rivers and not going to oceans. And governments provide financial support to the fishermen during such periods. These are done when the mother fishes discharge eggs in some specific areas. Secondly, the Bangladesh government is trying to clean the silted canals, dredging the rivers, etc. These are being done for some years, and the output was clearly observed by all.

Life on Land

The SDG goal is to protect, restore, and promote sustainable use of terrestrial ecosystems, sustainably manage forests, combat desertification, reverse land degradation, and halt biodiversity loss. And it seems the populations of this region are now aware of these factors very much. Hunting of the wild animals reduced. Catching or disturbing the seasonal birds arriving from the northern cold countries was reduced.

Most interestingly, Maldives is importing fertile soils from Bangladesh for cultivating different agricultural products.

Peace, Justice, and Strong Institutions

The SDG goal is to promote peaceful and inclusive societies for sustainable development, provide access to justice for all, and build effective, accountable, and inclusive institutions at all levels.

It was discussed at the beginning of this paper that the Panchayat or Saamaj, the indigenous informal community control system, yet prevails in all the countries of South Asia. And these indigenous bodies take care of their respective population from their own resources.

Partnerships to Achieve the Goal

Until now, all the South Asian countries are sincerely cooperating among themselves, as well as with UN organizations, toward achieving the goals. Significant outcomes were already observed in some countries, and it is expected that all the people of this region will benefit from the partnerships and cooperation.

15.3 Conclusions

Considering the existing global conflicts, urbanization patterns, pollution, racial discriminations, economic disparity, and many other drawbacks, it is concluded that the people of the South Asian countries are happy and having excellent subjective well-being. They sustained well, and if the existing family care systems continue, the people will benefit. The ancient community control bodies, the Panchayats or the Saamaj, succeeded well to manage the meaningful sustainability and changing care programs in the South Asian countries.

The psychology of sustainability and sustainable development (Di Fabio, 2017a, Di Fabio & Rosen, 2018) and the psychology of harmonization (Di Fabio & Tsuda, 2018) represent the current framework for sustainability and changing care programs using psychological lens. In particular, a positive sustainability perspective is introduced (Di Fabio, 2017b) paying attention to narratives and resources in terms of respect and regeneration on one side and on the other side to continuously renewing resources for promoting greater well-being for people and environments and their relationships. In this perspective, there is a call to take responsibility to renew and improve the flourishing of resources at any level (Di Fabio, 2017b).

Some recommendations are the following for people of the South Asian countries: advocating for the vulnerable, strengthening data and identifying the gaps, increased support for the persons with disabilities, increased support for the elderly persons, allowing the indigenous community control bodies to continue, enlightening the community leaders with data and facts, and overall awareness development.

References

Adhikari, P. (2017). Geriatric health care in India – Unmet needs and the way forward. *Archives of Medicine & Health Science [serial online].*, *5*, 112–114. Retrieved February 5, 2021; https://www.amhsjournal.org/text.asp?2017/5/1/112/208215

Asadullah, M. N., & Chaudhury, N. (2012). Subjective well-being and relative poverty in rural Bangladesh. *Journal of Economic Psychology, 33*(5), 940–950. https://doi.org/10.1016/j.joep.2012.05.003

Di Fabio, A. (2017a). Positive Healthy Organizations: Promoting well-being, meaningfulness, and sustainability in organizations. *Frontiers in Psychology. Organizational Psychology, 8*, 1938. https://doi.org/10.3389/fpsyg.2017.01938

Di Fabio, A. (2017b). The psychology of sustainability and sustainable development for well-being in organizations. *Frontiers in Psychology. Organizational Psychology, 8*, 1534. https://doi.org/10.3389/fpsyg.2017.01534

Di Fabio, A., & Rosen, M. A. (2018). Opening the black box of psychological processes in the science of sustainable development: A new frontier. *European Journal of Sustainable Development Research, 2*(4), 47. https://doi.org/10.20897/ejosdr/3933

Di Fabio, A., & Tsuda, A. (2018). The psychology of harmony and harmonization: Advancing the perspectives for the psychology of sustainability and sustainable development. *Sustainability, 10*(12), 4726. https://doi.org/10.3390/su10124726

Farooqi, M. R., Anwar, H. N., & Sher, F. (2019). A comparative study of Panchayat system and state's judicial system In Punjab Pakistan. *Journal of Organizational Behavior Research, 4*(2), 142–149.

Gopalan, S., & Shiva, M. (Eds.). (2000). *National profile on women, health and development: Country profile – India*. Voluntary Health Association of India and World Health Organization.

Gorman, M., Jones, S., & Turner, J. (2019). Older people, mobility and transport in low-and middle-income countries: A review of the research. *Sustainability, 11*(21), 6157. https://doi.org/10.3390/su11216157

Hossain, I., Akhtar, T., & Uddin, T. (2006). The elderly care services and their current situation in Bangladesh: An understanding from theoretical perspective. *Journal of Medical Sciences, 6*, 131–138. https://doi.org/10.3923/jms.2006.131.138

Islam, M. A., & Rahman, M. F. (2017). Health problems of elderly people in Bangladesh. *Journal of Armed Forces Medical College, Bangladesh, 13(1), 1–2.* https://doi.org/10.3329/jafmc.v13i1

Johnson, C. (2003). *Decentralisation in India: Poverty, politics and Panchayati Raj* (Vol. 199). Overseas Development Institute.

Kumar, S. (1964). The panchayat constitution of Nepal and its operation. *International Studies, 6*(2), 133–152. https://doi.org/10.1177/002088176400600202

Mcpherson, B. (1991). *Aging: The middle and later years. An introduction to sociology: The social order* (2nd ed.). Mcgraw Hill Ryerson.

Morgan, L., & Kunkel, S. (2001). *Aging: The social context*. Pine Forge Press.
Mullick, R., & Raaj, N. (2007, September 9). Panchayats turn into kangaroo courts. *The Times of India. September, 9*, 2007.
Sufi, A. H., Yamashita, K., & Nazneen, S. (1996). *Virgin Village Bandaikhara – Its women and mental handicaps*. The SIVUS Bangladesh.
Sufi, A. H., & Tushar, T. A. (2017). *Exploring the status of developmentally disabled children of Sukhandighi village*. University Grants Commission Research Project, Faculty of Life and Earth Sciences, University of Rajshahi.
Sukumar, N., Lal, L. D., & Mishra, V. K. (2019). Inclusiveness in the Panchayati Raj Institutions. *Journal of Social Inclusion Studies, 5*(1), 72–88. https://doi.org/10.1177/2394481119859675
Shushant Parashar, V. C., Saxena, S., & Meenakshi. (2019). Local Government Institutions and Environment Management in South Asia: A Case Study of India and Bhutan. *International Journal of Innovative Technology and Exploring Engineering (IJITEE), 8*, 7C2.
Udgaonkar, P. B. (1986). *Political institutions & administration*. Motilal Banarasidass Publishers.

Chapter 16
Access to Work-Life Balance Practices and Well-Being for Tunisian Managers

Alma Hafsi, Amira Laroui, and Takwa Chammem

16.1 Introduction

Though the question of work-life balance has frequently been dealt with in the literature, it is still an up-to-date issue. Dealing with that issue leads to new research questions linked, at the same time, to deep changes in the world of labor and social and familial evolutions. All these deep social and economic changes have undoubtedly had a direct impact on social norms and roles in different spheres of life. Moreover, the rapid technological innovations have imposed new labor organization methods, new relationships with the work environment, new human relations, and a re-thinking of the position of work. In that context, so many new functioning rationales, henceforth based on the culture of urgency, participate in intensifying the rhythm of life and that of work (Aubert, 2011). Practically, these conditions of speeding rates lead most often to mental overloads and constant pressures causing the loss of work's spatial dimension trampling upon the border of the family sphere (Bruchon-Schweitzer, 2002; Vézina et al., 2001). Consequently, reconciling both spheres of life becomes more and more difficult and burdensome (Barrère-Maurisson, 2003; Fusulier & Tremblay, 2013). Many persons may suffer from the inability to satisfactorily combine the different professional and familial activities (De Terssac & Tremblay, 2000). Coping with organizational and familial constraints at the same time becomes a real problem that significantly disrupts male and female employees' health and the purpose of work.

A. Hafsi (✉)
University of Tunis and Cnam-Paris, Tunis, Tunisia
e-mail: Alma.Hafsi@fshst.u-tunis.tn

A. Laroui
Ministry of Education, Tunis, Tunisia

T. Chammem
GIZ-German Society for international cooperation-(Gmbh), Tunis, Tunisia

A. Di Fabio (ed.), *Cross-cultural Perspectives on Well-Being and Sustainability in Organizations*, https://doi.org/10.1007/978-3-030-86709-6_16

In the academic literature, reconciling the professional activity and personal commitments is currently recognized as a potential source of stress likely to affect the employees' well-being and their life quality at work (Bustreel et al., 2012; Curie & Hajjar, 1987; Lallement et al., 2011; Loriol, 2006; Morin, 2008; Vézina et al., 2011). Thereon, many writers were challenged by the deleterious impact of the organizational constraints and the working conditions on the mental health of the employees that arise in the form of breakdown, psychological distress, and job burnout as specified by Karasek and Theorell (1990), Bruchon-Schweitzer (2002), Bakker and Demerouti (2007), and Lapointe et al. (2011). These constraints have a real cost for organizations, particularly in terms of absenteeism for health reasons and distress, dissatisfaction, lack of commitment, and departures (Cooper, 2006). Therefore, the organization should respond by implementing a set of measures likely to harmonize both spheres of life that are obviously in close interaction (Curie & Hajjar, 1987).

Though the question of work-life balance and the well-being at work at both the social and economic levels are currently important, hardly any research on the topic has so far been conducted in Tunisia, to our best knowledge. Henceforth, our study seeks to examine the working conditions of a proportion of Tunisian managers by trying to highlight the implications of access to work-life balance practices for the mental health, the centrality of the purpose of work, and the corollaries of measures taken by a number of public and private businesses. For this, we will first present the theoretical framework of this research and focus mainly on both major concepts examined, namely, work-life balance, health, and well-being. Consequently, we will present our research model, the methodology of conducting an electronic survey, the major tools of measurement, the data analysis method, and the main findings.

16.2 Work-Life Balance: A Dilatory Compromise

Work has to be placed in the changing trajectory of modernity which goes together with the socioeconomic, scientific, and technological advancement. In fact, work has a predominant position today in individuals' lives which is only second to family (Garner & Méda, 2006; Méda, 2010). Work has a pivotal position in our society. It is perhaps a source of self-sufficiency for women as much as for men. According to Arendt (1988) and Garner and Méda (2006), it fosters identity building, self-realization, and self- fulfillment.

The mass influx of women in the labor market which is a vector for social integration seems indeed as an essential component of modern social transformation and emancipation. However, social tribulations and gender stereotypes often make sure that women undertake different responsibilities in addition to professional ones (OIT, 2017). They find themselves immediately overwhelmed by social and familial burdens that could have repercussions not only on family life but also on work (Barrère-Maurisson, 2010; St-Onge et al., 2002; Thoemmes et al., 2011; Tremblay, 2012). With regard to this, public and managerial policies seek, on the one side, to keep skilled male and female employees and to improve the work-life balance, on the other, to improve the reconciliation between the two spheres of life in such a way to promote the well-being of parent-employees as well as the commitment to work (Baudry & Fillion, 2014; Di Fabio & Rosen, 2018; Pailhé & Solaz, 2010a; Tremblay, 2003; Vandenberghe, 2004). The work-life balance for Tremblay (2004) refers to the way individuals succeed in combining harmoniously their familial and professional responsibilities. This combination is not easy yet for managers due to various dependent variables that have to be taken into account, particularly the inherent characteristics of their job, the working conditions, and factors outside work (Guilbert & Lancry, 2009; St-Onge et al., 2002).

Being Manager Means No Longer Counting Working Hours

Being a manager in a private or a public organization certainly allows you to enjoy a high social and hierarchical position. The managers' daily task is to lead, to manage, and to direct staff. In other words, their organizational responsibility is important given the strategic position they hold in the firm (Desmarais & privé, 2010). That is the profitable representation we make of this job. Yet with the recent organizational changes, managers have started to re-define their roles and missions inside the firms. Thence, the volatility of the work environment and the economic uncertainty oblige managers to deploy efforts of adaptability and adjustment that could, in the long term, become a source of stress. Furthermore, the managers' role is considered as pivotal in the organization since it consists in steering projects, attaining the organization's objectives, and customer satisfaction. Consequently, they are always subject to accountability for results and often face pressures from work environment, market pressures, heightened competition from other skilled managers (Dagenais-Desmarais & Savoie, 2011), and the lack of peer solidarity (Monneuse, 2014; Thoemmes, 2012). Henceforth, the managers' work organization becomes more restraining at the psychological level (Dagenais-Desmarais & Privé, 2010; Dagenais-Desmarais & Savoie, 2011).

A manager is asked to accomplish simultaneously different tasks of various levels of technicality. He manages human resources, plans projects, develops strategies, solves problems, etc. The variety of the roles a manager is expected to play aggravates the health risks and disrupts the well-being at work.

Furthermore, the use of the technological innovations like network connection, laptops, cellphones, and connected watches imposes new links with the information based on the principle of urgency on managers (Thoemmes, 2012). In these conditions, meeting everyday pressing demands leads managers to live the working time in an oppressive way (Thoemmes et al., 2011). Thereon, work could be accomplished either inside the organization or by telecommuting that is why time management becomes harder (Di Fabio & Rosen, 2018). Thoemmes et al. (2011) insist that managers claim that they have difficulties in limiting working hours due to high professional demands. We often say that a manager never counts his working hours.

Therefore, the burden of the professional responsibility increases the workload for that social group and urges them to work more inside the organization as much as they do outside of working hours. Thus, they have difficulties in reconciling the two spheres of life and to balance between family commitments and professional duties. In other words, managers' commitment in one role makes it difficult to be committed in the other role, namely, the familial one.

Being a Parent Magnifies the Work-Family Conflict

In that issue of reconciliation, being a parent accentuates the scale of time management conflicts. In fact, the birth of a child causes changes in reconciling both spheres of life in a given way. Young parents of today find it difficult to undertake parental and professional tasks at the same time due to the lack of time (Tremblay & Vaillancourt-Laflamme, 2000). New time-related constraints emerge and lead to a new way of living private and working time, mainly for women. Undoubtedly, the time an individual devotes to work does not allow him to devote himself entirely to family life and play fully his role which creates a state of tension. Moreover, Tremblay (2012), Guérin et al. (1997), and Boulet (2013) reveal differences between men and women regarding the time they devote to the familial sphere and to the professional sphere.

In France, the survey led by Pailhé and Solaz (2010b) illustrates these differences. Out of 9547 persons aged between 20 and 49 working in 2673 private and public companies, the results show that births caused departures, switching to part time, important reorganizations of working time, and position changes for women. After a second birth, mothers switch to part time due to difficulties to reconcile a full-time job and a family life (Rivaud & Ulrich, 2007). In that way, according to Bèque (2019), a part-time job makes reconciliation easy. 23% of women who are in charge of at least one child, under the age of 18, thus answered that they are working part time to look after their children compared to less than 1% of men. The discrepancy is visible here between men and women and can be seen through parental activities. According to Pailhé and Solaz (2010b), still, fathers are blamed for being less present with their children on Wednesdays or at lunch during the week. Although fathers fully perform the tasks of accompanying their children to school and coaching them in their schoolwork, they are certainly doing so less than women. Moreover, it should be noted that when handling unexpected situations, it is always women who leave work for children's medical care and not men even though their jobs allow them to do so.

Furthermore, taking into consideration the level of education, the fair distribution of resources leads to a less gender-specific division of tasks within the couple. The involvement of fathers in the parental tasks seems, de facto, more active. In this respect, Bèque (2019) insists that managers acknowledge that they have difficulties with their relatives (18% men and 17% women) more than other employees (workers) due to intense working time. Being a manager could, nevertheless, give women some privileges, like hiring housemaids, thanks to their financial resources, and be therefore less blamed by their families.

This quest for balancing both spheres of life allows the individual to reach coherence in his link with work associated with a feeling of well-being and peace of mind that could enable him to deal with his functions.

The Impact of Work-Life Conflict on Health

In the last decades, researchers are more interested in the work-like balance highlighting its deleterious repercussions on mental health and the well-being at work (Demerouti & Geurts, 2004; Guérin et al., 1997; Paquet & Najem, 2005; St-Amour & Bourque, 2013; St-Onge et al., 2002; Tremblay, 2004; Tremblay & De Sève, 2002). The well-being at work is a subjective experience that is positive to express the best of oneself which is built through oneself and in interaction both with the self and with the organization (Dagenais-Desmarais & Privé, 2010). If this well-being is impaired and the health condition is perceived as influenced, that means that the individual has no more resources necessary to make him deal with his duties. That is why some individuals may, under the magnitude of the work-life conflict, fall into substance abuse, smoking, and even addiction to prescription drugs (St-Amour & Bourque, 2013), whereas other individuals specifically women would suffer from a psychological depression, a breakdown, or guilt-stress (Marchand et al., 2015; Tremblay & Vaillancourt-Laflamme, 2000). The situation could worsen and result in a state of depression and job burnout (Guérin et al., 1997; Marchand et al., 2015). The desire to permanently reconcile the requirements of both spheres remains costly for health, private, and professional lives (Lavoie, 2016).

Various recent studies and surveys have constructively clarified the representation of the working conditions by emphasizing the Job-Demand-Resource (JD-R) which was developed by Bakker and Demerouti (2007). This model was inspired by the Job Strain Model of Karasek (1979) and by Hobfoll's Conservation of Resources Theory (2001). The model sheds light on the concurrent interactions between work demands and resources. In this model, the demands include the characteristics of the working environment that are likely to produce professional tensions and stress and harm the employees' well-being. They are of different physical, psychological, social, or organizational character and refer to perceived strains such as work pressures and workloads, emotional requirements, organizational changes, and the re-organization of work (Demerouti et al., 2001).

Resources refer to physical, psychological, social, and organizational aspects of work that could facilitate reaching objectives, foster self-development and training, or decrease the adverse effects of demands and their costs (Bakker & Demerouti, 2007). Definitely, resources could either be specific to the individual (traits of personality, family context, etc.), or they could be provided by work itself and are therefore organizational. The social resources correspond to the hierarchical and collective support that are part of collective strategies of adaptation aiming at assisting the employee to deal with stressful situations. Moreover, taking into account the nature of the position and its characteristics (autonomy, decision-making authority, etc.) are resources available for the employee to enable him to control his work and resist against demands (Bakker & Demerouti, 2007). In this model, stress and burnout would result from the interaction between an excessive workload and weak personal resources. The resources would, therefore, play the role of a buffer. Their presence would facilitate answering the encountered demands which are no longer threats, but they become challenges. Their absence does not, nevertheless, allow the employee to meet the requirements of work and lead to an attitude of withdrawal and perhaps disengagement (Bakker & Demerouti, 2007; Demerouti et al., 2001). The study conducted by the European Agency for Health and Safety at Work (2012) describes significant links between psychosocial threats and work stress on the one side and the negative repercussions on the mental and physical health (depression and cardiovascular diseases) on the other side.

In the framework of the fifth European survey on working conditions (Eurofound, 2012), 45% of workers, among respondents, declared having experienced organizational changes in the previous 3 years that had an impact on their work environments, and 62 % acknowledged having worked with tight deadlines. Furthermore, the results of the European survey of companies on the new and emerging risks (ESENER, EU-OSHA, 2010) show that 79% of the European managers state that they were concerned by stress at work. Certainly, the psychological load of work is often associated with work partners and leads to further demands in the processing of clients' files and summons emergency work (Molinier, 2009). All these cited studies emphasize the influence of the working conditions on the well-being and health of workers. Their impact differs, however, according to individual characteristics, namely, sex (Ferguson et al., 2012). For example, in the survey on the working and employment conditions and occupational health and safety (EQCOTESST) conducted by Vézina et al. (2011), it is shown that women (21.7%) are more affected by psychological distress than men (15%) particularly due to the accumulation of constraints and psychological demands. For some women, combining career and private life generates difficulties in managing personal and professional responsibilities. This fact is equally clear in a study conducted by Tremblay et al. (2006) which proves that around a quarter of women surveyed (24%) conveyed a higher level of stress than 10% of men. Furthermore, regarding the work-life unbalance, researches show connections between the difficulties of reconciliation and depression or psychological distress (Gilbert-Ouimet et al., 2016; Vézina et al., 2011) and underline that workers affected by that distress constitute more than 40% of the rate of general absenteeism in the firm (Vézina et al., 2011). Yet other slighter differences are emphasized between the higher rate of absenteeism among men and women with children than among childless men and women. The rate of absenteeism is likely to decrease as long as children are growing older.

According to Lhuilier (2010), because of the porosity of the bordering lines between both professional and private spheres, the conflict could seem like a very noticeable drudgery originating directly from what seems to raise the invisibility of work. In that respect, the study conducted by St-Onge et al. (2002) on the bidirectionality of the work-family conflict reveals that the latter is more perceived among respondents than the family-work type. This situation becomes more accentuated when professional problems interfere in the personal life. For Ballesteros et al. (2016), the use of new technologies namely, the mobile ones, makes the employee remain permanently connected to the company and work everywhere, which is one of the drawbacks of that use. The informational opulence, the immediacy of answers, and reactivity are strongly accompanied by pressure and tension (Bobillier-Chaumon, 2012). Henceforth, work is overlapping, and settling the space of work and that of life is no longer possible from now on (Guilbert & Lancry, 2005). Sometimes the situation leads to a state of ill-being, causes the turnover of employees (Pierre & Tremblay, 2012), and increases the rate of absenteeism caused by psychological problems, job burnout (Guérin et al., 1997; Tremblay et al., 2006), or work accident (Vézina et al., 2011).

On the contrary, the supports of the organization and the work-life balance practices have beneficial effects on the organization itself, the employees' health, and their well-being at work.

16.3 The Access to Work-Life Balance Practices Guarantees Well-Being at Work

Definitely, work organization is at the heart of the challenging reconciliation of private and professional lives. Several researches were interested in adaptation measures taken by employers concerned by the well-being of their employees. In that respect, the majority of analyses show that these measures have beneficial effects on the organization and on the health of the employees as well. In fact, these measures act on the working conditions of employees in order to lighten the psychological load caused by the overlap of professional and familial responsibilities. Nevertheless, the access to work-life balance practices depends equally on the business culture, the position of the employee, trade union, and the quality of these practices themselves. Lero et al. (2009), like Boulet (2013), underline that access to work-life balance improves the mental health by decreasing the psychological distress of employees. For example, a study conducted by the Quebec Statistical Institute (ISQ) (2013) on the effects of implementing the reconciliation practices shows that the distress levels among employees who have the possibility to have leaves for familial reasons are lower than those among their colleagues who do not have access to these practices. This finding is consolidated by Boulet (2013) who reports that illness leaves, the shortened working week, and telecommuting have a protective effect on the health. Yet, according to the writer, the access to these practices conceals significant gender and work status differences. Consequently, women have access to a bigger number reconciliation practices than men mainly among women who have a high level of education and a permanent job. As it was underlined by Boyer and Nicolas (2009), once they become parents, women enjoy employer's sympathy and have easier access to part time than men since companies seem to comply with gender stereotypes related to parental roles. Others still identify working time management as the measure of reconciliation most sought by parents who have young children since that practice could best contribute to the well-being of employees (Pailhé & Solaz, 2009; Vézina et al., 2011). Furthermore, there are disparities of measures offered between private and public sectors. The employees of the public sector benefit from a favorable environment for the family, thanks to a social dialogue and which enables them to have an easier access to part time than those of the private sector (Pailhé & Solaz, 2010a). These latter have access to flextime, but practices remain so heterogeneous depending on the type of the activity and the will of the employer. Therefore, thanks to all these valuable measures, that the well-being of the employees is in line with the best functioning of the organization. It creates a decrease of absenteeism and the durability of workers on the one side and favors the establishment of the meaning of work on the other (Di Fabio & Rosen, 2018; Karatepe, 2012; Morin, 2010). This shows that the effects of work organization are noticeable at the level of the workers' mental health and are expressed in their psychological well-being (Morin & Aranha, 2008). According to Morin (2010), work has negative and positive effects on health, but the rectitude of practices which favors coherence between the person and professional relations in an environment endowed with moral values, equity, justice, gratitude, learning, autonomy, and human dignity influences that relation and encourages the worker to give value, meaning, and importance to what he is doing (May et al., 2004) by committing himself emotionally (Morin, 2008). In the same vein, in reference to the psychology of sustainability and sustainable development, a healthy organization is the one in the dynamic regulation between the functional process committed in an organization and the style of leadership in managing the human capital to reach a social cohesion and well-being at work (Di Fabio & Peiro, 2018; Di Fabio & Rosen, 2018). The operating mode centered, at the same time, on organizational sources and the positive inter- and intra-personal experiences promotes the significance of personal and professional lives (Di Fabio, 2017a).

To conclude, the literature review makes us believe that the access to work-life balance practices as organizational resources represents a lever for the health of the employees according to the type of job and sex. The factor of the centrality of work seems important and should be taken into consideration. In this way, the objectives of the current research are:

1. To better understand the impact of access to practices of CTF on the mental health of Tunisian managers and the meaning they give to their work
2. To identify the factors that influence the level of psychological distress among these managers

We thus make a first hypothesis according to which the access to reconciliation practices, the meaning of work, and the psychological distress could differ according to sex and age, the number of children, and the type of business organization. The second hypothesis states that psychological distress could become harder by the access to practices than by the meaning of work, personal characteristics, and the type of business organization.

16.4 Methodology

Participants

The questionnaire was put online via Google Forms platform. Two-hundred and four managers took part in that study including 118 women (57.8%) and 86 men (42.2%). Among these managers, 155 are in the private sector, and 49 have senior positions in the public sector. Out of 204 participants, 136 were between 25 and 35 years old (66.7%), 49 were between 35 and 45 years old (24%), and 19 were between 45 and 55 years old (9%).

Measures

In that study, three questionnaires are deployed to collect information on the accessibility to work-life balance practices provided by the company, the meaning of work, and the psychological well-being.

The Access to Work-Life Balance Practices

The questionnaire used originates from a Quebec survey on the working conditions, employment, health, and job security (EQCOTESST) (Boulet & Le Bourdais, 2016; Vézina et al., 2011). That tool is made of 10 items starting by: to your knowledge, are the following measures accessible to you in your working environment?

- A flextime which refers to the freedom of choice of the working time
- The possibility to swap working time with a colleague
- A time schedule accessible when needed
- Paid sick leaves
- Paid sick leaves for familial reasons (such as the death of a relative, the illness of a child, etc.)
- Access to shorter working week on voluntary bases
- Telecommuting
- The gradual return to work after maternity or paternity leaves
- Access to unpaid leave or deferred salary leave
- Access to nursery in the workplace

The participant responds whether the practice exists or not in his/her workplace by "yes" or "no." That survey allows one to assess the managers' access to ten practices according to Likert scale with three levels (weak, average, or high). If the number of positive answers ranges from 0 to 3 practices, the score is in the "weak" category; between 4 and 6, the score is in the "average" category; and finally between 7 and 10 practices, the score is in the "high" category. This scale enables analyzing the impact of access to practices of CTF on psychological distress. Nevertheless, the questionnaire does not highlight the use of the access to work-life balance practices. The internal consistency of this scale tested on the participants in the study was acceptable with a coefficient alpha of Cronbach $\alpha = 0.7$.

Kessler Scale of Psychological Distress Scale (K6)

Psychological distress is measured by K6 scale, the short version made of six items. This tool has been made by Kessler and his collaborators (Kessler et al., 2002) and used in many surveys. The choice of that tool is justified by its easy administration, concision, and validity. The objective of the K6, made of six items, is to evaluate the level and frequency of feeling nervous, desperate, restless, depressed, and good for nothing or have the impression that every task is exhausting during the last month. Respondents answer each item according to the 5-point Likert scale: never (0 point), rarely (1 point), sometimes (2 points), frequently (3 points), and always (4 points). The coefficient alpha of Cronbach of that scale was 0.79.

The score of the psychological distress is the total of points obtained for every question. That score ranges from 0 to 24. Managers who have a global score lower or equal to 1 are in the "weak" category, those who have a score superior to 1 and inferior to 7 are in the "average" category, and those whose score is equal or above 7 are in the "high "category displaying a high psychological distress.

May's Scale of the Meaning of Work

This scale measures the level of the work meaning that workers give to their professional activity. This scale has been built by May, Gilson and Harter (2004). It is made of six items, and answers are completed according to Likert 6-point scale examining the level of agreement coded from 0 to 6. The total of points indicates the total scores of the meaning of work. A high score represents a high level of the meaning of work for workers. The coefficient alpha of Cronbach of that scale was equal to 0.867.

Procedure

We have basically chosen LinkedIn which is a professional network to get access easily to the targeted population. We have written a publication in the form of a short text presenting the objective of our study and the link redirecting to the questionnaire. The latter was diffused online via "Google Forms" platform.

Data Analysis

We have first proceeded by evaluating the internal consistency of each scale by calculating the alpha coefficient (α) of Cronbach. Then, the descriptive analyses of the sum of variables were carried out. The links between the variables were estimated by using the chi-square test of interdependence since our variables are qualitative.

Finally, multivariate and predictive analyses by logistic regressions were applied to study the relation between the explanatory variables (the access to reconciliation practices and the meaning of work) and an explained variable (weak psychological distress).

16.5 Results

The distribution of our sample is made on the basis of their individual characteristics (sex, age range, dependent children), and organizational characteristics too (type of organization: private sector/public sector) are shown in Table 16.1. The table informs us on the distribution of the proportions of respondents in the three scales (access to reconciliation practices, the meaning of work, and the psychological distress).

To check the link between the three dependent variables and the personal and organizational characteristics, chi-square tests were carried out. Results show a significant link between the type of organization variable (private-public) and the variables of access to reconciliation practices (χ^2 (2) = 6.347, p=0.042<.05).

Significant links between the type of organization and psychological distress were equally observed (χ^2 (2) = 7.954, p=0.019<.05). Managers working in the public sector claim that they have a higher access to reconciliation practices and express a higher level of the meaning of work. Moreover, the same managers believe they have a low level of psychological distress when compared to their colleagues who work in the private sector. Among all managers in that study, those who have two dependent children express a low psychological distress level against the other categories (childless managers or those who have one child or three children).

Chi-square test of interdependence shows a significant link between these variables (number of dependent children and distress level) (χ^2 (6) = 12.835, p=0.046<.05).

However, the chi-square of interdependence shows that there is no link between the sex variable and the three dependent variables: the access to reconciliation practices (χ^2 (2)=1.781, p=0.410>.05), the meaning of work(χ^2 (2)=2.274, p=0.321>.05), and psychological distress (χ^2 (2)=, 647, p=0.724>.05). The chi-square of interdependence does not show any link between the age variable and reconciliation (χ^2 (4)=8.802), age and the meaning of work (χ^2 (4)=7.356, p=0.118.>05), and age and psychological distress (χ^2 (4)=8.554, p=0.073.>05).

To develop the analyses further and identify the major factors of the psychological distress of Tunisian managers, ordinal logistic regression, appropriate for the

Table 16.1 Distribution of employees according to personal and organizational characteristics on the three scales

		Access to reconciliation practices				Meaning of work				Psychological distress			
		Low	Medium	High	chi 2	Low	Medium	High	chi 2	Low	Medium	High	**chi 2**
Sex	**Women**	38 22.03%	54 45.76%	26 32.2%	1,781 (.410)	36 30.50%	56 47.45%	26 22.03%	2,274 (.321)	33 3837%	35 40,7%	18 23,72%	,647 (.724)
	Men	21 27.90%	41 47.67%	24 24.41%		26 12.25%	26 46.51%	52 30.23%		39 33,05%	51 43,22%	28 20,93%	
Age	**- 35**	42 30,88%	68 28,57%	26 (15,78%)	8,802 (.066)	40 29.41%	68 50%	28 20.58%	7,356 (.118)	40 29,41%	65 47,8%	31 22,79%	8,554 (.073)
	35–45	14 50%	20 40,81%	15 36,84%		13 26.53%	21 42.85%	15 30.61%		21 42,85%	16 32,65%	12 24,48%	
	45 +	3 19.11%	7 30,61%	9 47,36%		3 15.78%	7 36.84%	9 47.36%		11 57,89%	5 10,20%	3 15,78%	
Children	**0**	43 32.33%	65 48,87%	25 18,79%	9,582 (.143)	42 31.57%	63 47.36%	28 21.0%	8,319 (.216)	38 28,57%	63 47,36%	32 24,06%	12,835 **(.046)**
	1	7 28%	10 40%	8 32%		5 20%	13 52%	7 28%		9 36%	10 40%	6 24%	
	2	6 18.18%	13 39,39%	14 42,42%		5 15.15%	14 42.42%	14 42.42%		20 60.60%	7 21.21%	6 18,18%	
	3	3 23.07%	7 53,84%	33 23,07%		4 30.76%	6 46.15%	3 23.07%		5 38,46%	6 46.15%	2 15,38%	
Type of organization	**Private**	51 32,90%	71 45,80%	33 21%	6,347 **(.042)**	48 30.96%	73 47.09%	34 21.93%	6,106 **(.047)**	50 32,25%	63 40.6 %	42 27,09%	7,954 **(.019)**
	Public	8 16%	24 48,97%	17 34,69%		8 16.32%	23 46.93%	18 36.73%		22 44,89%	23 46.93%	4 8,16%	

study data, was carried out. The obtained logit model presents a significant likelihood ratio (χ^2 (22) = 204.395, p = 0.00) enabling the affirmation that an influence exists. Furthermore, Nagelkerke R^2 index (0.718) suggests that the validity of the model is important and enables us to explain 71.8% of the variance of the model.

In the regression analysis, the impact differs from an explanatory variable to another. In this way, results show that the access to reconciliation practices and the meaning of work are negatively linked to the psychological distress. Managers having a low or a medium access have thus much distress than those having a high access. In fact, managers having a low access have 69.504 times much psychological distress than their fellows who have a high access. Likewise, those having a medium access to reconciliation practices have 63.231 times much psychological distress than those having a high access. Moreover, it appears that managers having a low access to reconciliation practices and those having a high access have significant differences regarding the levels of psychological distress (β=4.241 (p=0.006)). It is the same for those having a medium and those who have a high access (β =4.147 (p=0.001)). It could be said that the increase in the access to reconciliation practices decreases the psychological distress. In addition, the results of logit show that managers having a low meaning of work have much distress than those having a high meaning of work. They are 704965.246 times more likely to have a higher level of distress than their colleagues who have a high meaning. Consequently, having a low level of the meaning of work is more associated with a high psychological distress (β = 13.466, p =.000< 0.05). This enables us to say that giving a meaning to work decreases psychological distress.

To conclude, managers having a low and medium access to reconciliation practices, in our study, seem to be more vulnerable to a medium level of psychological distress. Managers having a low level of the meaning of work are more vulnerable to a high distress. We could deduce that what distinguishes managers having a medium level of distress from those who have a high level of distress is the low level of the meaning of work.

16.6 Discussion

The main objective the research has aimed for is to better understand the impact of access to practices of CTF on the mental health of Tunisian managers and the meaning they give to their work. We have formulated a hypothesis according to which the access to reconciliation practices, the meaning of work, and the psychological distress could differ according to sex and age of the manager, the number of dependent children, and the type of business organization public/private. The findings we reached by the interdependence tests do not show any significant link between sex and access to practices of reconciliation, the meaning of work, and the psychological distress, even less for age, though many works have highlighted them (Bèque, 2019; Pailhé & Solaz, 2010a; Tremblay et al., 2006; Vézina et al., 2011). The same observation could be made about the meaning of work and the psychological distress. Nevertheless, when we compare the access to reconciliation practices, the meaning of work, and the psychological distress, according to the types of business organization, public and private sectors, the findings highlight differences of proportions between managers working in the public and the private sectors. Managers in the public sector seem more privileged than their colleagues in the private sector in terms of organizational resources which, generate motivation, engagement, and bear therefore a meaning for work (Bakker & Demerouti, 2007; Guilbert & Lancry, 2009; St-Amour & Bourque, 2013; St-Onge et al., 2002; Vandenberghe, 2004). Moreover, these same managers believe they have a low level of psychological distress when compared with those working in the private sector and the fact of having dependent children. At this level, we could think of the impact of the organization of work and the psychosocial climate within the organization as it was noted by Morin (2010) and Karatepe (2012). The characteristics of the work environment, the social relations, the management style, as well as the culture of the organization could promote assigning a meaning to work and therefore further the psychological well-being of employees (May et al., 2004; Di Fabio & Rosen, 2018). We should also mention that even if the reconciliation practices exist in the organization, the access to these practices remains, nonetheless, limited due to its negative consequences on the career trajectories and advancement to higher positions, mainly among managers (Tremblay, 2012). Hence, our first hypothesis could only be partly confirmed.

Furthermore, the current study equally suggests identifying the factors that influence the level of psychological distress among managers. For this purpose, we have carried out a logistic regression consisting in simultaneously assessing several explanatory variables in the same model. The analyses corroborate the findings of other studies concerning solely the contribution of the access to the practices of reconciliation and the meaning of work to explain the psychological distress. This is equally in accordance with the works of Morin (2010) which show the impact of the meaning of work on the mental health of workers. In fact, if a person finds a meaning in his work, he is likely to be healthier than a person who works with no meaning (Morin & Aranha, 2008). The access to reconciliation practices also significantly reduces the psychological distress and forms organizational resources capable of alleviating the probability of developing psychological disorders (Bakker & Demerouti, 2007; Boulet, 2013; Lero et al., 2009). Similar to the literature, these findings conclude that working in an environment that offers optimum conditions for the well-being, away from pressures to succeed in juggling personal and professional responsibilities

, is an environment where practices generate a collective solidarity and organizational fairness, sources of positive attitudes, and a meaning for work (Bakker & Demerouti, 2007; Di Fabio, 2017a ; Di Fabio & Rosen, 2018; Tremblay, 2012).

16.7 Conclusion

We could say that the data of the current research confirm the relation between the access to reconciliation practices and the well-being at work on the one side and the centrality of the meaning of work on the other. Nevertheless, within the issue of health at work, other aspects in relation to the working conditions and the organizational practices need further examination. Although we highlighted the impact of the type of business organization (public/private) of managers, particularly in terms of the characteristics of organizational functioning on the psychological distress, we still do not know whether we would find this relation in other types of jobs. Likewise, although we have examined the impact of the access to many practices of reconciliation work-family on the health, we could not have suggested courses of action. That is why other researches inscribed in the psychology of sustainability and sustainable development will be necessary to consider action strategies oriented toward promoting healthy organizations in which working conditions contribute to a better quality of life for employees (Di Fabio, 2017b) (Di Fabio & Rosen, 2018). The results of research can not only direct reflection toward theoretical orientations but also nurture awareness of the theme by the system of managing human capital which interacts with its environment while reconsidering the psychological dimensions at work.

That research shows, nevertheless, limitations which represent themselves perspectives for future researches. The first one consists in considering work-life reconciliation without taking into account other variables involved in the person's life, namely, life outside work. A second limitation is about the choice of quantitative method without completing it with data obtained by qualitative method.

We think that deeper researches on the qualitative plan will equally be relevant. These works will be continued in that respect with more important samples of participants in order to enlighten more the links between the difficulties of reconciliation and the job longevity. In fact, a methodological diversification and a bigger representativeness of the population of the study that will provide more elements will nonetheless be a future path of research.

References

Arendt, H. (1988). *Condition de l'homme moderne*. Pocket.

Aubert, N. (2011). La société hyper moderne, l'individu hyper moderne: Ruptures et contradictions. 3ème séminaire AFAPP. http://www.afapp.org/images/docs/110530%20Nicole%20Aubert(2).pdf

Bakker, A. B., & Demerouti, E. (2007). The Job Demands-Resources model: state of the art. *Journal of Managerial Psychology, 22*(3), 309–328. https://doi.org/10.1108/02683940710733115

Ballesteros Leiva, F., Poilpot-Rocaboy, G., & St-Onge, S. (2016). Les relations entre les perceptions des conflits et desenrichissements travail-famille et le bien-être. *Revue de gestion des ressources humaines, 100*, 5–20. http://www.cairn.info/revue-de-gestion-des-ressources-humaines-2016-2-page-5.htm

Barrère-Maurisson, M.-A. (2003). *Travail, famille : Le nouveau contrat*. Folio, Actuel inédit.

Barrère-Maurisson, M. -A. (2010). Conclusion: Quel avenir pour la conciliation travail-famille ? Nouveaux enjeux, nouveaux outils. *Revue Interventions économiques*. http://journals.openedition.org/interventionseconomiques/453

Baudry, S., & Fillion, F. (2014). La conciliation vie professionnelle / vie privée : quels outils pour les politiques publiques? *La découverte : Regards croisés sur l'économie, 15*, 245–250. https://www.cairn.info/revue-regards-croises-sur-l-economie-2014-2-page-245.htm

Bèque, M. (2019). Conciliation difficile entre vie familiale et vie professionnelle. Quels sont les salariés les plus concernés ? *DARES Analyses*, 45, 8p.

Bobillier-Chaumon, M. –E.. (2012). Nomadisme et dépendance : des pratiques professionnelles reconfigurées. In : Cadres CFDT. 2012. n° 448. https://www.cadrescfdt.fr/sites/default/files/cadres_cfdt_ndeg448_marc-eric_bobillier_chaumon_nomadisme_et_dependance.pdf

Boulet, M. (2013). *Les pratiques de conciliation travail et vie personnelle : un outil pour atténuer la détresse psychologique des salariés du Québec, Québec : Institut de la statistique du Québec* (32 pages).

Boulet, M., & Le Bourdais, C. (2016). Pratiques de conciliation travail-famille et détresse psychologique des salariés québécois : une comparaison selon le genre. *Relations industrielles, 71(3)*, 442–467. http://id.erudit.org/iderudit/1037660ar

Bruchon-Schweitzer, M. (2002). *Psychologie de la santé*. In *Modèles, concepts et méthodes*. Dunod.

Bustreel, A., Cornuau, F., & Pernod-Lemattre, M. (2012). Concilier vie familiale et vie professionnelle en France : les disparités d'horaires de travail. *Relations industrielles / Industrial Relations, 67*(4), 681–702. https://doi.org/10.7202/1013200ar

Cooper, C. L. (2006). The challenge of HRM in managing employee stress and improving well-being. In R. J. Burke & C. L. Cooper (Eds.), *The human resources revolution* (pp. 139–146). Elsevier.

Curie, J., & Hajjar, V. (1987). *Vie de travail – vie hors travail. La vie en temps partagé*. In C. Lévy-Leboyer & J. C. Spérandio (Eds.), *Traité de psychologie du travail* (pp. 37–55). PUF.

Dagenais-Desmarais, V., & Privé, C. (2010). Comment améliorer le bien-être psychologique au travail ? *Gestion, 3*(35), 69–77. https://www.cairn.info/revue-gestion-2010-3-page-69.htm

Dagenais-Desmarais, V., & Savoie, A. (2011). What is psychological well-being really? A grassroots approach from the organizational sciences. *Journal of Happiness Studies*. https://doi.org/10.1007/s10902-011-9285-3

Demerouti et al. (2001), Demerouti, Evangelia & Nachreiner, Friedhelm & Schaufeli, Wilmar. (2001). The Job Demands–Resources Model of Burnout. *The Journal of applied psychology. 86*, 499–512. https://doi.org/10.1037/0021-9010.86.3.499

Demerouti, E., & Geurts, S. (2004). Towards a Typology of Work-Home Interaction. *Community, Work and Family, 7*(3), 285–309.

De Terssac, G. & D.-G. Tremblay (2000). *Où va le temps de travail*?, Toulouse, éditions Octares.

Di Fabio, A. (2017a). Positive healthy organizations: promoting well-being, meaningfulness, and sustainability in organizations. *Frontiers in Psychology, 8*, 1938. https://doi.org/10.3389/fpsyg.2017.01938

Di Fabio, A. (2017b). The psychology of sustainability and sustainable development for well-being in organizations. *Frontiers in Psychology, 8*, 1534. https://doi.org/10.3389/fpsyg.2017.01534

Di Fabio, A., & Peiro, J.-M. (2018). Human capital sustainability leadership to promote sustainable development and healthy organizations: A new scale. *Sustainability, 10*, 2413. https://doi.org/10.3390/su10072413

Di Fabio, A., & Rosen, M. A. (2018). Opening the black box of psychological processes in the science of sustainable development: A new frontier. *European Journal of Sustainable Development Research, 2*(4), 47. https://doi.org/10.20897/ejosdr/3933

EU-OSHA. (2010). *European survey of enterprises on new and emerging risks: managing safety and health at work*. Communautés européennes. https://osha.europa.eu/en/publications/reports/esener1oshmanagement

Eurofound. (2012). *Fifth European working conditions survey*. Office des publications de l'Union européenne. http://www.eurofound.europa.eu/surveys/ewcs/2010/

European Agency for Safety and Health at Work (2012). Annual Report 2011.

Ferguson, M., Carlson, D., Hunter, E.M., & Whitten, D. (2012). A two-study examination of work– family conflict, production deviance and gender, *Journal of Vocational Behavior, 81*, n°2, p. 245–258.

Fusulier, B., & Tremblay, D.-G. (2013). Appartenances professionnelles et vie familiale : une entrée analytique particulière. *Revue Internationale, Enfances, Familles, Générations, 18*, 1–16. https://doi.org/10.7202/1014571ar

Garner, H., & Méda, D. (2006). La place du travail dans l'identité des personnes. *Données sociales, Insee*, 623–630. https://www.insee.fr/fr/statistiques/fichier/1371981/donsoc06zq.pdf

Guérin, G., St-Onge, S., Chenevier, L., Denault, K., & Deschamps, M. (1997). Le conflit emploi famille: Ses causes et ses conséquences: Résultats d'enquête. Montréal : Université de Montréal, École de relations industrielles, 23 p. https://papyrus.bib.umontreal.ca/xmlui/bitstream/handle/1866/1869/A1.328%20WP%209705.pdf

Guilbert, L., & Lancry, A. (2005). Les activités, temps et lieux de vie des cadres : Un système de déterminants individuels, contextuels et technologiques, @ctivités, vol. 2, n°2, 25–42.

Guilbert, L., & Lancry, A. (2009). La conciliation des sphères de vie des cadres. *Etudes de cas. Pratiques psychologiques, 15*, 271–285. https://doi.org/10.1016/j.prps.2008.09.008

Gilbert-Ouimet, M., Brisson, C., & Vézina, M. (2016). P344 Effet de la double exposition aux facteurs psychosociaux du travail et aux responsabilités familiales sur la détresse psychologique : une étude prospective de 5 ans chez des cols blancs, *Médecine du travail et de l'environnement* 73, A237.

Hobfoll, S.E. (2001). "The Influence of culture, community, and the nested-self in the stress process: Advancing conservation of resources theory". Applied Psychology. 50 (2): 337–421. https://doi.org/10.1111/1464-0597.00062

Karasek, R. A. (1979). Job demands, job decision latitude, and mental strain: Implications for job redesign. *Administrative Science Quarterly, 24*, 285–308. https://doi.org/10.2307/2392498

Karasek, R. A., & Theorell, T. (1990). *Healthy work: Stress, productivity, and the reconstruction of working life*. Basic Books.

Karatépé, O, M. (2012). Les effets du collègue et du soutien organisationnel perçu sur les résultats des employés de l'hôtel : le rôle modérateur de l'intégration au travail. *Journal of Hospitality & Tourism Research, 36*(4), 495–516. https://doi.org/10.1177/1096348011413592

Kessler, R., AndrewS, G., Colpe, L., Hiripi, E., Mroczek, D., Normand, S., Walters, E.E., & Zaslavsky, A. (2002). Échelles de dépistage courtes pour surveiller les prévalences de la population et les tendances de la détresse psychologique non spécifique. *Médecine psychologique, 32*(6), 959–976. https://doi.org/10.1017/S0033291702006074

Lallement, M., Mary, C., Loriol, M., Molinier, P., Gollac, M., Marichalar, P., & Martin, E. (2011). Dossier-débat « Maux du travail : dégradation, recomposition ou illusion ? ». *Sociologie du travail, vol., 53*(1). https://doi.org/10.1016/j.soctra.2010.12.001

Lapointe, D., Boudrias, J. S., & Brunet, L. (2011). Vérification de l'effet modérateur des ressources sur la relation entre les demandes et la tension psychologique au travail. In P. Desrumeaux, A. M. Vonthron, & S. Pohl. *Qualité de vie, risque et santé au travail*. Paris: Edition l'Harmattan.

Lavoie, A. (2016). Les défis de la conciliation travail-famille chez les parents salariés. In *Un portrait à partir de l'Enquête québécoise sur l'expérience des parents d'enfants de 0 à 5 ans 2015* (59 p). Institut de la statistique du Québec.

Lero, D. –S., Richardson, J., & Korabik, K. (2009). Examen des coûts et des avantages des pratiques de conciliation travail-vie. Document présenté à l'Association canadienne des administrateurs de la législation ouvrière (ACALO), University of Guelph : Centre for Families, Work and Well-Being, 138 pages.

Lhuilier, D. (2010). L'invisibilité du travail réel et l'opacité des liens santé-travail. *Sciences sociales et santé, 28*(2), 31–63. https://www.cairn.info/revue-sciences-sociales-et-sante-2010-2-page-31.htm
Loriol, M.(2006). *Je stresse donc je suis : comment bien dire son mal-être,* Paris, Mots et Cie.
Marchand, A., Bilodeau, J., Demers, A., Beauregard, N., Durant, P., & Haines, V. (2015). Gendered depression: Vulnerability or exposure to work and family stressors? *Social Science &Medecine, 166*, 160–168. https://doi.org/10.1016/j.socscimed.2016.08.021
May, D. R., Gilson, R. L., & Harter, L. M. (2004). The psychological conditions of meaningfulness, safety and availability and the engagement of the human spirit at work. *Journal of Occupational and Organizational Psychology, 77*, 11–37. https://doi.org/10.1348/096317904322915892
Méda, D. (2010). Travail : la révolution nécessaire. La Tour d'Aigues, Editions de l'Aube.
Molinier, P. (2009). Temps professionnel et temps personnel des travailleuses du care : perméabilité ou clivage ? , *Temporalités* [Online], 9 | 2009, Online since 30 September 2009. http://journals.openedition.org/temporalites/988; https://doi.org/10.4000/temporalites.988
Monneuse, D. (2014). Le silence des cadres, enquête sur un malaise.
Morin, E. (2008). Qu'est-ce qui donne un sens au travail ? *Objectif prévention, 31*(2), 10–14. http://www.ohdq.com/Ressources/Document s/SensAuTravail.pdf
Morin, E. (2010). La santé mentale au travail : une question de gros bon sens. *Gestion, 35*, 34–40. https://doi.org/10.3917/riges.353.0034
Morin, E., & Aranha, F. (2008). Sens du travail, santé mentale et engagement organisationnel (Rapport n° R-543). Montréal, Québec. : *Institut national de santé publique, Institut de recherche Robert-Sauvé en santé et en sécurité du travai*l .http://www.deslibris.ca/ID/252536
OIT. (2017). *Emploi et questions sociales dans le monde*. Tendances de l'emploi des femmes.
Pailhé, A. & Solaz, A. (2009). *Entre famille et travail: Des arrangements de couple aux pratiques des employeurs*. Paris: La Découverte. https://doi.org/10.3917/dec.pailh.2009.01
Pailhé, A., & Solaz, A. (2010a). Concilier, organiser, renoncer : quel genre d'arrangements ? *Travail, genre et sociétés*, n° 24, 29–46. https://doi.org/10.3917/tgs.0240029
Pailhé, A., & Solaz, A. (2010b). L'implication des entreprises dans l'aide à la parentalité en France : Une initiative bienvenue mais source d'inégalités entre salariés,. *Revue Interventions économiques*. http://journals.openedition.org/interventionséconomiques/412
Paquet, R., & Najem, E. (2005). L'évolution récente des pratiques de conciliation travail-famille dans les entreprises canadiennes. *Actes du Congrès relations de travail et organisations, 3*(2), 1028–1047.
Pierre, X., & Tremblay, D. (2012). Améliorer l'implication et la rétention des agents dans les centres d'appels : agir sur les pratiques de gestion. *Recherches en Sciences de Gestion, 4*(4), 17–38. https://doi.org/10.3917/resg.091.0017
Rivaud, S., & Ulrich, V. (2007). Le poids du temps partiel dans les trajectoires professionnelles des femmes, *Dares, Document d'études*, n° 127. https://dares.travail-emploi.gouv.fr/sites/default/files/pdf/de127_ulrichfemmestempspartiel-2
St-Amour, & Bourque, M. (2013). *Conciliation travail-famille et santé : Le Québec peut-il s'inspirer des politiques gouvernementales mises en place dans d'autres pays?, Institut national de santé publique du Québec* (161 pages). *Collection Politiques publiques et santé*.
St-Onge, S., Renaud, S., Guérin, G., & Caussignac, É. (2002). Vérification d'un modèle structurel à l'égard du conflit travail-famille. *Relations industrielles/ Industrial relations, 57*(3), 491–516. https://doi.org/10.7202/006887ar
Thoemmes, J. (2012). La fabrique des normes temporelles du travail. *La nouvelle revue du travail, 1*. http://nrt.revues.org.gate3.inist.fr/153
Thoemmes, J., Kanzari, R., & Escarboutel, M. (2011). Temporalités des cadres et malaise au travail. *Revue Interventions-économiques, 43*. http://journals.openedition.org/interventionseconomiques/1401;DOI:10.4000
Tremblay, D.-G. (2003). Articulation emploi-famille : Comment les pères voient-ils les choses? *Politiques sociales Bruxelles et Madrid., 63*(3-4), 70–86.

Tremblay, D.-G. (2004). *Conciliation emploi-famille et temps sociaux* (340 p). Presses de l'Université du Québec et Octares.
Tremblay, D.-G. (2012). *Articuler emploi et famille : le rôle du soutien organisationnel au coeur de trois professions. Québec : Québec : Presses de l'Université du Québec.* https://doi.org/10.7202/1028116ar
Tremblay, D.-G. & De Sève, M. (2002). Articulation emploi-famille et temps de travail ; résultats de l'analyse qualitative dans les secteurs de la santé et de l'éducation. Rapport de recherche.
Tremblay, D.-G., Renaud Paquet & Elmustapha Najem (2006). Telework: a way to balance work and family or an increase in work-family conflict? *Canadian Journal of communication. 31*, 2. octobre. 2006.
Tremblay, D.-G., & Vaillancourt-Laflamme, C. (2000). *La conciliation des responsabilités parentales et professionnelles chez les employés de bureau : Résultats d'une recherche menée au Québec.* Montréal: Chaire de recherche sur les enjeux socio-organisationnels de l'économie du savoir, Télé-université, . www.teluq.uquebec.ca/chaireecosavoir
Vandenberghe, C. (2004). Conserver ses employés productifs: Nature du problème et stratégies d'intervention. *Gestion, 3*(3), 64–72. https://doi.org/10.3917/riges.293.0064
Vézina, M., Derriennic, F., & Monfort, C. (2001). L'impact de l'organisation du travail sur l'isolement social. *Travailler, 5*, 101–117. https://www.cairn.info/revue-travailler-2001-1-page-101.htm
Vézina, M., Cloutier, E., Stock, S., Lipel, K., Fortin, E., Delisle, A., St-Vincent, M., Funes, A., Duguay, P., Vézina, S., & Prud'homme, P.(2011). *Enquête québécoise sur des conditions de travail, d'emploi et de santé et de sécurité au travail (EQCOTESST)*. Études et recherches/Rapport R-691. Montréal : Institut de recherche Robert-Sauvé en santé et en sécurité du travail (IRSST).

Chapter 17
When a Balanced Life Leads to a Better Life: Reflections on Innovation, Health, and Wealth

Yomna Sameer and Nada Rostom

17.1 Introduction

As much as the twenty-first century is characterized by change, complexity, and development (Di Fabio, 2017a), the labor market is characterized by insecurity, economic instability, and ambiguity, especially in the era of the COVID-19. At the same time, the world has never been concerned about sustainable well-being as much as the current era. The link between well-being and sustainability is still inadequately explored (Veenhoven, 2004; Di Fabio & Kenny, 2018). The question of sustainability has always revolved around the well-being of nations. However, we cannot achieve sustainability without considering the actions and behaviors of individuals. This brought the attention of psychologists, who made an effort to address the sustainability query from a psychological view (Di Fabio, 2017b). The psychology of sustainability and sustainable development focuses on the well-being of the individual and the environment across different dimensions, such as the natural, personal, and cross-cultural ones (Di Fabio & Rosen, 2018).

In the new framework of psychology of sustainability and sustainable development and highlighting on the importance of human factor and psychology on the welfare of nations, this paper addresses the link between a balanced life and health, wealth, and innovation, assuming that employees could balance between their work commitments and life without compromising innovations and wealth and while maintaining their health. With a special emphasis on the current COVID-19 situation, we argue in this paper that the number of hours employees have to work does not go hand in hand with innovation and wealth. So, we argue that the less "working

Y. Sameer (✉)
Abu Dhabi University, Dubai, UAE
e-mail: yomna.morsy@adu.ac.ae

N. Rostom
British University in Egypt, Cairo, Egypt

A. Di Fabio (ed.), *Cross-cultural Perspectives on Well-Being and Sustainability in Organizations*, https://doi.org/10.1007/978-3-030-86709-6_17

hours" the employees work, the better their health, and also the more innovative and wealthy they are. Of course, this is not to assume that "the less, the better" in general. However, it is to prove that it is not "the more, the better." Previous research on work-life balance focused mainly on the different tools employers offer to make employees' lives easier, and the link between work-life balance and stress has been evident. The lack of work-life balance affects one's stress level which may also translate to weaker physical health. Chronic stressors may also affect how people take decisions (Beehr & Newman 1978; see also Leslie et al., 2019). With the spread of the pandemic, people are worried that the hours they devote to their work may not be enough. Therefore, in this paper, we focus on the number of working hours and its relationship to health, wealth, and innovation across different countries.

17.2 Theoretical Framework

Sustainability and Work-Life Balance

Sustainability was introduced to the international agenda by the Brundtland Report (WCED, 1987) and was defined as the development that "meets the needs of the present without compromising the ability of future generations to meet their own needs." More generally, sustainability can be defined as the ability to maintain a source undiminished over a period of time (Lélé & Norgaard, 1996). It involves achieving sustainable production and consumption at both the macro (societal) and micro (individual) levels (Moll et al., 2008). However, the two levels interact with each other, and they may be incompatible, giving rise to negative psychological consequences, such as discomfort and a feeling of sacrifice (Verdugo, 2012). Also, it should not be considered sustainable if one generation overwork to achieve sustainability. So, sustainability from the new perspective of psychology has a lot of reflections on the labor market. The link between sustainability and well-being has been limited to looking at well-being as an outcome and not as a major sustainability component. Micro-level studies document a positive association between well-being and sustainable behavior (Bechtel & Corral-Verdugo, 2010). The adoption of the Sustainable Development Goals (SDGs) in 2015 signaled a worldwide commitment to pursue a more sustainable path toward comprehensive and impartial growth. Known also as Agenda 2030, the 17 SDGs have covered a broad range of issues related to development and comprise 304 indicators to monitor 169 targets. SDGs build on various advancements resulting from SD efforts over the past three decades and on from the Millennium Development Goals (MDGs), which were the first initiative of establishing global goals, during 2000–2015. There are so many links between sustainability and the labor market. At the core of sustainability are the security of jobs and the well-being of employees, that is, having a decent job and working in good working conditions. The core years of the Great Recession (2008–2010) and its aftershock resulted in broader labor market inequalities and more discriminated employment conditions in most European countries (Lewis et al., 2016).

As mentioned earlier, one main aim for this paper is to shed light on the link between working hours and positive outcomes. We look at these outcomes as "sustainability" outcomes in the sense that nations can still survive while working less hours. This is in addition to the normal benefits of working for a reasonable amount of time rather than overworking. In her book *The Overworked American*, Schor looked at work-life balance mainly from "working hours" perspective (Jacobs & Greens, 1998; Roberts, 2007). The term "work-life" balance then included many other forms that aim at helping employees balance between their work and personal life. Work-life balance measures and practices refer to "initiatives voluntarily introduced by firms which facilitate the reconciliation of employees' work and personal lives" (McCarthy et al., 2010). Such initiatives could be *temporal arrangements* that give employees room to reduce the number of hours they work (e.g., job sharing where two employees share one job, part-time working where an employee works less than a full-time equivalent) or *flexible working arrangements* such as flexi-time or flexible working hours which is working with flexibility and is mainly about managing one's own work time (Chung & Lippe, 2020). It also includes teleworking/home-working/e-working where employees have locational flexibility in completing their work. Finally, *work-life balance could take the form of support programs* such as employee counselling, employee assistance programs, time management training, stress management training, and *childcare* facilities on-site or financial support for childcare off-site (e.g., through subsidized childcare) (Darcy et al., 2012). In the current study, we focus mainly on working hours.

To date, much of the research in the work-life balance field has examined individual-level work-life balance factors such as employee demands for flexible working practices and discussions on the support of the public sector (Den Dulk & Groeneveld, 2013; Lewis et al., 2009), employee satisfaction with work-life or work-family policies and plans (Lewis et al., 2016), and impact of work-life balance programs on a number of employee-level outcomes such as stress, commitment, and productivity (Darcy & McCarthy, 2007; Lewis et al., 2016). However, there has been rare, if any, research on work-life balance on the country level and its impact on macro outcomes.

Work-Life Balance, Health, Wealth, and Innovation

Considered as a national crisis, the USA nation spends almost two trillion dollars annually on healthcare (Porter & Teisberg, 2006). Decades of reforms failed to satisfy anyone, either physicians, governments, or the patients themselves (Porter, 2010). However, all solutions aimed at targeting one or two aspects of the system believing they were the core problems, for example, advancements in technology or purchasing drugs from Canada (Porter & Teisberg, 2006), and none of the solutions worked. Several studies showed how a healthy lifestyle and having positive emotions, which is in a way the opposite to burnout, could lead to positive health outcomes. Kubzansky et al. (2001) followed 1306 men who were classified as optimists and pessimists. In a 10-year follow-up, frequency of coronary heart disease (CHD), non-fatal myocardial infarction, fatal CHD, and angina pectoris were recorded. Surprisingly, a strong positive correlation was found between increasingly high levels of optimism and increased protection against each of the cardiovascular events. In the same way, Kubzansky and Thurston (2007) found a strong positive relationship between emotional vitality and lack of CVD. Another similar study showed high loneliness among women was associated with increased risk of CHD (Thurston & Kbzansky, 2009). Also, a main component of the well-being theory (Seligman, 2012), purpose in life, or the sense that life has meaning and direction is associated with reduced risks of adverse health (Yu et al., 2015). It is also worth mentioning that positive affect has been found to be protective against stroke, whereas depression increases the risk of it (Ostir et al., 2001; Steptoe et al., 2005). Moreover, Cohen et al. (2006) found that the tendency to express positive emotions was associated with greater resistance to developing a cold (see also Cohen, 2014).

Better states of mind are also associated with higher mental resilience, recovery, and longevity. Taylor et al. (2000) not only associate this physical and mental resilience to personal control, a sense of meaning, or psychological beliefs such as optimism but also argue that these beliefs are effective even if not realistic. Scheier et al. (1989) found that dispositional optimism was associated with faster recovery rates during hospitalization in cases of coronary artery bypass surgery. All of these studies and much more shows how positive emotions and a happy life could lead to better health outcomes. The link between work-life balance and positive emotions has been discussed in the first section. Therefore, we can conclude here that it is likely to assume that work-life balance should lead to a happier life which would lead to a healthier one. Happier and more optimistic individuals are also more likely to deal with work-life imbalances. Indeed, stress level may depend on one's perception of a certain incidence. This perception is influenced by one's surrounding, personality traits, and ability to cope with the stressors (Cohen et al., 1983). According to Greenhaus and Powell (2006), persons with positive attitudes (PAs) are normally able to perceive or lever pressures of opposing demands of family and work in a balanced manner. In addition, positive psychology suggests that individuals with positive stance about self and others in their environment are likely to be optimistic (Seligman & Csikszentmihalyi, 2000). This can occur through positive spillover from the work domain into the home domain (Culbertson et al., 2012). Therefore, it is assumed that having a PA and an ability to maintain a positive outlook is the first coping strategy for individuals to develop their capacity to reduce work-family conflict which would help enhance health and well-being (Rotondo & Kincaid, 2008).

The link between work-life balance in general and health outcomes has been documented, specially the link between work-life balance and stress. Job stress has been conceptualized as a state in which job-related factors induce a person to change his/her physiological or psychological conditions extensively that the person is made to deviate from the normal functioning (Beehr & Newman, 1978). The lack of work-life balance affects one's stress level which may also translate to weaker physical health. Magee et al. (2012) indicate that work-family conflict strongly predicted poor physical and mental health, leading to employee stress. While stress is a part of everyday life, high imbalances between work and family life can cause excessive amounts of stress. These imbalances are linked to impaired functioning of individual employees (Ilies et al., 2007). This means that work-life imbalance may cause a vicious cycle of high stress levels and lower mental and physical health that may negatively affect work performance. Weakening work performance increases stress levels further, and the cycle goes on.

Contrarily, Hecht and Boies (2009) found that people engaged in recreation and volunteering or fitness activities have less work-life imbalances through a fulfillment of life goals in externally oriented activities. Therefore, it is likely that abilities to meet lifestyle (i.e., sporting, recreational, and social) and other community activities (i.e., voluntary work and club membership) are important individual coping approaches. These strategies contribute to enrich life outside work, which in turn help achieve physical and mental health, to obtain employee well-being.

Furthermore, chronic stressors may also affect how people take decisions. Since women are likely to be affected from the work-life imbalances more than men and in the same time are responsible for raising children, their stress-driven attitude and decision-making can also affect how their children will make decisions. That is, stress caused by work-life imbalances may result in intergenerational transfer of stress and pressure.

Happier and more optimistic individuals are also more likely to deal with work-life imbalances. Indeed, stress level may depend on one's perception of a certain incidence. This perception is influenced by one's surrounding, personality traits, and ability to cope with the stressors (Cohen et al., 1983). According to Greenhaus and Powell (2006), persons with positive attitudes (PAs) are normally able to perceive or lessen pressures of opposing demands of family and work in a balanced manner. In addition, positive psychology suggests that individuals with positive stance about self and others in their environment are likely to be optimistic (Seligman & Csikszentmihalyi, 2000). This can occur through positive spillover from the work domain into the home domain (Culbertson et al., 2012). Therefore, it is assumed that having a PA and an ability to maintain a positive outlook is the first coping strategy for individuals to develop their capacity to reduce work-family conflict which would help enhance health and well-being (Rotondo & Kincaid, 2008). Therefore, it is likely to assume that people who enjoy work-life balance, that is, in our case "less hours," are more likely to be happy and so to enjoy better health.

H 1: Work-life balance relates positively and significantly to better health.

H 2: The relationship between work-life balance and health is mediated by happiness.

Innovation behaviors have become increasingly enviable in today's unpredictable fast-moving and competitive work environment though been seen previously as out of place or insolent topic (Anderson et al., 2003; De Jong & Kemp, 2003). Innovative behavior is not limited to developing new product ideas and new technologies, but it also includes initiating new ideas or changes in managerial procedures that aims at improving work relationships or the submission of new ideas or technologies to work processes aiming at improving the value of work (Amabile, 1996, 1998; Kleysen & Street, 2001). It consists of various practices and behaviors such as opportunity discovery, idea generation, significant research and investigation, championing, and application (Kleysen & Street, 2001, De Jong & Kemp, 2003; Anderson et al., 2003). It is widely known in psychology that there is a link between flow and creativity. That is, artists and creative people get into the creativity mode when they are in the flow mood. Burnout at work could prevent this "flow" state. Happy people are likely to be creative too. Just by observing the world statistics, we can see that the happiest nations also rank at the top of the chart of the work-life balance statistics. The question is, will this lead to more creative ideas and implementation of it? We assume that the more work-life balance, the more happy nations will be, and so the more innovative.

H 3: Work-life balance relates positively and significantly to more innovation.
H 4: The relationship between work-life balance and innovation is mediated by happiness.

The link between work-life balance and wealth has not been examined before. The question whether working less could lead to more fortune is indeed a unique one. Drawing from the above discussion, we assume that people who enjoy a balanced happy life are more likely to be productive and so to have high income. That is, the less working hours does not necessarily mean less income.

H 5: Work-life balance relates positively and significantly to better income (wealth).
H 6: The relationship between work-life balance and wealth is mediated by happiness.

17.3 Methodology

Participants and Procedure

Our dataset is a balanced panel composed of 48 countries with 143 observations, combining developing and developed countries and spanning from 2016 to 2018. Only for health variable, data for 2018 is not published yet, and hence, analysis for health includes only 96 observations.

Measures

Work-life balance is approximated by average work hours per capita data obtained from the International Labour Organization. Happiness is obtained from the World Poll Survey. The happiness index is estimated from nationally representative samples of respondents who are asked to rate their level of happiness on a 0–10 scale. Finally, income per capita and health conditions, approximated with life expectancy at birth, are retrieved from the World Bank. Income per capita is measured in constant $US 2010 terms. An approximation of health condition with life expectancy at birth can be explained by the strong correlation between the two variables as shown in previous research. Life expectancy at birth reflects the mean number of years a newborn is expected to live if the current mortality rate continues to apply. Innovation level is approximated by the Global Innovation Index which is estimated by the World Intellectual Property Organization.

Data Analysis

Means and standard deviations for descriptive statistics, correlation analysis, and mediation analysis using seemingly unrelated regression were carried out to verify the hypotheses of the present study.

17.4 Results

Descriptive Statistics

The main aim of this research is to observe the impact of work-life balance on health, wealth, and innovation, hypothesizing that happiness can be a main mediator. Precisely, we propose that greater work-life balance induces greater levels of happiness and that happiness can increase creative thinking. Also, happier people are also likely to be healthier on the other hand.

Greater balance in work and leisure is also expected to have a positive impact on wealth. Hence, we hypothesize that an increase in the number of hours worked and a reduction in work-life balance will be negatively related to both health outcomes (H1, 2), innovation (H3, 4), and wealth (H5, 6). Figure 17.1 illustrates the hypotheses in a schematic model.

Table 17.1 shows the summary statistics of our dataset. The average hours worked in our dataset is around 40 hours per week, with a standard deviation of almost 3 hours above or below the mean. The average life expectancy at birth is around 77 years, with 5-year deviations around this mean. Happiness is an index that takes the value from 0 to 10, with 10 being the highest happiness level that can

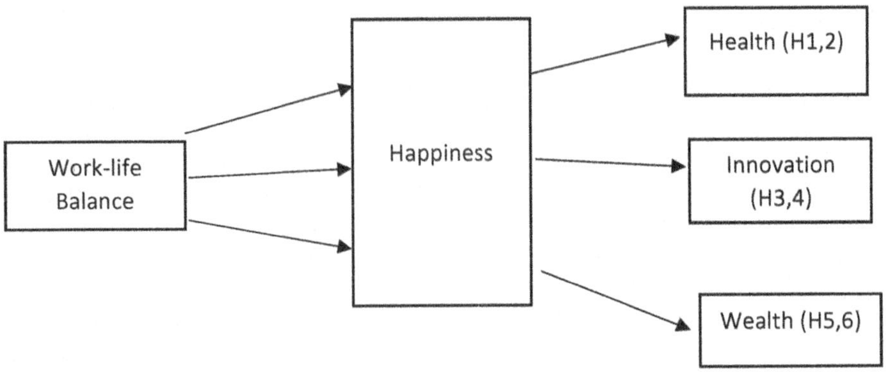

Fig. 17.1 Hypothesized model

Table 17.1 Descriptive statistics

	Mean	Standard deviation
Innovation	42.303	11.961
Happiness	6.050	0.967
Work-life balance	39.327	3.358
Health	77.372	4.687
Income	27662.040	19439.860

be reached. On average, all countries in the dataset report perceiving themselves as being in the a bit happier than the average. Also, most countries are within a 1 score more or less than the average in terms of happiness perception. Income per capita shows a very wide discrepancy, and this is because developing and developed countries are included in the dataset. Innovation index is a simple average of the score of two sub-indices designed by the WIPO which include the innovation input and innovation output index. The average score of innovation in our dataset is 42, with 11 points deviation from the mean, while the maximum is 68, and the minimum is 23 points. Likewise, the minimum and maximum boundaries for all the variables show that the dataset is representative to different countries in the world and does not focus on a particular group of country. Analysis was conducted using Stata15.

Correlation Analysis

The correlation analysis shows that all variables are strongly correlated to each other since correlation ranges between 0.6 and 0.8. Only the correlation between happiness and work-life balance is negative and moderate (Table 17.2).

Table 17.2 Correlations

	Innovation	Happiness	Hours worked	Health	Income
Innovation	1				
Happiness	0.745	1			
Work-life balance	−0.778	−0.546	1		
Health	0.786	0.719	−0.674	1	
Income	0.872	0.728	−0.677	0.757	1

17.5 Hypothesized Mediation Model Using Seemingly Unrelated Regression

Seemingly unrelated regression is used to conduct the mediation analysis. Through this method, several linear regression models are estimated together. Each linear equation is a valid regression on its own and can be estimated separately with different dependent and explanatory variables. This method however assumes exogeneity of the explanatory variables. Reverse causality between the different estimates presented here is likely to be a problem. For instance, wealthier countries also have better institutions and better working hour regulations. Thus, as a country gets richer, it may have better work-life balance and not the opposite. Therefore, causality cannot be claimed, and these estimates only show correlations. Post-estimation using non-linear combinations of estimators is then used to estimate the indirect impact or the mediation effect of happiness on the relationship between work-life balance and innovation. Hence, as per the hypothesis provided above, three seemingly unrelated regressions are estimated with happiness as a mediator and work-life balance as an independent variable.

From the innovation, happiness, and work-life balance regression, it is observed that higher work hours decrease happiness score (b=−0.160, p-value<0.01). A greater imbalance between the hours worked and the time dedicated for leisure is also correlated with the innovation score (b=1.837, p-value <0.01). This is the direct effect of work-life balance on innovation. Also, a higher happiness index is associated with a greater level of innovation (b=5.76, p-value<0.01). The indirect effect of work-life balance through happiness can be found in the bottom of the table where the post-estimation is provided. First, happiness seems to be an adequate mediator since the results are significant. The result (b=−0.921, p-value=0.01) also shows that the work-life balance affects happiness negatively and that, eventually, lower happiness level may lead to lower innovation score.

Similar results can also be observed for health and wealth. In both cases, the direct impact of work-life balance is negative and significant (for health, b=−0.557, p-value<0.01, and for wealth, b=−0.107, p-value). The indirect effect using the post-estimation is negative and significant which implies that a work-life imbalance is correlated with lower happiness levels and that less happy nations are likely to suffer from health issues that affect their longevity. A significant negative association can also be observed between wealth and work-life balance. It seems that populations that work less hours tend to be richer. Again happiness plays a significant mediator role (Tables 17.3, 17.4, and 17.5).

17.6 Discussion

Working hours and the ability to balance between life and work have been under research for years due to its significant implications on health and productivity (Kossek & Ozeki, 1999). The COVID-19 situation has made it even more critical. This study examined the links between working hours, health, wealth, and innovation. Happiness as a mediator for such relationships was examined too. Happiness can be an important channel that can explain this relationship since overworking will likely affect one's life balance and causes less enjoyment of life, which would eventually negatively affect one's psychological and physical health. Income per capita on the other hand shows that the link between working hours and wealth is not necessarily positive, that is, it is not the number of hours that determine how much money you make. A population that on average work more hours but is also likely to earn on average more income is not expected to have the same health and sustainability outcomes as one that works the same number of hours but earns less. Indeed, a person is expected to be happier if they work more hours but also earn more.

Literature has shown the positive impact of work-life balance interventions like new procedures or programs such as flexi-time or developing a family-friendly culture (see Prottas & Kopelman, 2009). However, the outcomes have been narrowed to job satisfaction, commitment, and engagement (Lambert, 2016). Macro-level outcomes specifically sustainable development and health (taken as a sustainable development outcome too) have not been linked to work-life balance before,

Table 17.3 Regressions

Seemingly unrelated regression		
Innovation, happiness, and work-life balance		
	Happiness	Innovation
Happiness		5.762***
		(0.623)
Work-life balance	−0.160***	−1.837***
	(0.02)	(0.179)
Constant	12.334***	79.677***
	(0.79)	(9.679)
Indirect effect of happiness		−0.921***
		(0.152)

Standard error reported in parentheses
*Significant at the 10% level
**Significant at the 5% level
***Significant at the 1% level

Table 17.4 Regressions

Seemingly unrelated regression		
Health, happiness, and work-life balance		
	Happiness	Health
		2.406***
		(0.359)
Work-life balance	−0.159***	−0.557***
	(0.024)	(0.103)
Constant	12.313***	84.712***
	(0.966)	(5.578)
Indirect impact of happiness		−0.383***
		(0.082)

Standard error reported in parentheses
*Significant at the 10% level
**Significant at the 5% level
***Significant at the 1% level

Table 17.5 Regressions

Seemingly unrelated regression		
Wealth, happiness, and work-life balance		
	Happiness	Wealth
Happiness		0.411***
		(0.044)
Work-life balance	−0.160***	−0.107***
	(0.02)	(0.013)
Constant	12.330***	11.679***
	(0.793)	(0.689)
Indirect impact of happiness		−0.66***
		(0.109)

Standard error reported in parentheses
*Significant at the 10% level
**Significant at the 5% level
***Significant at the 1% level

especially in the form of "less working hours." There is indeed research that links work-life balance and health outcomes (Eby et al., 2005). However, not on a macro level and not in terms of working hours. The current study aimed at examining this relationship on a macro-level. Besides, work-life balance has been studied more from a conflict perspective as mentioned earlier, whether it is "life interfering with work" or "work interfering with life" (Frone, 1997).

The mediating role of happiness is also very significant as it explains the reason behind the more positive health outcomes. Previous research showed the relationship between work-life conflict and burnout and absenteeism (see Allen, 2001) which in turn could lead to poor health. The link between work-life conflict and physical and mental health outcomes on the micro-level has been investigated too where it was found to predict depression, mood disorders, anxiety, and physical health complains (Frone, 1997, 2000; Parasuaman et al., 1996). The current study completed the dots by exploring the link across countries. Besides, this link has not been studied where work-life balance is measured by working hours. This is consistent with a previous study by Parasuaman and colleagues (1996) where they found that workers who spent more time at work reported more work overload and so greater parental demands and less family involvement which in turn leads to more stress. So, the more working hours, the less happy or satisfied one is and so the less healthy (see also Parasuraman & Simmers, 2001).

From a sustainability perspective, research has not paid enough attention to the link between work-life balance and sustainability. In this paper, we looked at health, wealth, and innovation as critical success factors for the economic and human development of any nation. Besides, there are other indirect links, like how working less could lead to better sustainable outcomes in the form of less energy used. It could also be that when people are in a good mood and not pressured and so balancing life and work, they are more likely to be responsible toward the environment. Therefore, new perspectives in the framework of the psychology of sustainability and sustainable development (Di Fabio, 2017a, b; Di Fabio & Rosen, 2018) and of the Psychology of Harmonization (Di Fabio & Tsuda, 2018) could be opened.

Finally, like any study, this study suffered from potential limitations. First, happiness is subjectively measured. Its meaning can be interpreted differently depending on cultural factors. The approximation of work-life balance is also flawed. However, it could still be considered an adequate approximation since with a higher number of hours, one will have less time for leisure and will face family and life conflict. Although there is no perfect relationship between less leisure time, health outcomes, or economic development, the relationship is still expected to be statistically significant. Furthermore, the sample collected from developing countries may not be representative of the poorest groups in the society. This can cause a measurement error bias since it is likely to be a systematic problem to developing countries.

To conclude, the link between work-life balance as measured by less working hours and its relationships to innovation, health, and wealth has not been examined before. Though previous research linked work-life balance to stress, family conflicts, and satisfaction (see Hirschi et al., 2019), research has been rare on country-level macro analysis like innovation on a country level, health, and wealth. The positive psychology movement too has been focused on micro- rather than macro-level outcomes (see Seligman, 2008). This paper is considered to be the first to focus on living a happy balanced life and its implications on health, wealth, and innovation. Previous research focused on happiness from an individual perspective and an institutional level, while the country level has been neglected. Innovation is key to development, and discovering that less working hours is not necessarily affecting it is indeed worth more investigation and is a good start for more research on the topic. Health and the costs associated with the healthcare have been under investigation for years. This study gives an indirect solution that is relevant to the issue by focusing on the healthy lifestyle and its implications on health. Finally, the link between income and work-life balance is indeed a finding that is worth investigating more and brings a lot of implications whether practical or theoretical.

References

Allen, T. D. (2001). Family-supportive work environments: The role of organizational perceptions. *Journal of Vocational Behavior, 58*(3), 414–435. https://doi.org/10.1006/jvbe.2000.1774

Amabile, T. M. (1996). *Creativity in context*. Westview Press.

Amabile, T. M. (1998). From individual creativity to organizational innovation. In K. Gronhaug & G. Kaufmann (Eds.), *Innovation: A cross-disciplinary perspective* (pp. 139–166). Oxford University Press.

Anderson, N., De Dreu, C. K. W., & Nijstad, B. A. (2003). The Routinization of Innovation Research: A constructively critical review of the state-of-the-science. *Journal of Organizational behavior, 25*, 147–173. https://doi.org/10.1002/job.236

Bechtel, R., & Corral-Verdugo, V. (2010). Happiness and sustainable behavior. In V. Corral-Verdugo, C. García, & M. Frías (Eds.), *Psychological approaches to sustainability* (pp. 185–204). Nova Science Publishers.

Beehr, T. A., & Newman, J. E. (1978). Job stress, employee health, and organizational effectiveness: A facet analysis, model, and literature review 1. *Personnel Psychology, 31*(4), 665–699. https://doi.org/10.1111/j.1744-6570.1978.tb02118.x

Chung, H., & Van der Lippe, T. (2020). Flexible working, work–life balance, and gender equality: Introduction. *Social Indicators Research, 151*, 1–17. https://doi.org/10.1007/s11205-018-2025-x

Cohen, S. (2014). A global measure of perceived stress. *Journal of Health and Social Behavior, 24*(4), 385–396. https://doi.org/10.2307/2136404

Cohen, S., Alper, C. M., Doyle, W. J., Treanor, J. J., & Turner, R. B. (2006). Positive emotional style predicts resistance to illness after experimental exposure to rhinovirus or influenza A virus. *Psychosomatic Medicine, 68*(6), 809–815. https://doi.org/10.1097/01.psy.0000245867.92364.3c

Culbertson, S. S., Mills, M. J., & Fullagar, C. J. (2012). Work engagement and work-family facilitation: Making homes happier through positive affective spillover. *Human Relations, 65*(9), 1155–1177. https://doi.org/10.1177/0018726712440295

Darcy, C., & McCarthy, A. (2007). Work-family conflict. *Journal of European Industrial Training, 31*(7), 530–549. https://doi.org/10.1037/0021-9010.81.4.400

Darcy, C., McCarthy, A., Hill, J., & Grady, G. (2012). Work–life balance: One size fits all? An exploratory analysis of the differential effects of career stage. *European Management Journal, 30*(2), 111–120. https://doi.org/10.1016/j.emj.2011.11.001

De Jong, J. P. J., & Kemp, R. (2003). Determinants of co-workers' innovative behavior: An investigation into knowledge intensive services. *International Journal of Innovation Management, 7*(2), 189–212. https://doi.org/10.1142/S1363919603000787

Den Dulk, L., & Groeneveld, S. (2013). Work–life balance support in the public sector in Europe. *Review of Public Personnel Administration, 33*(4), 384–405. https://doi.org/10.1177/0734371X12449024

Di Fabio, A. (2017a). Positive healthy organizations: Promoting well-being, meaningfulness, and sustainability in organizations. *Frontiers in Psychology. Organizational Psychology, 8*, 1938. https://doi.org/10.3389/fpsyg.2017.01938

Di Fabio, A. (2017b). The psychology of sustainability and sustainable development for well-being in organizations. *Frontiers in Psychology. Organizational Psychology, 8*, 1534. https://doi.org/10.3389/fpsyg.2017.01534

Di Fabio, A., & Kenny, M. E. (2018). Intrapreneurial self-capital: A key resource for promoting well-being in a shifting work landscape. *Sustainability, 10*(9), 3035. https://doi.org/10.3390/su10093035

Di Fabio, A., & Rosen, M. A. (2018). Opening the Black Box of psychological processes in the science of sustainable development: A new frontier. *European Journal of Sustainable Development Research, 2*(4), 47. https://doi.org/10.20897/ejosdr/3933

Di Fabio, A., & Tsuda, A. (2018). The psychology of harmony and harmonization: Advancing the perspectives for the psychology of sustainability and sustainable development. *Sustainability, 10*(12), 4726. https://doi.org/10.3390/su10124726

Eby, L. T., Casper, W. J., Lockwood, A., Bordeaux, C., & Brinley, A. (2005). Work and family research in IO/OB: Content analysis and review of the literature (1980–2002). *Journal of Vocational Behavior, 66*(1), 124–197. https://doi.org/10.1016/j.jvb.2003.11.003

Frone, M. R. (1997). Developing and testing an integrative model of the work – Family interface. *Journal of Vocational Behavior, 50*(2), 145–167.

Frone, M. R. (2000). Work-family conflict and employee psychiatric disorders: The national comorbidity survey. *Journal of Applied Psychology, 85*(6), 888–895. https://doi.org/10.1037/0021-90I0.85.6.888

Greenhaus, J. H., & Powell, G. N. (2006). When work and family are allies: A theory of work-family enrichment. *Academy of Management Review, 31*(1), 77–92. https://doi.org/10.5465/amr.2006.19379625

Hecht, T. D., & Boies, K. (2009). Structure and correlates of spillover from nonwork to work: An examination of nonwork activities, well-being, and work outcomes. *Journal of Occupational Health Psychology, 14*(4), 414–426. https://doi.org/10.1037/a0015981

Hirschi, A., Shockley, K. M., & Zacher, H. (2019). Achieving work-family balance: An action regulation model. *Academy of Management Review, 44*(1), 150–171. https://doi.org/10.5465/amr.2016.0409

Ilies, R., Schwind, K. M., Wagner, D. T., Johnson, M. D., DeRue, D. S., & Ilgen, D. R. (2007). When can employees have a family life? The effects of daily workload and affect on work-family conflict and social behaviors at home. *Journal of applied psychology, 92*(5), 1368.

Jacobs, J. A., & Green, K. (1998). Who are the overworked Americans? *Review of social Economy, 56*(4), 442–459. https://doi.org/10.1080/00346769800000044

Kleysen, R. F., & Street, C. T. (2001). Towards a Multi-dimensional measure of individual innovative behavior. *Journal of Intellectual Capital, 2*(3), 284–296. https://doi.org/10.1108/EUM0000000005660

Kossek, E. E., & Ozeki, C. (1999). Bridging the work-family policy and productivity gap: A literature review. *Community Work Family, 2*(1), 7–32. https://doi.org/10.1080/13668809908414247

Kubzansky, L. D., & Thurston, R. C. (2007). Emotional vitality and incident coronary heart disease: Benefits of healthy psychological functioning. *Archives of General Psychiatry, 64*(12), 1393–1401. https://doi.org/10.1001/archpsyc.64.12.1393

Kubzansky, L. D., Sparrow, D., Vokonas, P., & Kawachi, I. (2001). Is the glass half empty or half full? A prospective study of optimism and coronary heart disease in the normative aging study. *Psychosomatic Medicine, 63*(6), 910–916.

Lambert, S. J. (2016). Added benefits: The link between work-life benefits and organizational citizenship behavior. *Academy of Management, 43*(5), 801–815. https://doi.org/10.5465/1556411

Lélé, S., & Norgaard, R. B. (1996). Sustainability and the Scientist's Burden. *Conservative Biology, 10*(2), 354–365. https://doi.org/10.1046/j.1523-1739.1996.10020354.x

Leslie, L. M., King, E. B., & Clair, J. A. (2019). Work-life ideologies: The contextual basis and consequences of beliefs about work and life. *Academy of Management Review, 44*(1), 72–98. https://doi.org/10.5465/amr.2016.0410

Lewis, S., Brannen, J., & Nilsen, A. (Eds.). (2009). *Work, families and organisations in transition: European perspectives*. Policy Press.

Lewis, S., Anderson, D., Lyonette, C., Payne, N., & Wood, S. (Eds.). (2016). *Work-life balance in times of recession, austerity and beyond: Meeting the needs of employees, organizations and social justice*. Taylor & Francis.

Magee, C. A., Stefanic, N., Caputi, P., & Iverson, D. C. (2012). The association between job demands/control and health in employed parents: The mediating role of work-to-family interference and enhancement. *Journal of Occupational Health Psychology, 17*(2), 196–205. https://doi.org/10.1037/a0027050

McCarthy, A., Darcy, C., & Grady, G. (2010). Work-life balance policy and practice: Understanding line manager attitudes and behaviors. *Human Resource Management Review, 20*(2), 158–167.

Moll, H. C., Noorman, K. J., Kok, R., Engström, R., Throne-Holst, H., & Clark, C. (2008). Pursuing more sustainable consumption by analyzing household metabolism in european countries and cities. *Journal of Industrial Ecology, 9*(1-2), 259–275. https://doi.org/10.1162/1088198054084662

Ostir, G. V., Markides, K. S., Peek, M. K., & Goodwin, J. S. (2001). The association between emotional well-being and the incidence of stroke in older adults. *Psychosomatic Medicine, 63*(2), 210–215.

Parasuraman, S., & Simmers, C. A. (2001). Type of employment, work–family conflict and well-being: A comparative study. *Journal of Organizational Behavior: The International Journal of Industrial, Occupational and Organizational Psychology and Behavior, 22*(5), 551–568. https://doi.org/10.1002/job.102

Parasuraman, S., Purohit, Y. S., Godshalk, V. M., & Beutell, N. J. (1996). Work and family variables, entrepreneurial career success, and psychological well-being. *Journal of Vocational Behavior, 48*(3), 275–300.

Porter, M. E. (2010). What is value in health care? *New England Journal of Medicine, 363*(26), 2477–2481. https://doi.org/10.1056/NEJMp1011024

Porter, M. E., & Teisberg, E. O. (2006). *Redefining health care: Creating value-based competition on results*. Harvard Business Press.

Prottas, D. J., & Kopelman, R. E. (2009). Comparative work-family practice availability and employee attitudes. *Psychologist-Manager Journal, 12*(2), 79–96. https://doi.org/10.1080/10887150802665380

Roberts, K. (2007). Work-life balance-the sources of the contemporary problem and the probable outcomes: A review and interpretation of the evidence. *Employee Relations, 29*(4), 334–351. https://doi.org/10.1108/01425450710759181

Rotondo, D. M., & Kincaid, J. F. (2008). Conflict, facilitation, and individual coping styles across the work and family domains. *Journal of Managerial Psychology, 23*(5), 484–506. https://doi.org/10.1108/02683940810884504

Scheier, M. F., Matthews, K. A., Owens, J. F., Magovern, G. J., Lefebvre, R. C., Abbott, R. A., & Carver, C. S. (1989). Dispositional optimism and recovery from coronary artery bypass surgery: The beneficial effects on physical and psychological well-being. *Journal of Personality and Social Psychology, 57*(6), 1024–1040. https://doi.org/10.1037/0022-3514.57.6.1024

Seligman, M. E. (2008). Positive health. *Applied Psychology, 57*(s1), 3–18. https://doi.org/10.1111/j.1464-0597.2008.00351.x

Seligman, M. E. (2012). *Flourish: A visionary new understanding of happiness and well-being*. Simon and Schuster.

Seligman, M. E. P., & Csikszentmihalyi, M. (2000). Positive psychology. An introduction. *The American Psychologist, 55*(1), 5–14. https://doi.org/10.1037/0003-066X.55.1.5

Steptoe, A., Wardle, J., & Marmot, M. (2005). Positive affect and health-related neuroendocrine, cardiovascular, and inflammatory processes. *Proceedings of the National academy of Sciences of the United States of America, 102*(18), 6508–6512.

Taylor, S. E., Kemeny, M. E., Reed, G. M., Bower, J. E., & Gruenewald, T. L. (2000). Psychological resources, positive illusions, and health. *American psychologist, 55*(1), 99–109. https://doi.org/10.1037//0003-066x.55.1.99

Thurston, R. C., & Kubzansky, L. D. (2009). Women, loneliness, and incident coronary heart disease. *Psychosomatic Medicine, 71*(8), 836–842. https://doi.org/10.1097/PSY.0b013e3181b40efc
Veenhoven, R. (2004). *Sustainable consumption and happiness*. MPRA.
Verdugo, V. C. (2012). The positive psychology of sustainability. *Environment, Development and Sustainability, 14*(5), 651–666. https://doi.org/10.1007/s10668-012-9346-8
WCED (World Commission on Environment and Development) (1987). Our common future, Oxford: Oxford University Press.
Yu, L., Boyle, P. A., Wilson, R. S., Levine, S. R., Schneider, J. A., & Bennett, D. A. (2015). Purpose in life and cerebral infarcts in community-dwelling older people. *Stroke, 46*(4), 1071–1076. https://doi.org/10.1161/STROKEAHA.114.008010

Index

A. Di Fabio (ed.), *Cross-cultural Perspectives on Well-Being and Sustainability in Organizations*, https://doi.org/10.1007/978-3-030-86709-6